Performance and Application of Novel Biocomposites

Performance and Application of Novel Biocomposites

Editor

Oisik Das

MDPI • Basel • Beijing • Wuhan • Barcelona • Belgrade • Manchester • Tokyo • Cluj • Tianjin

Editor
Oisik Das
Luleå University of Technology
Sweden

Editorial Office
MDPI
St. Alban-Anlage 66
4052 Basel, Switzerland

This is a reprint of articles from the Special Issue published online in the open access journal *Polymers* (ISSN 2073-4360) (available at: https://www.mdpi.com/journal/polymers/special_issues/Per_App_Nov_Bio).

For citation purposes, cite each article independently as indicated on the article page online and as indicated below:

LastName, A.A.; LastName, B.B.; LastName, C.C. Article Title. *Journal Name* **Year**, *Volume Number*, Page Range.

ISBN 978-3-0365-0312-7 (Hbk)
ISBN 978-3-0365-0313-4 (PDF)

Contents

About the Editor

Oisik Das research activities pertain to carbon-based materials and polymeric composites, specifically improvement of their performance properties (e.g., mechanical, flammability, dimensional) through physical and chemical means. Of particular interest is the production and characterisation of biochar (i.e., bio-based carbon materials) for composite applications. Oisik has extensive experience in determining the material properties of numerous types of biochars through nanoindentation. Additionally, Oisik is interested in enhancing the fire-resistant properties of polymeric composites by using both conventional and natural fire retardants. Oisik teaches courses related to material science and fire engineering and supervises students. Oisik worked at the KTH Royal Institute of Technology, Stockholm, Sweden for two years as a post-doctoral fellow conducting research on bio-based polymers and their composites. Oisik completed his PhD at the Centre for Advanced Composite Materials (CACM) at the University of Auckland, New Zealand. His research was focused on the utilisation of biochar (obtained from the pyrolysis/thermochemical conversion of lignocellulosic wastes) in areas of biocomposite development. Oisik's master's degree is from Washington State University, Pullman, USA where he worked on the thermochemical conversion of lignocellulosic biomass to produce value-added products (e.g., biocarbon and bio-oil). In the past, Oisik served Maharishi Markandeshwar (M.M.) University in India as an Assistant Professor where he worked with students regarding various applications of biocarbon/biochar.

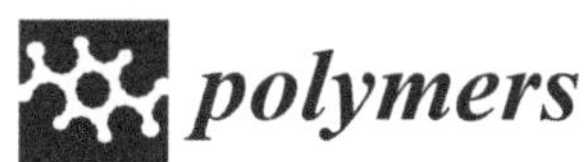 *polymers*

Editorial

Education and Research during Pandemics: Illustrated by the Example of Experimental Biocomposites Research

Oisik Das [1],* and Seeram Ramakrishna [2],*

[1]   Department of Engineering Sciences and Mathematics, Luleå University of Technology, 97187 Luleå, Sweden
[2]   Department of Mechanical Engineering, National University of Singapore, Singapore 117575, Singapore
*   Correspondence: oisik.das@ltu.se (O.D.); seeram@nus.edu.sg (S.R.)

Received: 5 August 2020; Accepted: 17 August 2020; Published: 18 August 2020

In late 2019, a novel Coronavirus was detected in Wuhan city of China, giving rise to the catastrophic pandemic that is still rampant today. Initially, the worst-hit districts were put under lockdown, which then extended to cities and eventually whole countries. Travel of people, along with logistics of goods and services, were (and still are) severely affected. Most nations of the world urged their citizens to stay indoors so as to avoid exposure to the virus, and thus remain infection-free. One of the demographics that are negatively affected by the lockdown measures is the students and researchers. Numerous universities around the world had to shut their premises at short notice, thus prompting a rapid shift from in-classroom education to online education, a transition that normally would take decades to happen. In particular, students received their classes through digital platforms, which included Zoom, Microsoft Teams, Skype, etc., whereas the researchers adopted tele-working. Although this strategy employed by universities is effective in curbing the further spread of the virus, it has some unintended consequences. Firstly, owing to the uncertainly regarding the end date for the current coronavirus pandemic, millennials and freshmen are unsure about their immediate enrolment in their chosen courses and programmes. For example, the University of Ohio in USA and the University of Cambridge in the United Kingdom will hold online classes for the upcoming fall and until the summer of 2021, respectively. This is particularly disheartening for international students, who are anticipating an active academic experience that includes campus life, engagement in classrooms, obtaining in-person feedback from lecturers, bonding and networking in cafes, etc. Secondly, and more importantly, students whose programmes warrant undertaking a significant amount of laboratory work are stressed about the stagnant nature of their research. While a few fields of study can be conducted on a digital platform, experimental research requires the presence of the person in laboratories for a substantial amount of time. Biocomposite education is at its core an experimental one, which includes the design of the biocomposite, preparation of raw materials, fabrication and manufacturing, prototyping, and finally testing and characterisation. Therefore, it is critical to identify some effective means to propagate biocomposites education during pandemics, wherein students and researchers are confined to quarantines. In other words, educators should create paths for effective learning in the biocomposite field in a distanced education system via alternative routes and remote controlled laboratories and equipment.

In light of the aforementioned, five strategies could be adopted by the students and researchers to sustain biocomposites education and learning during viral outbreaks and disruptions. The first strategy, which is one of the most obvious ones, is to bolster the theoretical knowledge regarding composite science and technology. Often, a student or a researcher learns on the job, i.e., learning by doing. While this is imperative to activate the psychomotor taxonomic domain, the cognitive domain can be made robust by indulging in the comprehension of background knowledge regarding various scientific phenomena and engineering concepts [1]. Although a student can progress through his/her academic career and reach higher positions of lecturer or assistant professor by relying solely on the 'working knowledge' of biocomposites, an in-depth understanding of concepts like micromechanics, macromechanics, laminate theory, structural mechanics, analytical modelling and finite element modelling will make them reflective practitioners [2]. Additionally, these academics will be intrinsically motivated [3] to conduct effective teaching and ground-breaking research. Therefore, the imparting of theoretical knowledge on biocomposites will garner self-regulation [4], confidence and self-efficacy [5] in the students and researchers.

In the second strategy, the students and researchers can devote their time to preparing comprehensive and critical review articles meant for beginners and experienced researchers, respectively. Not only does the preparation of review articles inadvertently facilitate the absorbance of overall knowledge, but also their eventual publication in peer-reviewed journals attracts more citations (compared to the narrowly focused research articles), which will boost the person's academic career and visibility. The writing of review articles enables the author to develop a holistic overview regarding specific aspects of the biocomposite field. Additionally, the author becomes aware of the latest developments in the state-of-the-art research, and is able to critically analyse and well position his/her own research so as to address specific scientific and technological challenges and needs. Thus, the above-mentioned facets of writing a review article are conducive for the development of biocomposites education because students/researchers will learn by immersing themselves in loops of experience, theories and practice, as specified by Boyatzis and Kolb, 1995 [6].

In the third strategy, the students and researchers can perform life cycle analyses (LCA) of various biocomposite products. LCA does not require access to laboratories, and thus can be performed from the safety of one's home. Through LCA analysis, the student/researcher will be able to grasp the importance of manufacturing and environmental sustainability, and attaining a circular economy mind-set. It is critical to reduce greenhouse gas (GHG) emissions and wastage at every stage of the biocomposites' life cycle, and LCA will shine light into the environmental impact of sourcing raw materials and feedstock, processing, manufacture, distribution, use, repair, maintenance and disposal or recycling, i.e., the cradle-to-grave life of the product. The performing of LCA studies will not only create opportunities for journal publications, but also encourage the student/researcher to undertake industry-facing and market-oriented sustainable design and re-design of biocomposites in the future. This will lead to the academic being environmentally conscious and striving towards waste minimisation and pollution reduction during the biocomposite's development and life cycle.

The fourth strategy is related to simulation studies of various aspects of biocomposites. Simulation studies can be related to the determination of process feasibility parameters, its lifetime prediction, failure mechanisms, etc. Although simulation without experimental validation could be futile, students/researchers can delve into the modelling world, which can enable process optimisation and effective product life cycle engineering. Furthermore, the students/researchers can visualise the performance of the biocomposite without having to actually manufacture the product. Therefore, simulation studies will not only enhance one's theoretical understanding of composite science, but also prepare one to tailor the design in order to have desirable performance properties and functionalities. Simulation studies will be the closest thing for the students/researchers to experimentally designing and developing biocomposites, and characterising their various properties in a manner akin to a real-life laboratory session.

If performing real-world experiments is unavoidable, maybe the students/researchers can do so in a simulated laboratory environment of virtual reality (VR), which is the fifth strategy. Nevertheless, VR technology would not be accessible to all the students, especially in developing nations where such technologies could be non-existent. VR technology can potentially allow students/researchers to collaborate and interact with the artificially created biocomposite laboratory by moving through its spaces and experiencing visual and auditory feedback from common instruments, such as injection moulding machines, Instron Universal testing machines, cone calorimetry equipment, etc. Since VR has been used in medicine in a way that has allowed the trainee doctors to rectify errors [7], the same can be emulated in biocomposite education. VR in biocomposite education will be beneficial in enabling the student/researcher to develop his/her experimental skills, and will reduce the total cost of the programme, since raw materials will not be expended.

In summary, there are several ways by which a student or a researcher can be immersed in continuing biocomposites education during pandemics and massive disruptions. Adherence to the aforementioned strategies will ensure that students/researchers can come back with a strong foundation once the pandemic ends and the laboratories reopen. The following Figure 1 depicts the ideas put forward in this article. An ideal solution for maintaining the flow of biocomposites research and education is the combination of all the five strategies in some form or another.

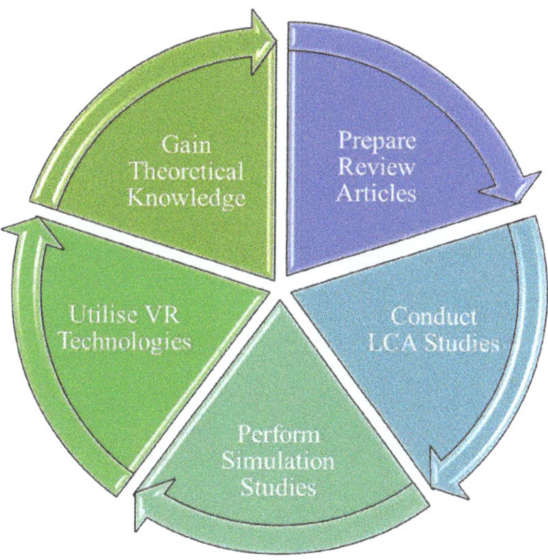

Figure 1. The five strategies for students and researchers to adopt in order to maintain the continuity of biocomposites education during a pandemic.

References

1. Adesoji, F.A. Bloom taxonomy of educational objectives and the modification of cognitive levels. *Adv. Soc. Sci.* **2018**, *5*, 5. [CrossRef]
2. Schön, D.A. *Educating the Reflective Practitioner*; Jossey-Bass: San Francisco, CA, USA, 1987.
3. Rust, C. The impact of assessment on student learning: How can the research literature practically help to inform the development of departmental assessment strategies and learner-centred assessment practices? *Act. Learn. High. Educ.* **2002**, *3*, 145–158. [CrossRef]
4. Ng, E.M. Integrating self-regulation principles with flipped classroom pedagogy for first year university students. *Comput. Educ.* **2018**, *126*, 65–74. [CrossRef]
5. Baker, D. What works: Using curriculum and pedagogy to increase girls' interest and participation in science. *Theory Pract.* **2013**, *52*, 14–20. [CrossRef]
6. Boyatzis, R.E.; Kolb, D.A. From learning styles to learning skills: The executive skills profile. *J. Manag Psychol.* **1995**, *10*, 3–17. [CrossRef]
7. Li, L.; Yu, F.; Shi, D.; Shi, J.; Tian, Z.; Yang, J.; Wang, X.; Jiang, Q. Application of virtual reality technology in clinical medicine. *Am. J. Transl.* **2017**, *9*, 3867.

Article

Study of the Compatibilization Effect of Different Reactive Agents in PHB/Natural Fiber-Based Composites

Estefanía Lidón Sánchez-Safont [1], Abdulaziz Aldureid [1], José María Lagarón [2], Luis Cabedo [1] and José Gámez-Pérez [1,*]

[1] Polymers and Advanced Materials Group (PIMA), Universitat Jaume I (UJI), Av. Vicent Sos Baynat s/n, 12071 Castelló de la Plana, Spain; esafont@uji.es (E.L.S.-S.); aldureid@uji.es (A.A.); lcabedo@uji.es (L.C.)

[2] Novel Materials and Nanotechnology Group, Institute of Agrochemistry and Food Technology (IATA), Spanish National Research Council (CSIC), Calle Catedrático Agustín Escardino Benlloch 7, 46980 Paterna, Spain; lagaron@iata.csic.es

[*] Correspondence: gamez@uji.es

Received: 5 August 2020; Accepted: 26 August 2020; Published: 30 August 2020

Abstract: Fiber–matrix interfacial adhesion is one of the key factors governing the final properties of natural fiber-based polymer composites. In this work, four extrusion reactive agents were tested as potential compatibilizers in polyhydroxylbutyrate (PHB)/cellulose composites: dicumyl peroxide (DCP), hexamethylene diisocyanate (HMDI), resorcinol diglycidyl ether (RDGE), and triglycidyl isocyanurate (TGIC). The influence of the fibers and the different reactive agents on the mechanical properties, physical aging, and crystallization behavior were assessed. To evaluate the compatibilization effectiveness of each reactive agent, highly purified commercial cellulose fibers (TC90) were used as reference filler. Then, the influence of fiber purity on the compatibilization effect of the reactive agent HMDI was evaluated using untreated (U_RH) and chemically purified (T_RH) rice husk fibers, comparing the results with the ones using TC90 fibers. The results show that reactive agents interact with the polymer matrix at different levels, but all compositions showed a drastic embrittlement due to the aging of PHB. No clear compatibilization effect was found using DCP, RDGE, or TGIC reactive agents. On the other hand, the fiber–polymer interfacial adhesion was enhanced with HMDI. The purity of the fiber played an important role in the effectiveness of HMDI as a compatibilizer, since composites with highly purified fibers showed the greatest improvements in tensile strength and the most favorable morphology. None of the reactive agents negatively affected the compostability of PHB. Finally, thermoformed trays with good mold reproducibility were successfully obtained for PHB/T_RH/HMDI composition.

Keywords: PHB; natural fiber; compatibilizer; cellulose; biocomposite

---

1. Introduction

The development of biobased biodegradable thermoplastic materials is a topic research of special interest because it can represent a cost-effective and environmental-friendly alternative to commodities [1]. Among the different biopolymers, polyhydroxylbutyrate (PHB), a bacterial origin biopolyester from the polyhydroxyalcanoates family (PHAs), has attracted a lot of attention. The applicability fields where the PHB-based material results are more interesting are those in which biodegradability is desired either because composting could be a viable option for their waste management or because they can potentially end up in the environment. Among those applications, we can highlight food packaging or disposable products such as single-use tableware, hygiene-related single-use products, straws, etc. [1–4]. The main strengths of the PHB that make it suitable for this type of application are its natural origin, its biodegradability, the absence of toxicity, and the high service temperature [5]. Indeed, PHB presents mechanical properties in terms of a stiffness and strength that is similar to PP, good barrier properties, which are comparable or even superior to PET [6–10], and it is biodegradable in different environments, such as soil and marine [7,11,12], and compostable at lab-scale, industrial, and home composting conditions [13].

However, PHB presents some shortcomings that limit its industrial applicability. PHB is a semicrystalline polymer that is capable of a high degree of crystallinity but has a relatively low crystallization rate. Hence, PHB suffers an appreciable embrittlement with time due to secondary crystallization and physical aging [14–17], and its long-term mechanical properties are characterized by low ductility and toughness. Indeed, the processing temperature window of PHB is very narrow: the lower limit is relatively high due to its high crystallinity, and the upper limit is relatively low because of its poor thermal stability in molten state (the degradation temperature is close to the melting temperature [7]). Altogether, these factors make PHB quite difficult to process, especially in the case of thermoforming [18]. In addition, one of the main limiting factors is its current high price. In this sense, the development of PHB-based composites using lignocellulosic fibers as fillers could contribute to a large extent to overcome the cost drawback maintaining the biodegradability and even improving the mechanical performance of PHB, allowing the valorization of vegetal wastes contributing to the circular economy.

Lignocellulosic fibers are hydrophilic materials composed by bundles of cellulose fibers embedded in a matrix of other non-cellulosic materials such as lignin, hemicelluloses, pectin, waxes, and other minor components [19]. The advantages of use lignocellulosic fibers as fillers are their availability, low cost, biodegradability, low density, high stiffness, and acceptable specific strength [20]. However, they also present shortcomings related to their thermal sensitivity and hydrophilic nature. In addition, depending on the vegetal source and/or the plant location and time of harvest, the composition, properties, morphology, and surface characteristics of different lignocellulosic fibers may differ significantly [21].

It is well known that the resultant properties of fiber-based composites depend not only on the properties of the constituents but are also determined by the fiber–matrix adhesion. The hydrophilic nature of the lignocellulosic fibers lowers the compatibility with the hydrophobic polymer. Nevertheless, according to Bhardwaj et al. [22], the relatively polar nature and presence of carbonyl groups ($-C=O$) in PHB as compared with other nonpolar matrices such as PP might cause a hydrogen-bonding-type interaction with the cellulosic fibers and relative better compatibility, as it has been also noticed by others in PHA/lignocellulosic composites [23,24]. However, these interactions are not enough to provide strong adhesion of PHB with lignocellulosic fibers, as it has been shown previously in PHA-based composites, which are filled with untreated lignocellulosic fibers [25–27]. Thus, the enhancement of fiber–matrix adhesion may be a key factor to exploit the full capabilities of these composites.

Some attempts to improve interfacial adhesion are physical treatments (plasma or corona discharge), chemical purification treatments (dewaxing and delignifying treatments) of the fibers, grafting, or the use of additives such as compatibilizers or coupling agents [19,28,29]. Reactive compatibilization is an interesting cost-effective one-step strategy consisting of the use of small amounts of reactive agents that possess functional groups with a tendency to react with the –OH groups of the fibers and with the carboxylic end groups from polyesters by covalent bond interactions. Thus, the most popular reactive agents used include maleic anhydride groups, epoxy groups, or isocyanate groups [30,31]. Several examples of the use of reactive agents in polyester/fiber-based composites can be found in the literature. Diisocyanates have been used in PHBV/bamboo fibers [32] or poly(3-hydroxybutyrate-co-3-hydroxyvalerate) (PHBV)/poly(butylene adipate-co-terephthalate) (PBAT) /Switchgrass systems [33]. Epoxy-based reactive agents have been used in PLA/sisal fiber composites [34,35].

Another strategy could be by using radical generators that could arouse random linkages between the matrix and the reinforcement via radical intermediate species, such as peroxides. Dycumil peroxide (DCP) has been used to compatibilize PHBV/Miscanthus fiber composites [31] or PHB and PHBV/α-cellulose composites [36].

In this work, the efficiency as compatiblizers of four different reactive agents in fiber-based PHB composites was tested. The reactive agents used were dicumyl peroxide (DCP), hexamethylene diisocyanate (HMDI), resorcinol diglycidyl ether (RDGE), and tryglicidyl isocyanurate (TGIC). The chemical structures of them are shown in Figure 1. In order to reduce variables and better understand the role of each reactive agent in this study, a high purified commercial cellulose fiber (TC90) with an α-cellulose content >99.5 % was selected, being the filler load set at 10 phr (i.e., per hundred mass of resin) for all compositions.

The effect of the different reactive agents on the PHB/cellulose interfacial interactions was studied by scanning electron microscopy (SEM), tensile tests, and dynamic mechanical analysis (DMA). Indeed, the effect of aging was assessed for all compositions. As maintained biodegradability is an important requirement for the applicability of these systems, the effect of the different reactive agents on the biodisintegration under standard composting conditions (ISO 20200) was also evaluated.

With the aim of analyzing the influence of fiber purity on the compatibilization efficiency, untreated rice fibers (U_RH) and chemically purified rice husk fibers (T_RH) according to a previous work [37] were used using HMDI as a compatibilizer. The mechanical performance and the morphology were analyzed, and the results were compared with the use of the commercial cellulose.

Finally, since packaging is one of the potential application fields for these composites, the suitability of PHB/T_RH/HMDI composites to be processed by thermoforming was tested. This process has been chosen for both its difficulty and for being one of the most popular forming techniques used in the packaging industry.

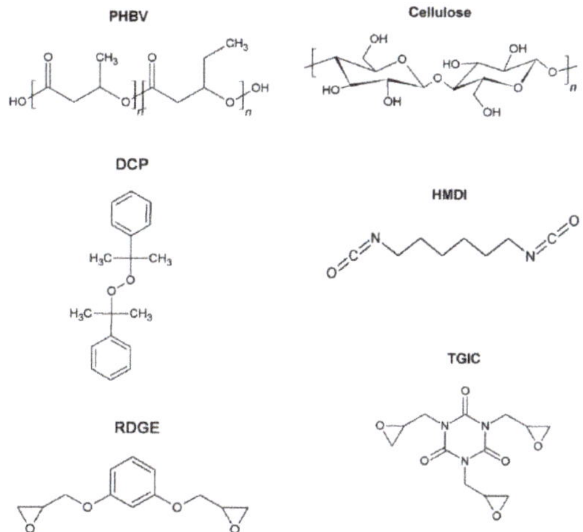

Figure 1. Chemical structures of polyhydroxylbutyrate (PHB), cellulose, and the reactive agents.

2. Materials and Methods

2.1. Materials

Poly(3-hydroxybutyrate) was supplied by Biomer® (Schwalbach, Germany) in pellet form (P309). Purified alpha-cellulose fiber grade with an alpha-cellulose content >99.5% (TC90) was purchased from CreaFill Fibers Corp. (Chestertown, MD, USA). Rice husk (RH) by-product from the rice production process was kindly provided by Herba Ingredients (Valencia, Spain). The four reactive agents used (dicumyl peroxide (DCP), hexamethylene diisocyanate (HMDI), resorcinol diglycidyl ether (RDGE), and triglycidyl isocyanurate (TGIC)) were purchased from Sigma Aldrich (Madrid, Spain). Sodium hydroxide (NaOH, 98%), hydrogen peroxide (H_2O_2, 30%), glacial acetic acid (CH_3COOH, 99%), and sulfuric acid (H_2SO_4, 98%) were purchased from Sigma Aldrich (Madrid, Spain).

2.2. Rice Husk Fibers Preparation

RH fibers were ground in a mechanical knife mill and then sieved in 140 µm mesh. These untreated RH fibers were named U_RH. A fraction of the ground and sieved RH fibers were subjected to a two-stage purification treatment in order to remove the major parts of impurities and non-cellulosic components such as waxes, lignin, and hemicelluloses. The first stage consisted of an alkaline attack with NaOH (5% *wt/v*, fiber/liquid ratio of 1:20, 80 °C, 2 h). This treatment was applied twice. The second stage consisted of an oxidative attack with peracetic acid (PAA) (fiber/liquid ratio of 1:20, 80 °C, 4 h). The peracetic acid was prepared by the mixing of 30%(*v/v*) hydrogen peroxide and acetic acid in the reaction medium with a volume ratio of 3:1 at room temperature and 1% (*w/w*) of sulfuric acid as catalyzer. This procedure was adapted from the literature [38,39]. After each stage, the fibers were filtered and washed repeatedly in distilled water until neutral pH was reached. The purified powder was dried at 60 °C for at least 24 hours and ground again to break the aggregates formed during the filtration process and then sieved in a 140 µm mesh. The as-treated RH fibers were named as T_RH.

2.3. Composites Preparation

In order to assess the role of reactive agents as compatibilizers, compounds of purified commercial cellulose (TC90) were prepared with all reactive agents. The effect of the cellulose purity was studied on rice husk fibers, with and without chemical treatment, using HMDI as the reactive agent. For the sake of comparison and to evaluate the effects of the compatibilizers on the matrix, blank compounds (without cellulose) were prepared as controls. All the compositions studied are summarized in Table 1.

The compounds were prepared by melt extrusion in a twin-screw co-rotating extruder (DUPRA SL, Castalla, Spain) with an L/D ratio of 24 and a diameter of 2.5 cm. All the components were dried before extrusion; PHB pellets were dried in a dehumidifier Piovan DPA50 (Piovan, Maria di Sala VE, Italy) at 60 °C following the producer's drying recommendations and the fibers (TC90, U_RH and T_RH) were dried in an oven at 100 °C for at least 2 h. The formulations were manually premixed in zip-bags. The temperature profile of the extruder was set as follows: 165/170/175/180 °C (from the hopper to the extruder die), and the screw speed was kept constant at 40 rpm. The extrudate material was pelletized and dried following the same considerations as pure PHB.

Table 1. Summary of studied formulations. DCP: dicumyl peroxide, HMDI: hexamethylene diisocyanate, RDGE: resorcinol diglycidyl ether, TGIC: triglycidyl isocyanurate.

Sample	Component (phr)							
	PHB	**TC90**	**U_RH**	**T_RH**	**DCP**	**HMDI**	**RDGE**	**TGIC**
PHB	100			-	-	-	-	-
PHB/DCP	100			-	1	-	-	-
PHB/HMDI	100			-		1	-	-
PHB/RDGE	100			-	-	-	1	-
PHB/TGIC	100			-	-	-	-	1
PHB/TC90	100	10		-	-	-	-	-
PHB/TC90/DCP	100	10		-	1	-	-	-
PHB/TC90/HMDI	100	10		-	-	1	-	-
PHB/TC90/RDGE	100	10		-	-	-	1	-
PHB/TC90/TGIC	100	10		-	-	-	-	1
PHB/U_RH	100		10	-	-	-	-	-
PHB/U_RH/HMDI	100		10	-	-	-	-	-
PHB/T_RH	100			10	-	-	-	-
PHB/T_RH/HMDI	100			10	-	1	-	-

From the extruded pellets, different samples were obtained by compression molding in a parallel plate hot-press (180 °C, 2 min for premelting followed by 2 min at 3 bar): bars of 50 × 12.5 × 3.5 mm for dynamic mechanical analysis tests, films of 0.4 mm nominal thickness for uniaxial mechanical tests, films of 0.2 mm nominal thickness for composting tests, and films of 0.8 mm nominal thickness for thermoforming essays. Samples with neat PHB were processed and tested at the same conditions as the compounds.

2.4. Methods

The morphology of PHB/TC90, PHB/U_RH, and PHB/T_RH composites with and without reactive agents was examined by scanning electron microscopy (SEM), using a high-resolution field-emission microscope (JEOL 7001F, Tokyo, Japan). The samples were prepared by cryofracturing after immersion in liquid nitrogen and then coated by sputtering with a thin layer of Pt.

Differential scanning calorimetry (DSC) experiments were conducted on a DSC2 (Mettler Toledo, Columbus, OH, USA) with an intracooler Julabo FT900 (Julabo, Seelbach, Germany) calibrated with Indium standard before use. Samples were analyzed at 0 days (after hot-pressed films obtention) and after 100 days, to account for physical aging at room temperature. The samples weighing typically 6 mg were first heated from −20 °C to 200 °C at 10 °C/min, kept for 5 min to erase thermal history, and cooled down to −20 °C at 10 °C/min. Then, a second heating scan to 200 °C at 10 °C/min was performed. Crystallization temperatures (T_c), melting temperatures (T_m), and melting enthalpies (ΔH_m) were calculated from all respective heating/cooling scans. The crystallinity (X_c) of the PHB–reactive agent compositions was determined by applying the expression (1) [10]:

$$X_c(\%) = \frac{\Delta H_m}{w \cdot \Delta H_m^0} \times 100 \tag{1}$$

where ΔH_m (J/g) is the melting enthalpy of the polymer matrix, $\Delta H°_m$ is the melting enthalpy of 100% crystalline PHB (perfect crystal) (146 J/g) [16], and w is the PHB weight fraction in the blend.

Tensile tests were conducted in a universal testing machine Shimatzu AGS-X 500N (Shimatzu, Kyoto, Japan) at room temperature with a crosshead speed of 10 mm/min. Dumbbell 400 μm-thick samples were die-cut from the hot-pressed films and tested according to ASTM D638 (Type IV) standard. The samples were tested immediately after processing (0 days) and after 15 days of aging at room temperature. All the samples were stored in a vacuum desiccator at ambient temperature until tested.

Dynamic mechanical analysis (DMA) experiments were conducted on hot-pressed sample bars (55 × 12.5 × 3.5 mm) in an AR G2 oscillatory rheometer (TA Instruments, New Castle, DE, USA) equipped with a clamp system for solid samples (torsion mode). Samples were heated from −20 °C to melting temperature with a heating rate of 2 °C/min at a constant frequency of 1 Hz. The maximum deformation (γ) was set to 0.1%.

Disintegration tests under standard composting conditions (ISO 20200 [41]) were carried out with samples of (15 × 15 × 0.2 mm^3) obtained from hot-pressed plates. Solid synthetic waste was prepared by mixing 10% of activated mature compost (VIGORHUMUS H-00, purchased from Burás Profesional, S.A., Girona, Spain), 40% sawdust, 30% rabbit feed, 10% corn starch, 5% sugar, 4% corn seed oil, and 1% urea. The water content of the mixture was adjusted to 55%. The samples were placed inside mesh bags to simplify their extraction and allow the contact of the compost with the specimens; then, they were buried in compost bioreactors at 4–6 cm depth. Bioreactors were incubated at 58 °C. The aerobic conditions were guaranteed by mixing the synthetic waste periodically and adding water according to the standard requirements. Two replicates of each sample were removed from the boxes at different composting times for analysis. Samples were washed with water and dried under vacuum at 40 °C until reaching a constant mass. The disintegration degree was calculated by normalizing the sample weight to the initial weight with Equation (2):

$$D = \frac{m_i - m_f}{m_i} \times 100 \tag{2}$$

where m_i is the initial dry mass of the test material and m_f is the dry mass of the test material recovered at different incubation stages. The disintegration study was completed taking photographs for visual evaluation.

The thermoformability of PHB/T_RH/HMDI was tested by a vacuum-assisted thermoforming technique in a pilot plant (SB 53c, Illig, Helmut Roegele, Heilbronn, Germany) equipped with an infrared emitter heating device. The mold used was a female circular tray that was 55 mm in diameter and 15 mm in depth with an edge radium of 5 mm. Rectangular hot-pressed sheets of a typical thickness of 800 µm were used for this study. The sheets were stamped with a square grid pattern (0.5 × 0.5 cm) in order to track the deformation that occurred during their mold conformation. The infrared heater was set to 600 °C, whereas the heating and vacuum times (ranging between 20–45 s and 3–20 s, respectively) were optimized in each case to obtain the best results.

3. Results

3.1. Influence of Reactive Agents in PHB/Cellulose Composites

3.1.1. Morphological Analysis

In order to assess the role of the reactive agents, blends with TC90 were prepared as detailed in the experimental section. The morphology of the PHB/TC90 composites with and without the reactive agents has been analyzed by SEM. Low magnification images were used to study the distribution of the fibers within the polymer matrix, and high magnification ones were used to examine the fiber/matrix interface. The micrographs of the different composites are shown in Figure 2.

As it can be observed in Figure 2a,c,e,g,i, in general, the fibers are well distributed within the polymer matrix, and we do not detect the presence of fiber aggregates, indicating an effective compounding. Despite this well dispersed and distributed morphology points to some type of fiber/matrix interaction (probably hydrogen bonding), the presence of some voids and prints caused by detached fibers (Figure 2a) as well as the gap observed between the fiber and the matrix (Figure 2b) are indicative of a certain lack of adhesion. Then, it can be said that there are some interactions in the melt that favor homogeneous dispersion, but those are not strong enough to provide an effective interface between both components.

Regarding the reactive agents, no remarkable differences in morphology are detected in PHB/TC90 composites with reactive agents compared with the composite without them. Although the major part of the fibers seems to be well embedded into the polymer matrix, some pull-out and detached fibers are detected, as well as a small gap between the fibers and the matrix. With regard to the use of DCP, contrary to other works reported in literature [31,36,42], in our case, no clear enhancement of compatibilization between fibers and matrix can be appreciated by SEM. In the same way, any compatibilization effect was found for the RDGE. Similarly, the TGIC did not show any additional compatibilization effect, as this was unexpected [34]. However, in PHB/TC90/HMDI composites, there is an improvement of fiber–matrix adhesion, finding no fiber pull-outs or detachment in the micrographs (Figure 2e). In this case, the fibers seem to be well covered by the polymer, and fibers broken on their longitudinal direction can be observed (Figure 2f), thus indicating a cohesive failure. Thus, SEM observations would be in agreement with a strong adhesion between the fibers and the PHB matrix. This is probably due to the formation of urethane linkages between the isocyanate groups of HMDI and hydroxyl (–OH) groups from the fibers and/or hydroxyl or carboxylic chain ends of PHB, as it has been proposed in the literature for biopolyester/fiber systems compatibilized with isocyanates [33].

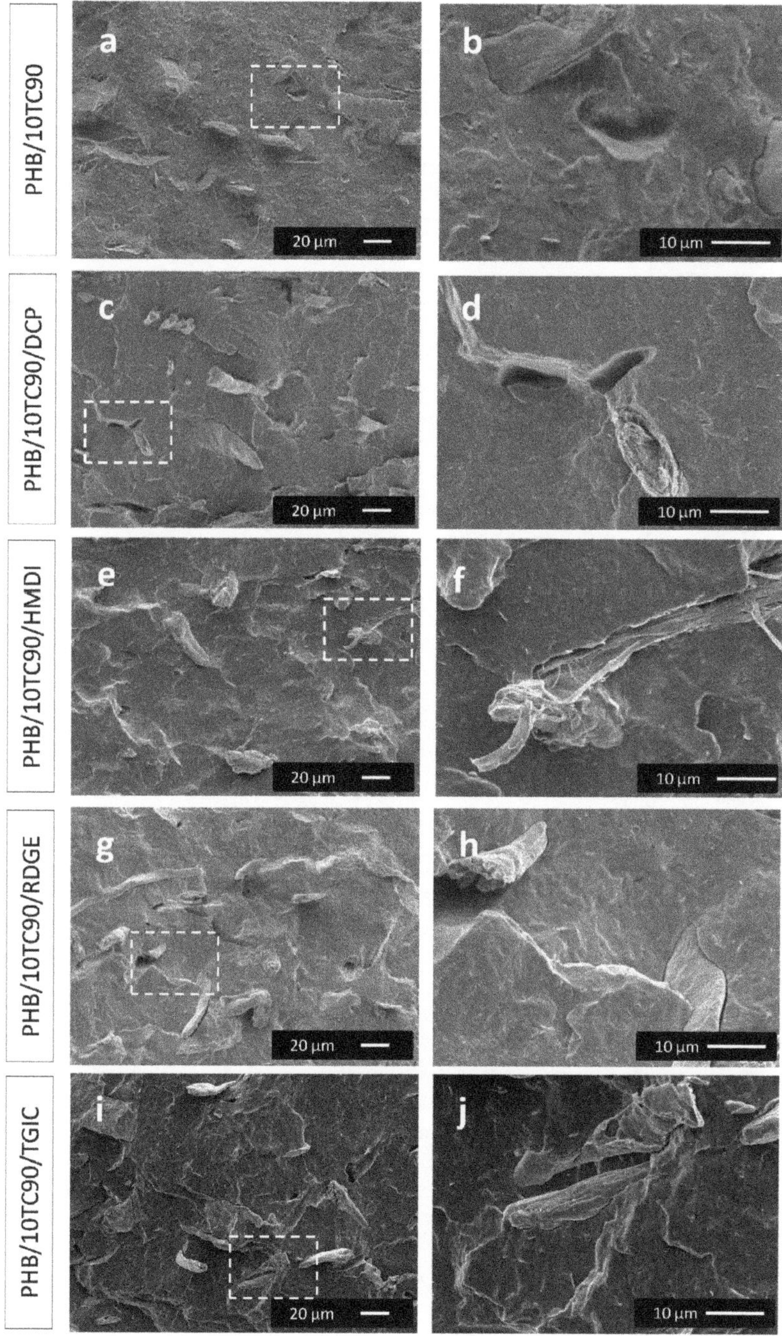

Figure 2. SEM micrographs of PHB/TC90 composites with and without reactive agents. The images in the right column (**b,d,f,h,j**) show higher magnifications of the areas indicated with a square in their corresponding images in the left column (**a,c,e,g,i**).

3.1.2. Thermal Properties

DSC experiments were run in all samples. The thermograms were obtained from the films recently processed (0 days) and after 100 days of storage at room temperature, when it is supposed that all secondary crystallization and physical aging phenomena has taken place [15,43]. It can be seen in the thermograms that aging only affected the first heating scans. The second heating scans, after erasing the thermal history and controlled cooling at 10 °C/min, were the same for 0 and 100 days, thus confirming that there were no significant structural changes during storage. Therefore, no signs of degradation were evidenced for this period. All the results are summarized in Table 2.

Table 2. Differential scanning calorimetry (DSC) parameters for neat PHB with and without reactive agents and PHB/TC90 composites with and without reactive agents.

| | 1st Heating Scan | | | | | | Cooling Scan | | 2nd Heating Scan | | |
| | 0 days | | | 100 days | | | T_c (°C) | ΔH_c (J/g) | T_m (°C) | ΔH_m (J/g) | X_c (%) |
	T_m (°C)	ΔH_m (J/g)	X_c (%)	T_m (°C)	ΔH_m (J/g)	X_c (%)					
PHB	175	72	49	170	86	59	117	91	170	94	64
PHB/DCP	167	71	49	167	73	50	115	81	160	84	58
PHB/HMDI	173	69	48	173	79	55	106	86	167	90	62
PHB/RDGE	172	75	52	173	81	56	115	88	168	91	63
PHB/TGIC	173	77	53	172	80	56	116	89	168	92	64
PHB/TC90	173	74	56	173	73	55	117	82	169	84	63
PHB/TC90/DCP	166	67	51	165	66	50	118	75	161	78	59
PHB/TC90/HMDI	173	71	49	173	72	55	111	78	169	81	62
PHB/TC90/RDGE	172	76	52	173	73	56	115	79	169	85	65
PHB/TC90/TGIC	167	78	54	172	73	55	118	79	166	85	64

The melting behavior and crystallinity index during first heating scans of the composites are affected in different extents by the different components. However, it must be considered that the crystal morphologies developed during cooling correspond to processing conditions, which implies higher cooling rates with respect to DSC-controlled cooling at 10 °C/min.

Neat processed PHB presents a melting peak temperature of 175 °C and a crystallinity index (X_c) of 49% at 0 days. After aging, X_c increases to 59%, and the melting peak temperature changes to 170 °C, due to secondary crystallization [44]. After erasing thermal history, the cooling of PHB yields a crystallization peak temperature of 117 °C and ΔH_c of 91 J/g, which corresponds to an X_c value of 63% (for either aged or unaged samples).

Regarding the influence of the reactive agents on X_c, at 0 days, PHB/DCP and PHB/HMDI show similar crystallinity indexes than neat PHB, whereas in PHB/RDGE and PHB/TGIC compounds, X_c is slightly higher. The melting temperatures, on the other hand, are similar to that corresponding to neat PHB, with the exception of DCP, which is slightly inferior. After 100 days of aging, the crystallinity index of PHB with the different reactive agents is in all cases inferior with respect to neat PHB, especially in case of DCP, for which melting parameters remain practically unchanged compared to the unaged sample. This can be related with the crosslinking effect of DCP, which generates free radicals and disrupts the linearity of the PHB chains, thus limiting the maximum crystallinity that can be developed [45].

The presence of the reactive agents (with or without fibers) seems to partially hinder the secondary crystallization [46], as the crystallinity index achieved at 100 days is in all cases inferior to that corresponding to neat PHB. Indeed, it seems that the sole presence of the fibers restricts the mobility of polymer chains, partly hindering the development of crystallization during aging, since the increase in the crystallinity index is not observed over time for the PHB/TC90 composites. According to the literature, fillers can restrict the polymer chain mobility [24], which could result in a stabilized crystallinity index over time.

After erasing the thermal history, no remarkable differences in crystallization temperatures or enthalpies were observed among the compounds, except in the case of HMDI addition. Compounds with HMDI showed lower T_c values than the other compositions, finding a reduction of T_c from 117 °C to 106 °C in PHB/HMDI and 111°C in PHB/TC90/HMDI samples. These findings can be related with some hindered motion of the polymer chains, thus suggesting the interaction of HMDI with the polymer matrix and the fibers [47].

With respect to melting in second heating scans, after low cooling rates where polymer chains had enough time and mobility to develop high crystallinity (10 °C/min during DSC test conditions), the crystallinity indexes and melting temperatures of the different compositions were similar to those corresponding to neat PHB, except for the compositions containing DCP. For DCP-containing compounds, significant reductions of T_m and X_c was detected, being in agreement with some crosslinking of the PHB matrix with the peroxide initiator DCP [45].

3.1.3. Mechanical Properties

PHB is known by its physical aging and secondary crystallization [15,16,43]. So, the mechanical performance depends on time after its processing. For such a reason, tensile properties have been assessed after processing (0 day) and after 15 days stored at room temperature. According to Corre et. al., after such a period of time, the variations on the mechanical properties are so small that it can be said that the properties are stabilized [16]. The mechanical properties of all compounds were determined by uniaxial tensile tests up to failure. Representative stress versus strain curves of the composites with TC90 are shown in Figure 3. The average parameters obtained from the curves are summarized in Table 3, and selected values are represented in Figure 4 to illustrate the trends observed.

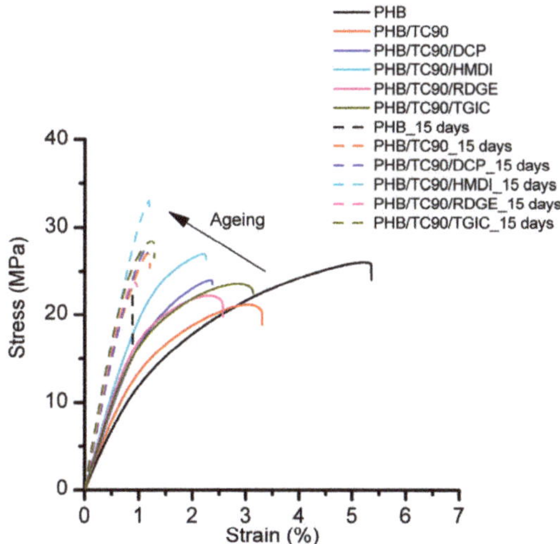

Figure 3. Representative stress–strain curves of neat PHB and PHB/TC90 composites at 0 days (solid lines) and after 15 days of aging (dashed lines).

Neat PHB shows a high variation in its tensile behavior due to aging. The elastic modulus increased more than 100% after 15 days, and the deformation at break decreased five times. In both cases, the failure mode is brittle, with unstable fracture and without showing necking nor evidence of shear yielding. The cause for such behavior is attributed to secondary crystallization and the physical aging of the amorphous region [15,16,43].

Table 3. Mechanical parameters corresponding to tensile tests.

	0 Days	15 Days	0 Days	15 Days	0 Days	15 Days	0 Days	15 Days
	Elastic Modulus (GPa)		Tensile Strength (MPa)		Elongation at Break (%)		Static Toughness (mJ/m^3)	
PHB	1.41	3.04	25.89	23.03	4.96	0.98	0.84	0.13
P309/DCP	1.54	2.75	24.88	29.77	3.87	1.55	0.62	0.27
P309/HMDI	1.82	3.15	23.36	24.36	2.50	0.86	0.38	0.10
P309/RDGE	1.54	2.90	23.42	29.32	3.62	1.55	0.57	0.27
P309/TGIC	1.48	3.03	24.76	30.95	4.28	1.43	0.72	0.25
P309/TC90	1.71	3.44	21.27	27.14	3.49	1.21	0.53	0.19
P309/TC90/DCP	1.92	3.19	23.64	26.83	2.23	1.15	0.34	0.18
P309/TC90/HMDI	2.31	3.49	26.75	32.11	2.04	1.16	0.35	0.20
P309/TC90/RDGE	2.10	3.40	21.71	25.06	2.42	1.07	0.37	0.16
P309/TC90/TGIC	2.02	3.59	23.89	28.06	3.22	1.22	0.56	0.21

When reactive agents are added to PHB, at 0 days, there is a minimum increase in elastic modulus in the case of addition of DCP, RDGE, and TGIC and a more pronounced one with HMDI (Table 3). However, after 15 days, PHB/DCP elastic modulus is 9% lower than neat PHB. The reactivity with the polymer matrix can account for this behavior. For DCP, the generation of some crosslinking is in agreement with lower development in crystallinity and hence a reduction on elastic modulus [45]. With respect to RDGE and TGIC, some reactivity is also evidenced, but in a way that it seems to affect the amorphous region, since after 15 days, they are able to withstand higher deformations prior to rupture and show higher tensile strength before a crack generates and propagates through the PHB. It could be hypothesized that more voluminous RDGE and TGIC disrupt the pseudo-order in the rigid amorphous phase region. This is not observed in PHB/HMDI compounds, probably because of the small and linear geometry of the HMDI molecule, which does not prevent the rearrangements of the amorphous phase that take place during physical aging [16,17,43,48].

When cellulose (TC90) is added to neat PHB (at 0 days), an increase in elastic modulus and a reduction of tensile strength and ductility is observed (Figure 3). The increase in modulus of elasticity is attributed to the reinforcement effect of the cellulose fibers. However, the addition of TC90 promotes rupture at even lower stress that in the case of Neat PHB (and therefore, much lower than yielding). Along with the increase of elastic modulus, this behavior suggest that cellulose acts as a reinforcement in PHB matrix at low strains, but after a certain point, it promotes the appearance of large defects that nucleate cracks that lead to brittle fracture, explaining the low values of tensile strength and elongation at break.

When reactive agents are added, the elastic modulus rises in all cases with respect to either cellulose without reactive agents or PHB with reactive agents. Similarly, the tensile strength also rises; both suggest an increase of affinity between the matrix and reinforcement. However, deformation at break does not increase, and it remains at a similar value as uncompatibilized TC90.

To better understand the role of TC90 and the different reactive agents on the mechanical behavior of PHB after aging (15 days), the elastic modulus, tensile strength, and elongation at break values of the composites are depicted in Figure 4.

After aging, the incorporation of TC90 fibers to PHB results in a slight improvement of elastic modulus and tensile strength, while showing a comparable elongation at break. In Figure 4a, it can be appreciated that the elastic modulus of all the composites is improved with respect to neat PHB (about 13% for PHB/TC90). This behavior can be reasonably ascribed to the affinity between the rigid cellulose fibers and the PHB matrix. The addition of the reactive agents to the PHB/TC90 compounds did not show any remarkable additional improvement of the elastic modulus at 15 days with respect to the composites without them. However, in the case of tensile strength values, an interesting increase of about 18% with respect to PHB/TC90 is detected in the PHB/TC90/HMDI compound. This rise suggests that HMDI had an effective compatibilizer role, strengthening the bonding of the cellulose fibers with the PHB matrix.

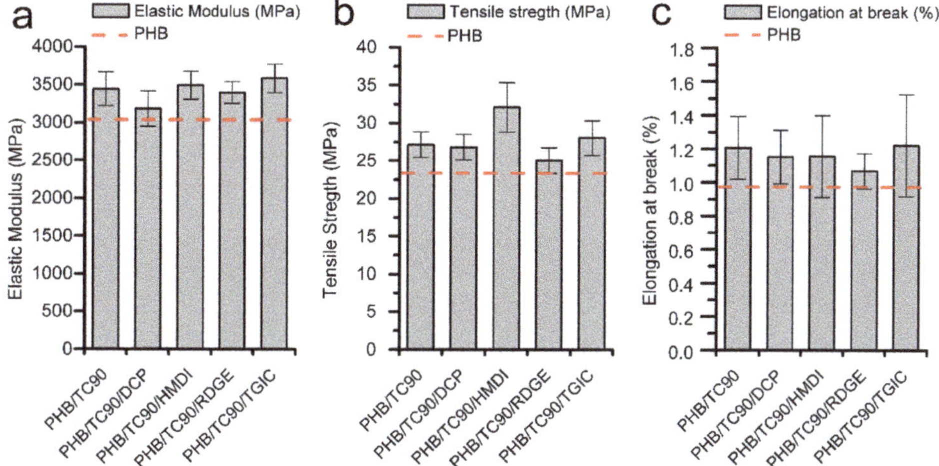

Figure 4. (**a**) Elastic modulus, (**b**) tensile strength, and (**c**) elongation at break of aged samples of neat PHB and PHB/TC90 composites with and without reactive agents.

The mechanical characterization of the composites was completed by DMA testing. Their storage modulus (G′) and the damping factor (tan δ) evolution with temperature are represented in Figure 5.

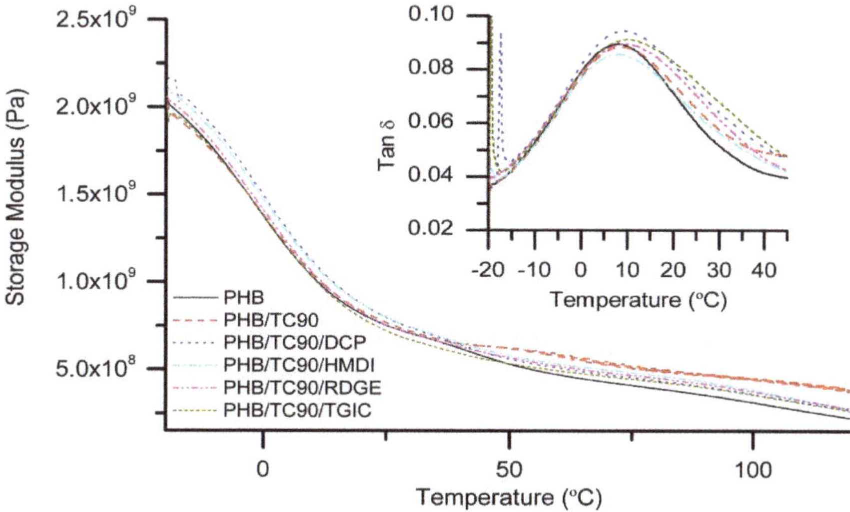

Figure 5. Storage modulus (G′) and tan δ (inset) evolution with temperature for neat PHB and PHB/TC90 composites with reactive agents.

No appreciable differences in storage modulus among the different samples are observed. These results are in agreement with the elastic modulus observed in mechanical tests. The tan δ represents the energy dissipated during the dynamic tests, and the tan δ peak is usually used to determine glass transition (T_g) in semicrystalline polymers [49]. In this case, all the samples present T_g

values around 10 °C. Nevertheless, the tan δ versus temperature plot suggest certain restricted mobility of the polymer chains in composites, since the tan δ peaks are broader than neat PHB [50]. Moreover, the height of the tan δ peak corresponding to PHB/TC90/HMDI is reduced compared to the rest of composites, indicating the further hindered motion of the polymer chains. This can be related with a better interaction between PHB and TC90 fibers due to the compatibilization effect of HMDI [30].

3.2. Influence of Fiber Purity

3.2.1. Morphological Analysis

As it has been discussed above, HMDI has demonstrated its efficiency at improving the interfacial adhesion between PHB and highly purified commercial cellulose fibers (TC90). In this section, the compatibilization ability of HMDI is tested using an unpurified rice husk fiber (U_RH) and a chemically treated rice husk fiber (T_RH) to study the combined effect of purification treatment and compatibilizer on the interfacial PHB/fiber interactions.

SEM micrographs (shown in Figure 6) allow visualizing qualitatively the matrix/fiber interface interactions of PHB/ U_RH and PHB/T_RH composites, with and without HMDI.

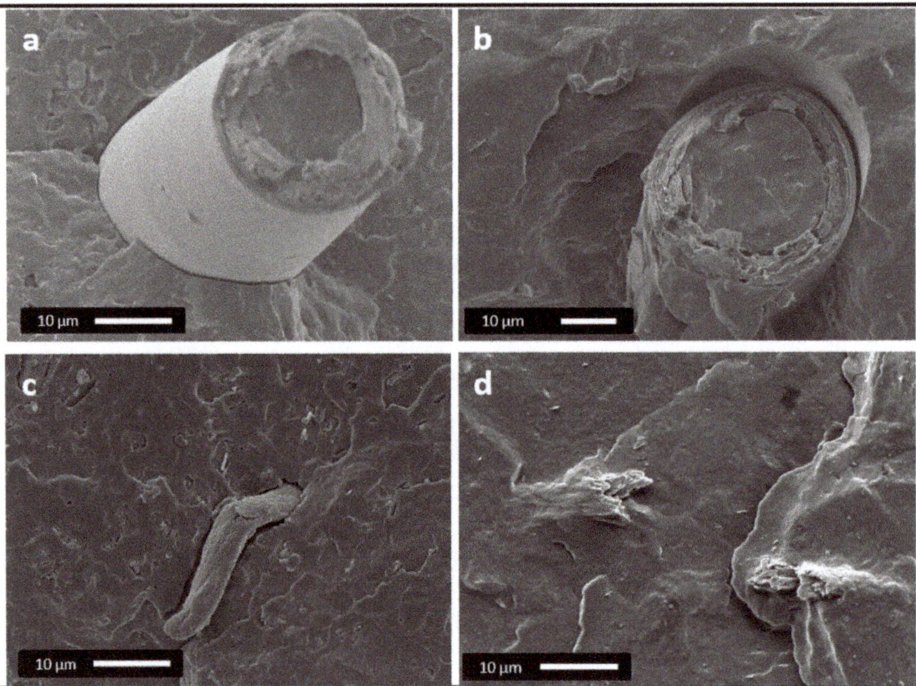

Figure 6. (a) SEM micrographs of PHB/U_RH, (b) PHB/U_RH/HMDI, (c) PHB/T_RH, and (d) PHB/T_RH/HMDI.

As it can be observed in Figure 6a, U_RH fibers within the compound present a smooth surface and show a gap between the fiber and the polymer matrix. A similar gap between these two components is also observed in PHB/U_RH/HMDI composition (Figure 6b). In case of treated fibers (PHB/T_RH), the fibers present a smaller diameter and a rougher surface than the untreated ones, but detachment of the fibers is also detected, thus indicating a certain lack of adhesion (Figure 6c). In the case of treated and compatibilized fibers (Figure 6d), they appear well covered by the polymer and no gap or signs of detachment at the interphase are detected, suggesting an improved adhesion.

3.2.2. Mechanical Properties

Uniaxial tensile tests of PHB/U_RH and PHB/T_RH with and without HMDI were performed on 15-day aged samples at room temperature. Elastic modulus, tensile strength, elongation at break, and the static toughness obtained from the area below the stress–strain representative curves for each composite are depicted in Figure 7. The mechanical parameters corresponding to neat PHB are also represented as a reference.

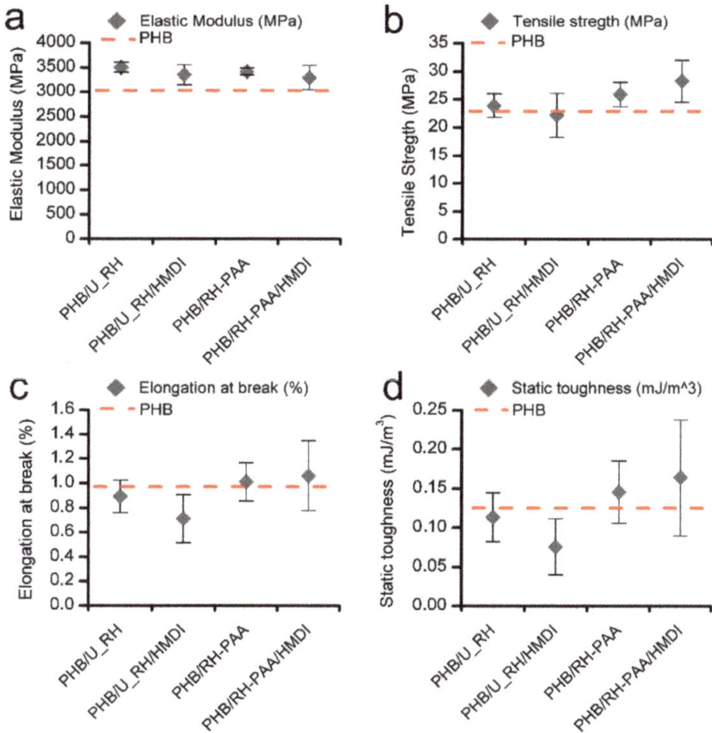

Figure 7. (**a**) Elastic modulus, (**b**) tensile strength, (**c**) elongation at break, and (**d**) static toughness of neat PHB, PHB/U_RH, and PHB/T_RH composites with and without reactive agents.

The addition of either U_RH or T_RH fibers produces a reinforcement effect with respect to neat PHB, since an increment of the modulus of elasticity is detected in all cases. On the other hand, the tensile strength seems to increase only when the treated fibers are the ones added, where the use of HMDI on such treated fibers produced an additional improvement of this parameter. For composites with treated fibers, the elongation at break remains simmilar to that of neat PHB, whereas for the untreated ones, this parameter is reduced (especially in the case with HMDI, in agreement with values reported for PHB/HMDI samples in Table 3). These results suggest that in the case of the untreated fibers, there is no interaction of HMDI between the polymer and the reinforcement.

Despite the fact that in all cases, the samples present a brittle behavior, the addition of the treated fibers leads to a visible trend of improvement of the static toughnes, compared with neat PHB, especially with the addition of HMDI. Nevertheless, the enhancement of the mechanical performance was not as pronounced as in the case of the compound prepared in the previous section, with high-purity commercial cellulose (PHB/TC90/HMDI).

3.2.3. Thermoforming Ability

PHB-based composites reinforced with fibers are considered an attractive alternative to commodities for short-term life applications such as packaging. For this reason, it is of particular interest to test their processability by thermoforming, which is a conventional technique that is usually applied in this industrial field. In this regard, PHB/T_RH/HMDI films have been thermoformed into trays following the procedure described in the experimental section, and the best results obtained are presented in Figure 8.

Figure 8. PHB/T_RH/HMDI thermoformed tray: (**a**) bottom-side view, (**b**) top-side view and (**c**) bottom view.

As it can be observed in Figure 8, thermoformed trays with good mold reproducibility and relatively good thickness distribution (punctual thickening is observed in Figure 8c) can be obtained by adjusting the operational parameters. In our case, the best results were obtained for a heat resistance temperature fixed at 600 °C, a heating time of 35 s, and 7 s applying vacuum. The difficulties of thermoforming semicrystalline polymers [18,51,52] and even more filled with fibers are well recognized, so the results obtained here are very promising.

3.2.4. Biodisintegration in Composting Conditions

One of the strengths of the PHB/fiber composites that make them especially attractive for packaging applications is their biodegradability and specifically their compostability. For this reason, it is of special interest to evaluate the effect of the reactive agents on the behavior under normalized composting conditions of the composites. Biodisintegration tests were conducted according to the ISO 20200 standard. The weight loss over time of the tested materials is represented in Figure 9, and pictures of the samples at different composting times are depicted in Figure 10.

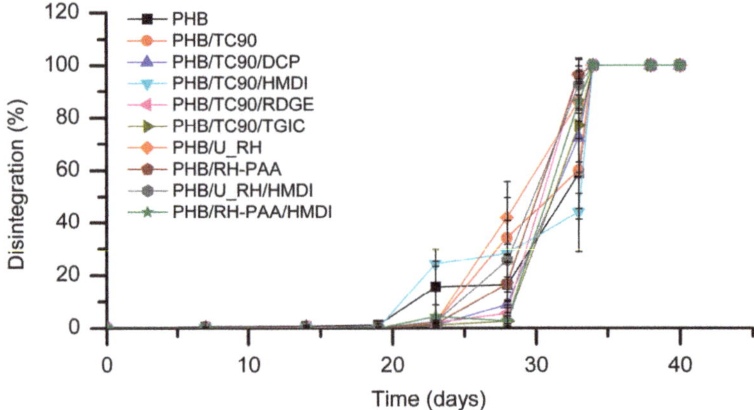

Figure 9. Disintegration of neat PHB and the studied composites over time under standard composting conditions (ISO 20200).

As shown in Figure 9, in neat PHB, the biodisintegration process occurs with an incubation period of about 23 days. From this time, appreciable weight loss is detected, and total disintegration (considered when fragments >2 mm are not detectable) is reached at 35 days of composting. For PHB/TC90 composites containing DCP, RDGE, and TGIC, the incubation period seems to be slightly longer. At intermediate composting times, some differences in weight loss among the samples were detected. However, these differences do not reveal a clear trend related with the presence or not of the different fibers or reactive agents. In any case, all the compositions reached total biodisintegration in the same period than neat PHB (35 days). Some authors have reported an accelerated biodegradation of biopolyesters with the incorporation of lignocellulosic fibers [23,25]. In our case, no remarkable differences were observed, as we had already previously noticed in PHBV/TC90 [53] and PHB/lignocellulosic composites [54].

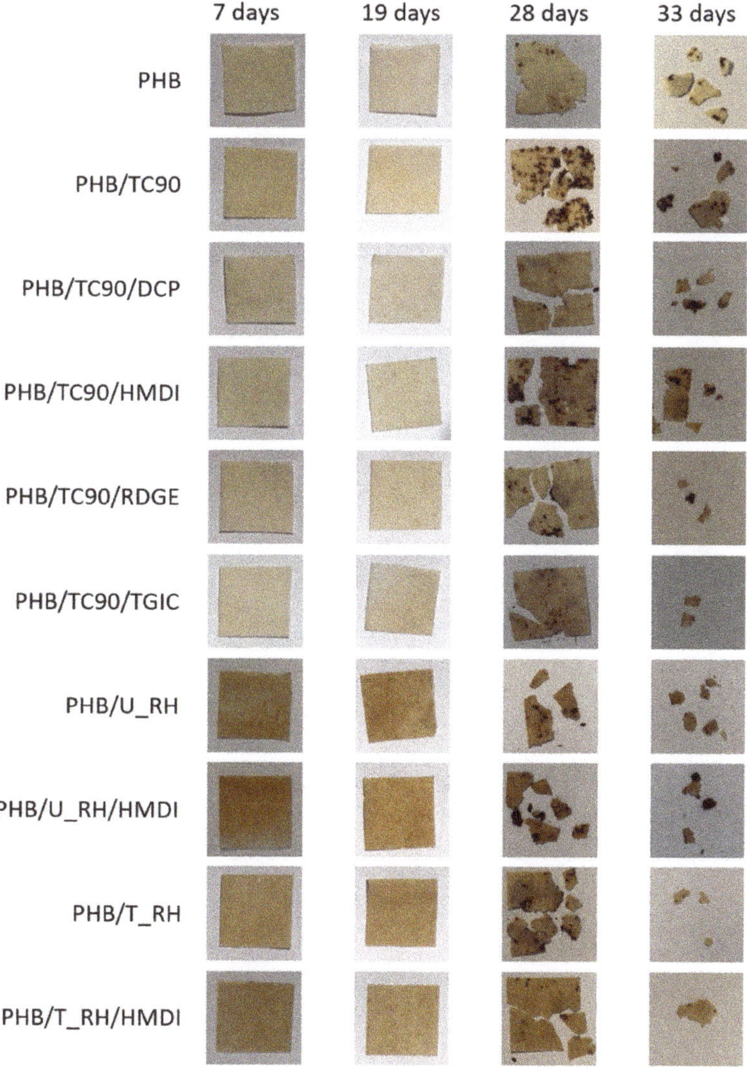

Figure 10. Pictures of neat PHB and the studied composites at different composting times.

As shown in Figure 10, no remarkable physical changes can be visually detected in the samples during the incubation period (pictures corresponding to 7 and 19 days of composting). At 28 days, clear deterioration of the samples is observed. The specimens are eroded and broken. As it has been reported in the literature, the biodisintegration process of PHB occurs by erosion from the surface to the bulk being the amorphous regions degraded in first place [55,56]. This leads to embrittlement and breakage of the samples. At 33 days of composting, all the samples appear broken into very small fragments.

According to these results, it can be concluded that neither the fibers nor the reactive agents have a negative effect on the biodisintegrability of PHB in composting conditions.

4. Discussion

Analyzing together the results obtained for the composites prepared with TC90, U_RH, and T_RH fibers with and without HMDI, it seems clear that the compatibility efficiency of the HMDI depends to a large extent on the purity of the fibers. Probably, this dependence is due to a greater presence of OH groups on the surface of the purified fibers, since it is assumed that compatibilization occurs through the formation of urethane bonds between the isocyanate groups of the HMDI and the hydroxyl groups of the fibers [33]. Tran et al. [57] also showed better results in PLA composites filled with rice and einkorn wheat husks using silane coupling agents, when fibers were previously submitted to an alkaline treatment.

As reported in previous works [54], U_RH fibers present waxes and impurities on their surface, which could be responsible for their poor adhesion with the PHB matrix. In addition, in lignocellulosic fibers, the cellulose is embedded in a matrix of non-cellulosic components (mainly composed by lignin, hemicelluloses, and pectin). All this limits the exposition of the reactive –OH groups of the cellulose for compatibilization. The purification treatment applied removed those waxes and impurities on the fibers surface, as well as most of the non-cellulosic components, leaving a greater amount of reactive –OH groups at the surface that could react with the compatibilizer. In addition, the treatment resulted in fibers with more favorable morphology and surface characteristics (i.e., a higher aspect ratio and roughness) that also contributed to improve the mechanical properties, even without the presence of a compatibilizer.

5. Conclusions

In this work, PHB/TC90 composites were obtained by reactive extrusion using four different reactive agents: dicumyl peroxide (DCP), hexamethylene diisocyanate (HMDI), resorcinol diglycidyl ether (RDGE), and triglycidyl isocyanurate (TGIC). The effect of aging attending on the influence of the reactive agents and the fibers on the mechanical and crystallization behavior of the composites was assessed. Aging produced a drastic embrittlement of the PHB characterized by increased elastic modulus and decreased tensile strength, elongation at break, and toughness. This embrittlement was attributed to secondary crystallization, which increased the value of X_c by about 17% in neat PHB samples, but also to the physical aging of the amorphous fraction. Both fibers and reactive agents partly hinder secondary crystallization, leading to lower crystallinity in aged composites with respect to neat PHB. The melting temperature and crystallinity index for the compositions containing DCP were reduced at high cooling rates (processing conditions) and low cooling rates (DSC conditions), suggesting crosslinking of the PHB matrix.

No clear compatibilization effect was found for the reactive agents DCP, RDGE, and TGIC. Elastic modulus, tensile strength, and elongation at break remained practically unchanged for PHB/TC90/DCP and PHB/TC90/TGIC compared with PHB/TC90. In the case of PHB/TC90/RDGE, those values were even lower than PHB/TC90. On the contrary, the tensile strength in PHB/TC90/HMDI was improved by about 40% with respect to neat PHB and around 18% with respect to the PHB/TC90 composite. The improved mechanical performance together with the SEM observations indicate enhanced interfacial adhesion between the TC90 fibers and PHB.

A clear influence of the purity of the fibers on the effectiveness of HMDI at improving interfacial adhesion was observed. HMDI only demonstrated effectiveness in purified cellulose fiber composites, the greatest improvements in the mechanical performance being the ones obtained with highly purified commercial cellulose fibers (TC90).

The different reactive agents did not negatively affect the compostability of PHB.

Finally, thermoformed trays with good mold reproducibility were successfully obtained for PHB/T_RH/HMDI composite.

Author Contributions: Conceptualization, J.G.-P., L.C. and J.M.L.; methodology, E.L.S.-S.; validation, J.G.-P., L.C. and J.M.L.; formal analysis, E.L.S.-S.; investigation, E.L.S.-S. and A.A.; resources, J.G.-P., L.C. and J.M.L.; data curation, E.L.S.-S. and A.A.; writing—original draft preparation, E.L.S.-S.; writing—review and editing, J.G.-P., L.C. and J.M.L.; supervision, J.G.-P., L.C. and J.M.L.; funding acquisition, J.G.-P., L.C. and J.M.L. All authors have read and agreed to the published version of the manuscript.

Funding: The authors would like to thank the financial support for this research from Ministerio de Ciencia e Innovación (RTI2018-097249-B-C22), Pla de Promoció de la Investigació de la Universitat Jaume I (UJI-B2019-44) and H2020 EU Project YPACK (H2020-SFS-2017-1, Reference 773872).

Acknowledgments: The authors would like to acknowledge the Instituto de Tecnología de Materiales of Universitat Politècnica de València-Campus de Alcoy, the Unidad Asociada IATA-UJI "Polymers Technology" and Servicios Centrales de Instrumentación Científica (SCIC) of Universitat Jaume I. We are also grateful to Raquel Oliver and Jose Ortega for experimental support.

Conflicts of Interest: The authors declare no conflict of interest.

References

1. Muneer, F.; Rasul, I.; Azeem, F.; Siddique, M.H.; Zubair, M.; Nadeem, H. Microbial Polyhydroxyalkanoates (PHAs): Efficient Replacement of Synthetic Polymers. *J. Polym. Environ.* **2020**, *28*, 2301–2323. [CrossRef]

2. Kumar, M.; Rathour, R.; Singh, R.; Sun, Y.; Pandey, A.; Gnansounou, E.; Andrew Lin, K.Y.; Tsang, D.C.W.; Thakur, I.S. Bacterial polyhydroxyalkanoates: Opportunities, challenges, and prospects. *J. Clean. Prod.* **2020**, *263*, 121500. [CrossRef]

3. Rujnić-Sokele, M.; Pilipović, A. Challenges and opportunities of biodegradable plastics: A mini review. *Waste Manag. Res.* **2017**, *35*, 132–140. [CrossRef] [PubMed]

4. Albuquerque, P.B.S.; Malafaia, C.B. Perspectives on the production, structural characteristics and potential applications of bioplastics derived from polyhydroxyalkanoates. *Int. J. Biol. Macromol.* **2018**, *107*, 615–625. [CrossRef] [PubMed]

5. Peelman, N.; Ragaert, P.; Ragaert, K.; De Meulenaer, B.; Devlieghere, F.; Cardon, L. Heat resistance of new biobased polymeric materials, focusing on starch, cellulose, PLA, and PHA. *J. Appl. Polym. Sci.* **2015**, *132*. [CrossRef]

6. Keskin, G.; Kızıl, G.; Bechelany, M.; Pochat-Bohatier, C.; Öner, M. Potential of polyhydroxyalkanoate (PHA) polymers family as substitutes of petroleum based polymers for packaging applications and solutions brought by their composites to form barrier materials. *Pure Appl. Chem.* **2017**, *89*, 1841–1848. [CrossRef]

7. Bugnicourt, E.; Cinelli, P.; Lazzeri, A.; Alvarez, V. Polyhydroxyalkanoate (PHA): Review of synthesis, characteristics, processing and potential applications in packaging. *Express Polym. Lett.* **2014**, *8*, 791–808. [CrossRef]

8. Wang, Y.; Yin, J.; Chen, G.Q. Polyhydroxyalkanoates, challenges and opportunities. *Curr. Opin. Biotechnol.* **2014**, *30*, 59–65. [CrossRef]

9. Keshavarz, T.; Roy, I. Polyhydroxyalkanoates: Bioplastics with a green agenda. *Curr. Opin. Microbiol.* **2010**, *13*, 321–326. [CrossRef]

10. Możejko-Ciesielska, J.; Kiewisz, R. Bacterial polyhydroxyalkanoates: Still fabulous? *Microbiol. Res.* **2016**, *192*, 271–282. [CrossRef]

11. Dilkes-Hoffman, L.S.; Lant, P.A.; Laycock, B.; Pratt, S. The rate of biodegradation of PHA bioplastics in the marine environment: A meta-study. *Mar. Pollut. Bull.* **2019**, *142*, 15–24. [CrossRef] [PubMed]

12. Arcos-Hernandez, M.V.; Laycock, B.; Pratt, S.; Donose, B.C.; Nikolić, M.A.; Luckman, P.; Werker, A.; Lant, P.A. Biodegradation in a soil environment of activated sludge derived polyhydroxyalkanoate (PHBV). *Polym. Degrad. Stab.* **2012**, *97*, 2301–2312. [CrossRef]

13. Hermann, B.G.; Debeer, L.; De Wilde, B.; Blok, K.; Patel, M.K. To compost or not to compost: Carbon and energy footprints of biodegradable materials' waste treatment. *Polym. Degrad. Stab.* **2011**, *96*, 1159–1171. [CrossRef]

14. De Koning, G.J.M.; Scheeren, A.H.C.; Lemstra, P.J.; Peeters, M.; Reynaers, H. Crystallization phenomena in bacterial poly[(R)-3-hydroxybutyrate]: 3. Toughening via texture changes. *Polymer* **1994**, *35*, 4598–4605. [CrossRef]

15. Alata, H.; Aoyama, T.; Inoue, Y. Effect of Aging on the Mechanical Properties of Poly(3-hydroxybutyrate-co-3-hydroxyhexanoate). *Macromolecules* **2007**, *40*, 4546–4551. [CrossRef]

16. Corre, Y.-M.; Bruzaud, S.; Audic, J.-L.; Grohens, Y. Morphology and functional properties of commercial polyhydroxyalkanoates: A comprehensive and comparative study. *Polym. Test.* **2012**, *31*, 226–235. [CrossRef]

17. Esposito, A.; Delpouve, N.; Causin, V.; Dhotel, A.; Delbreilh, L.; Dargent, E. From a Three-Phase Model to a Continuous Description of Molecular Mobility in Semicrystalline Poly(hydroxybutyrate-co-hydroxyvalerate). *Macromolecules* **2016**, *49*, 4850–4861. [CrossRef]

18. González-Ausejo, J.; Sanchez-Safont, E.; Lagaron, J.M.; Olsson, R.T.; Gamez-Perez, J.; Cabedo, L. Assessing the thermoformability of poly(3-hydroxybutyrate-co-3-hydroxyvalerate)/poly(acid lactic) blends compatibilized with diisocyanates. *Polym. Test.* **2017**, *62*, 235–245. [CrossRef]

19. Pereira, P.H.F.; de Freitas, M.R.; Cioffi, M.O.H.; de Caravalho, K.C.C.B.; Milanese, A.C.; Voorwald, H.J.C.; Mulinari, D.R. Vegetal fibers in polymeric composites: A review. *Polímeros* **2015**, *25*, 9–22. [CrossRef]

20. Mohanty, A.K.; Misra, M.; Hinrichsen, G. Biofibres, biodegradable polymers and biocomposites: An overview. *Macromol. Mater. Eng.* **2000**, *276–277*, 1–24. [CrossRef]

21. Lagarón, J.M.; Lopez-Rubio, A.; Fabra, M.J. Bio-based packaging. *J. Appl. Polym. Sci.* **2015**, *133*. [CrossRef]

22. Bhardwaj, R.; Mohanty, A.K.; Drzal, L.T.; Pourboghrat, F.; Misra, M. Renewable resource-based green composites from recycled cellulose fiber and poly(3-hydroxybutyrate-co-3-hydroxyvalerate) bioplastic. *Biomacromolecules* **2006**, *7*, 2044–2051. [CrossRef] [PubMed]

23. Gunning, M.A.; Geever, L.M.; Killion, J.A.; Lyons, J.G.; Higginbotham, C.L. Mechanical and biodegradation performance of short natural fibre polyhydroxybutyrate composites. *Polym. Test.* **2013**, *32*, 1603–1611. [CrossRef]

24. Torres-Tello, E.V.; Robledo-Ortíz, J.R.; González-García, Y.; Pérez-Fonseca, A.A.; Jasso-Gastinel, C.F.; Mendizábal, E. Effect of agave fiber content in the thermal and mechanical properties of green composites based on polyhydroxybutyrate or poly(hydroxybutyrate-co-hydroxyvalerate). *Ind. Crops Prod.* **2017**, *99*, 117–125. [CrossRef]

25. Batista, K.C.; Silva, D.A.K.; Coelho, L.A.F.; Pezzin, S.H.; Pezzin, A.P.T. Soil Biodegradation of PHBV/Peach Palm Particles Biocomposites. *J. Polym. Environ.* **2010**, *18*, 346–354. [CrossRef]

26. Barkoula, N.M.; Garkhail, S.K.; Peijs, T. Biodegradable composites based on flax/polyhydroxybutyrate and its copolymer with hydroxyvalerate. *Ind. Crops Prod.* **2010**, *31*, 34–42. [CrossRef]

27. Michel, A.T.; Billington, S.L. Characterization of poly-hydroxybutyrate films and hemp fiber reinforced composites exposed to accelerated weathering. *Polym. Degrad. Stab.* **2012**, *97*, 870–878. [CrossRef]

28. Rastogi, V.K.; Samyn, P. Novel processing of polyhydroxybutyrate with micro-to nanofibrillated cellulose and effect of fiber morphology on crystallization behaviour of composites. *Express Polym. Lett.* **2020**, *14*, 115–133. [CrossRef]

29. Wei, L.; McDonald, A. A Review on Grafting of Biofibers for Biocomposites. *Materials* **2016**, *9*, 303. [CrossRef]

30. Anderson, S.; Zhang, J.; Wolcott, M.P. Effect of Interfacial Modifiers on Mechanical and Physical Properties of the PHB Composite with High Wood Flour Content. *J. Polym. Environ.* **2013**, *21*, 631–639. [CrossRef]

31. Muthuraj, R.; Misra, M.; Mohanty, A.K. Reactive compatibilization and performance evaluation of miscanthus biofiber reinforced poly(hydroxybutyrate-co-hydroxyvalerate) biocomposites. *J. Appl. Polym. Sci.* **2017**, *134*. [CrossRef]

32. Jiang, L.; Chen, F.; Qian, J.; Huang, J.; Wolcott, M.; Liu, L.; Zhang, J. Reinforcing and Toughening Effects of Bamboo Pulp Fiber on Poly(3-hydroxybutyrate-co-3-hydroxyvalerate) Fiber Composites. *Ind. Eng. Chem. Res.* **2010**, *49*, 572–577. [CrossRef]

33. Nagarajan, V.; Misra, M.; Mohanty, A.K. New engineered biocomposites from poly(3-hydroxybutyrate-co-3-hydroxyvalerate) (PHBV)/poly(butylene adipate-co-terephthalate) (PBAT) blends and switchgrass: Fabrication and performance evaluation. *Ind. Crops Prod.* **2013**, *42*, 461–468. [CrossRef]

34. Hao, M.; Wu, H. Effect of in situ reactive interfacial compatibilization on structure and properties of polylactide/sisal fiber biocomposites. *Polym. Compos.* **2017**, *39*, E174–E187. [CrossRef]

35. Hao, M.; Wu, H.; Qiu, F.; Wang, X. Interface Bond Improvement of Sisal Fibre Reinforced Polylactide Composites with Added Epoxy Oligomer. *Materials* **2018**, *11*, 398. [CrossRef]

36. Wei, L.; Stark, N.M.; McDonald, A.G. Interfacial improvements in biocomposites based on poly(3-hydroxybutyrate) and poly(3-hydroxybutyrate-co-3-hydroxyvalerate) bioplastics reinforced and grafted with α-cellulose fibers. *Green Chem.* **2015**, *17*, 4800–4814. [CrossRef]

37. Sánchez-Safont, E.L.; Aldureid, A.; Lagarón, J.M.; Gamez-Perez, J.; Cabedo, L. Effect of the Purification Treatment on the Valorization of Natural Cellulosic Residues as Fillers in PHB-Based Composites for Short Shelf Life Applications. *Waste Biomass Valorization* **2020**, *1*, 3. [CrossRef]

38. Zhao, X.; van der Heide, E.; Zhang, T.; Liu, D. Delignification of sugarcane bagasse with alkali and peracetic acid and characterization of the pulp. *BioResources* **2010**, *5*, 1565–1580.

39. Mariano, M.; Cercená, R.; Soldi, V. Thermal characterization of cellulose nanocrystals isolated from sisal fibers using acid hydrolysis. *Ind. Crops Prod.* **2016**, *94*, 454–462. [CrossRef]

40. Carli, L.N.; Crespo, J.S.; Mauler, R.S. PHBV nanocomposites based on organomodified montmorillonite and halloysite: The effect of clay type on the morphology and thermal and mechanical properties. *Compos. Part A Appl. Sci. Manuf.* **2011**, *42*, 1601–1608. [CrossRef]

41. UNE-EN ISO UNE-EN ISO 20200. *Determinación del Grado de Desintegración de Materiales Plásticos Bajo Condiciones de Compostaje Simuladas en un Laboratorio*; Amigos Museo del Prado: Madrid, Spain, 2006.

42. Zhang, K.; Misra, M.; Mohanty, A.K. Toughened sustainable green composites from poly(3-hydroxybutyrate-co-3-hydroxyvalerate) based ternary blends and miscanthus biofiber. *ACS Sustain. Chem. Eng.* **2014**, *2*, 2345–2354. [CrossRef]

43. Crétois, R.; Chenal, J.M.; Sheibat-Othman, N.; Monnier, A.; Martin, C.; Astruz, O.; Kurusu, R.; Demarquette, N.R. Physical explanations about the improvement of PolyHydroxyButyrate ductility: Hidden effect of plasticizer on physical ageing. *Polymers* **2016**, *102*, 176–182. [CrossRef]

44. De Koning, G.J.M.; Lemstra, P.J. Crystallization phenomena in bacterial poly[(R)-3-hydroxybutyrate]: 2. Embrittlement and rejuvenation. *Polymer* **1993**, *34*, 4089–4094. [CrossRef]

45. Fei, B.; Chen, C.; Chen, S.; Peng, S.; Zhuang, Y.; An, Y.; Dong, L. Crosslinking of poly[(3-hydroxybutyrate)-co-(3-hydroxyvalerate)] using dicumyl peroxide as initiator. *Polym. Int.* **2004**, *53*, 937–943. [CrossRef]

46. Crist, B.; Schultz, J.M. Polymer spherulites: A critical review. *Prog. Polym. Sci.* **2016**, *56*, 1–63. [CrossRef]

47. Berthet, M.-A.; Angellier-Coussy, H.; Machado, D.; Hilliou, L.; Staebler, A.; Vicente, A.A.; Gontard, N. Exploring the potentialities of using lignocellulosic fibres derived from three food by-products as constituents of biocomposites for food packaging. *Ind. Crop. Prod.* **2015**, *69*, 110–122. [CrossRef]

48. Srubar, W.V.; Wright, Z.C.; Tsui, A.; Michel, A.T.; Billington, S.L.; Frank, C.W. Characterizing the effects of ambient aging on the mechanical and physical properties of two commercially available bacterial thermoplastics. In *Polymer Degradation and Stability, Proceedings of the 3rd International Conference on Biodegradable and Biobased Polymers (BIOPOL-2011), Strasbourg, Austria, 29–31 August 2011*; Elsevier: Amsterdam, The Netherlands, 2012; Volume 97, pp. 1922–1929.

49. Saba, N.; Jawaid, M.; Alothman, O.Y.; Paridah, M.T. A review on dynamic mechanical properties of natural fibre reinforced polymer composites. *Constr. Build. Mater.* **2016**, *106*, 149–159. [CrossRef]

50. Yu, L. *Biodegradable Polymer Blends and Composites from Renewable Resources*; John Wiley and Sons: Hoboken, NJ, USA, 2009; ISBN 9780470146835.

51. Giménez, E.; Lagarón, J.M.; Cabedo, L.; Gavara, R.; Saura, J.J. Study of the thermoformability of ethylene-vinyl alcohol copolymer based barrier blends of interest in food packaging applications. *J. Appl. Polym. Sci.* **2004**, *91*, 3851–3855. [CrossRef]

52. Giménez, E.; Lagarón, J.M.; Maspoch, M.L.; Cabedo, L.; Saura, J.J. Uniaxial tensile behavior and thermoforming characteristics of high barrier EVOH-based blends of interest in food packaging. *Polym. Eng. Sci.* **2004**, *44*, 598–608. [CrossRef]

53. Sánchez-Safont, E.L.; González-Ausejo, J.; Gámez-Pérez, J.; Lagarón, J.M.; Cabedo, L. Poly(3-Hydroxybutyrate-co-3-Hydroxyvalerate)/Purified Cellulose Fiber Composites by Melt Blending: Characterization and Degradation in Composting Conditions. *J. Renew. Mater.* **2016**, *4*, 123–132. [CrossRef]

54. Sánchez-Safont, E.L.; Aldureid, A.; Lagarón, J.M.; Gámez-Pérez, J.; Cabedo, L. Biocomposites of different lignocellulosic wastes for sustainable food packaging applications. *Compos. Part B Eng.* **2018**, *145*, 215–225. [CrossRef]

55. Weng, Y.-X.; Wang, Y.-Z.Y.; Wang, X.-L.; Wang, Y.-Z.Y. Biodegradation behavior of PHBV films in a pilot-scale composting condition. *Polym. Test.* **2010**, *29*, 579–587. [CrossRef]

56. Arrieta, M.P.; López, J.; Rayón, E.; Jiménez, A. Disintegrability under composting conditions of plasticized PLA–PHB blends. *Polym. Degrad. Stab.* **2014**, *108*, 307–318. [CrossRef]

57. Tran, T.P.T.; Bénézet, J.-C.; Bergeret, A. Rice and Einkorn wheat husks reinforced poly(lactic acid) (PLA) biocomposites: Effects of alkaline and silane surface treatments of husks. *Ind. Crops Prod.* **2014**, *58*, 111–124. [CrossRef]

Article

Injection Molding of Coir Coconut Fiber Reinforced Polyolefin Blends: Mechanical, Viscoelastic, Thermal Behavior and Three-Dimensional Microscopy Study

Miguel A. Hidalgo-Salazar [1,*], Juan P. Correa-Aguirre [1], Serafín García-Navarro [2] and Luis Roca-Blay [2]

[1]   Research Group for Manufacturing Technologies GITEM, Universidad Autónoma de Occidente, Cali, DC 760030, USA; jpcorrea@uao.edu.co
[2]   AIMPLAS, Gustave Eiffel 4 (València Parc Tecnològic), 46980 Paterna, Spain; sgarcia@aimplas.es (S.G.-N.); lroca@aimplas.es (L.R.-B.)
*   Correspondence: mahidalgo@uao.edu.co; Tel.: +57-2-3188000

Received: 14 May 2020; Accepted: 23 June 2020; Published: 7 July 2020

Abstract: In this study, the properties of a polyolefin blend matrix (PP-HDPE) were evaluated and modified through the addition of raw coir coconut fibers-(CCF). PP-HDPE-CCF biocomposites were prepared using melt blending processes with CCF loadings up to 30% (*w/w*). CCF addition generates an increase of the tensile and flexural modulus up to 78% and 99% compared to PP-HDPE blend. This stiffening effect is caused by a decrease in the polymeric chain mobility due to CCF, the higher mechanical properties of the CCF compared to the polymeric matrix and could be an advantage for some biocomposites applications. Thermal characterizations show that CCF incorporation increases the PP-HDPE thermal stability up to 63 °C, slightly affecting the melting behavior of the PP and HDPE matrix. DMA analysis shows that CCF improves the PP-HDPE blend capacity to absorb higher external loads while exhibiting elastic behavior maintaining its characteristics at higher temperatures. Also, the three-dimensional microscopy study showed that CCF particles enhance the dimensional stability of the PP-HDPE matrix and decrease manufacturing defects as shrinkage in injected specimens. This research opens a feasible opportunity for considering PP-HDPE-CCF biocomposites as alternative materials for the design and manufacturing of sustainable products by injection molding.

Keywords: biocomposites; mechanical properties; thermal properties; natural fibers; injection molding

---

1. Introduction

A polymeric blend is a material formed from the physical combination of at least two polymers. These materials are used in various technological applications due to the possibility of modifying several properties considering the main characteristics of the polymers in the blend and their mixing ratio. Some studies estimated that currently, a large quantity of the polymers in the global market are sold as polymeric blends [1]. In the research context, these materials have been studied by several authors in the last years [2–4]. According to these studies, polymeric blends can be classified as immiscible (heterogeneous), compatible, or miscible (homogeneous), been most of these materials (around 90%) immiscible or multiphase systems with partial miscibility [1,3,4].

According to a Colombian non-profit organization that represents the companies related to chemical sector including plastics, rubber, paints, inks (coatings), and fibers, ACOPLASTICOS, the Colombian production of plastic resins was about 1.34 million tons in 2017 and polyolefins (low density polyethylene—LDPE, high density polyethylene and polypropylene—PP) represent around 42% of this production capacity [5]. Also, those polymers represent most of the post-consumer plastic wastes and the separation into the individual polymer and complete sorting during mechanical recycling processes are expensive and sometimes impossible. However, valorization of these polyolefins is possible because they can be easily recycled by converting them into good performance polymer blends [6,7].

PP-PE blends have been studied for several researchers in the last years. Some of these studies show the limited miscibility of PP-LDPE blends [8]. However, most of the revised literature shows that PP and PE are immiscible, resulting in phase separation during melt blending, low adhesion between the constituents' phases, and poor mechanical properties [9–11]. Nonetheless, due to their availability, recyclability, sustainable character, and low-cost, PP-PE blends could become strategic materials for several industry applications facing a circular plastics economy [7,12–14].

The reinforcement of polymeric matrices with natural fibers as coir coconut fiber, hemp, sisal, pineapple, sugarcane bagasse, fique, wood flour and their combinations has been studied during the last years [15–22]. These materials are known as natural fiber reinforced polymer composites (NFRPCs) or biocomposites and have the potential to be used in several applications as automobile parts, construction, and furniture due to the lower cost of natural fibers in comparison with traditional fibers and the enhancement of the polymeric matrices properties induced by natural fibers incorporation [22,23]. Also, biocomposites provide some advantages such as reduction overweight, less dependence on oil resources, lower costs and CO_2 emissions, recycling, among others [24–26].

The coconut fruit constituents are the white meat (28% wt), which is protected by the shell (12% wt) and the coir (35% wt). Also, the raw coconut husk is formed by coir fibers (30% wt) and a cork-like material called pith (70% wt). Coir Coconut Fibers (CCF) main constituents are cellulose (42% wt), hemicellulose (0.25% wt), lignin (47% wt), ashes (2% wt), pectin (3% wt) and about 5% wt of moisture [17]. According to Alvarado [27], estimated coconut production in Colombia was around 139,000 metric tons, which generates at least 50,000 metric tons of coconut husks mainly used in the hydroponic industry, soil stabilization, compost, and fuel.

Several studies exploring the characterization of thermoplastic polymers-CCF biocomposites have been reported. The term biocomposite is often used to name polymeric reinforced composites, where the reinforcing phase and/or the matrix are derived from materials of biological origin [28,29]. According to the reviewed literature, several studies have reported the formulation and characterization of biocomposites, which have a status of renewable and sustainable materials since they are composed of natural fibers embedded in non-degradable and biodegradable polymeric matrices [30–33].

Mir et al. [33] studied the incorporation of CCF (up to 20% wt) in a PP matrix using thermocompression. The CCF were chemically treated with chromium sulfate and sodium bicarbonate in hydrochloric acid media to improve the compatibility between fibers and PP. Their results show that the tensile properties of PP-CCF biocomposites change with the fiber load. For the biocomposite PP-CCF 10% wt, the tensile strength increases 11% as compared with PP, whereas for PP-CCF 20% wt, the tensile strength drops to values lower than those of PP.

Haque et al. explored the effect of fiber content (up to 30% wt) and chemical modification of the fibers with sodium hydroxide and a benzene diazonium salt on the mechanical properties of biocomposites based on abaca fiber, CCF, and PP obtained by extrusion and injection molding [34]. The results showed that CCF generates better mechanical properties in biocomposites than abaca fiber. Tensile tests show that chemically modified CCF increases tensile strength up to 10% and tensile modulus up to 250% compared to PP. Finally, the authors conclude that based on fiber loading, biocomposites with 30% wt of fibers had the best set of mechanical properties among the materials studied. In another study, Perez-Fonseca et al. report the effect of the hybridization of CCF with agave fiber (up to 30%) and the addition of a coupling agent (maleated polyethylene, MAPE) on the water absorption and mechanical properties on HDPE based biocomposites obtained by extrusion and injection molding. Their results show that fibers and MAPE combination generates biocomposites with enhanced tensile and flexural strengths while lowering water absorption of the biocomposites.

The reviewed literature presents an overview of the characterization of biocomposites based on chemically modified CCF [17,35]. However, less effort has been focused on the study and production of biocomposites based on untreated CCF and polyolefin blends using high-volume manufacturing processes such as extrusion and injection molding, which could be an advantage for the development of these materials in conventional plastic processing companies. In the present study, biocomposites based on a PP-HDPE blend and untreated CCF were prepared using extrusion following by injection molding. The thermal, mechanical, viscoelastic and morphological properties of the obtained materials were studied in order to evaluate CCF addition effect on the PP-HDPE blend behavior. Also, the analysis of the shrinkage and the dimensional stability of the injected specimens were studied through a novel three-dimensional microscopy study.

2. Materials and Methods

2.1. Materials

The polymeric materials used were an injection grade PP reference 575P and an injection grade HDPE reference M80064s with melt flow indexes (MFI) of 4.8 and 8.8 g/10 min respectively (measured at 190 °C, 2.16 kg). Both polymeric materials were purchased from SABIC (Al-Jubail, Saudi Arabia). The raw CCF, shown in Figure 1, were kindly supplied by "Super de Alimentos" (Manizales, Colombia) and were generated in the coconut candies production process. Before PP-HDPE-CCF biocomposites formulation, the CCF was grinded and sieved through a 400 μm sieve.

Figure 1. Raw coir coconut.

2.2. Methodology

Neat Polymers and Biocomposites Processing

The extrusion process of the different materials was performed in a co-rotating twin-screw extruder with a L:D ratio 40:1 and a screw diameter of 22 mm, equipped with two volumetric feeders and a pelletizer located next to the die zone. During the extrusion, the following parameters were fixed:

- Neat PP, HDPE and a 50–50 (% *w/w*) PP-HDPE blend pellets were fed in the extruder feeding zone using a volumetric feeder
- For PP-HDPE-CCF biocomposites, the CCF was fed with another volumetric side feeder at L/D 20. The side feeder speed was variated to obtain PP-HDPE-CCF biocomposites with CCF loadings of 10%, 20%, and 30% (*w/w*)
- Temperature: 165 (feeding zone) to 185 °C (die)
- Twin-screw rotation speed: 50 rpm

Then the pellets of the different materials were dried in an X-DRY AIR mini dryer (Moretto, Massanzago, Italy) at 60 °C and a dew point of −52 °C. The specimens used for mechanical characterizations (described in Section 2.3.2) were injected in a microinjection molding machine BOY XS (BOY Machines Inc., Exton, PA, USA) with the following parameters:

- Temperature: 180 (feeding zone) and 185 °C (nozzle).

- Filling pressure: 60 to 80 bars
- Holding pressure: 60 bars
- Clamping force: 30 kN

Figure 2 shows the injected flexural specimens of each material. The processing temperatures were set below 200 °C to avoid thermal degradation of the CCF (see Section 3.1), the remaining processing parameters were set based on reviewed literature regarding the processing optimization of natural fiber-polyolefin biocomposites [16,36].

The injection molding process allows to obtain complex geometric parts with fast function elements and in large quantities [37]. It offers several advantages over other manufacturing process as compression molding. According to Pickering [37], the content of natural fibers that can be incorporated during the injection process is between 20 to 40% by weight. A higher content is not recommended since an increase in fiber clearly reduces the flow capacity of the melt, generating instabilities during the process. Besides this, natural fibers have a similar morphology, but differ from each other by factors such as the internal area of the lumens, the number of lumens, the number and size of the fiber cells, the thickness of the secondary cell walls and the actual cross section [38]. These characteristics influence different fiber properties such as mechanical properties and bulk density which is related to the packing capacity of the fiber and the maximum throughput of the fibers [36]. In our case, the CCF throughput within the extruder was optimized to obtain PP-CCF biocomposites with a maximum fiber weight percentage of 30% and considerable flow properties to successfully apply in the injection molding process.

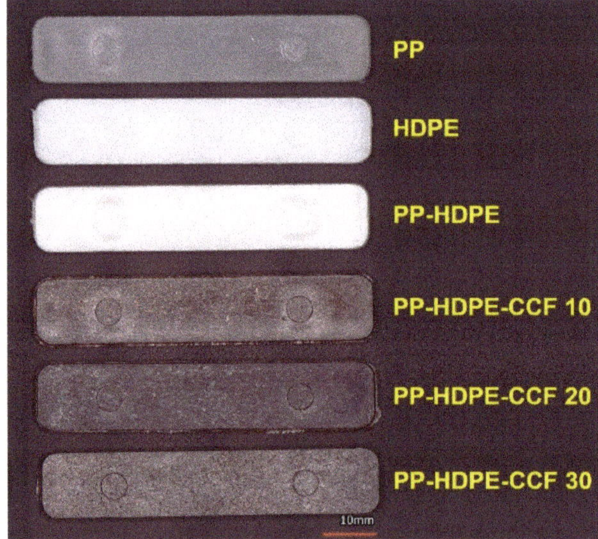

Figure 2. Injected specimens of neat PP, HDPE, PP-HDPE blend, and their biocomposites.

2.3. Materials Characterization

2.3.1. Melt Flow Index (MFI)

Melt flow index tests of neat PP and HDPE were performed at 190 °C and 2.16 kg using a plastics melt flow indexer.

2.3.2. Mechanical Properties

Tensile tests and three-point bending flexural tests were performed in a universal testing machine INSTRON Model 3366 (INSTRON, Norwood, MA, USA) equipped with an axial extensometer Epsilon model 3555 BP (Epsilon Tech, Jackson, WY, USA). Before being testing, the specimens were conditioned at 23 °C and 50% relative humidity for seven days. Tensile tests were performed on type V specimens (according to standard ASTM-D 638-14) at 23 °C, using a cross-head speed of 5 mm/min while flexural tests were carried out on bars with a rectangular cross-section at 23 °C, using a cross head speed of 1.3 mm/min, a distance between support spans of 50 mm and were performed up to 5% strain under standard ASTM D 790-17. The results were taken as the average of five samples. The impact strength of neat PP, HDPE, PP-HDPE blend, and their biocomposites was determined with an impact machine equipped with a 2.5 Joules pendulum. Notched IZOD impact tests were carried out at 23 °C and a starting angle of 150° under standard ASTM D 256-10 (2018). The results were taken as the average value of five samples.

2.3.3. Thermal Characterization

DSC tests were carried out using a TA Q2000 differential scanning calorimeter (Texas Instruments, Dallas, TX, USA) under nitrogen atmosphere at a scanning speed of 10 °C/min, with a sample of 10 mg in aluminum pans from 20 to 200 °C at a scanning speed of 10 °C/min, with a sample of 10 mg in aluminum pans. First, the samples were subjected to heating cycles at 10 °C/min from 20 to 200 °C to erase the thermal history related to processing events, following by cooling cycles at 10 °C/min from 200 to 0 °C. Finally, second heating cycles were performed at 10 °C/min from 0 to 200 °C. The samples were analyzed in aluminum crucibles under a N2 atmosphere. On the other hand, thermogravimetric analysis (TGA) tests were performed using a TA Q500 thermogravimeter (Texas Instruments, Dallas, TX, USA) from 25 to 600 °C at a heating rate of 10 °C/min. The samples were analyzed in aluminum crucibles under a N2 atmosphere.

2.3.4. Dynamic Mechanical Analysis (DMA)

DMA tests were performed using a DMA RSA-G2 (Texas Instruments, Dallas, TX, USA) in three-point bending mode from −80 to 150 °C, a frequency of 1 Hz, a constant heating rate of 3 °C/min and 0.01% of strain (taken from the linear viscoelastic domain of the plot E′ vs. strain reported earlier for PP and PE [31,32] and PP-natural fiber biocomposites [16,33]). Changes in storage modulus (E′), loss modulus (E″), and tan delta (loss factor) were recorded.

2.3.5. Morphology

Scanning electronic microscopy (SEM) of the PP, HDPE, PP-HDPE blend, and their biocomposites was performed using a Quanta FEG 250 microscope (ThermoFisher Scientific, Hillsboro, OR, USA) operating at a voltage of 10 kV. To obtain a brittle fracture on the visualized surfaces, the samples were immersed in a container with liquid nitrogen for 15 min. Later, the fracture was generated inside the container using two steel forceps. Finally, with the aim of increasing their electric conductivity, the samples were previously sputter-coated with gold.

2.3.6. Particle Size and Roughness Measurements

The particle size of the milled and sieved CCF was measured with a Three-dimensional microscope VR-3000 Series with a wide-area three-dimensional measurement system from Keyence (Keyence Corporation of America, Itasca, IL, USA). The roughness measurements were also performed twenty single milled fibers with this measurement system. The reported results were Ra (arithmetical mean height) and Rz (maximum height of profile).

2.3.7. Statistical Analysis

Tensile, flexural, and impact properties of the materials were subject to analyses of variance (ANOVA), and Tukey's test was applied at the 0.05 level of significance. All statistical analyses were performed using Minitab Statistical Software Release 14 (Minitab LLC, State College, PA, USA).

3. Results and Discussion

3.1. CCF Characterization

The milled CCF (Figure 3) contains fibers and cork-like particles with an average length (l) of and 0.94 ± 0.22 mm and 0.38 ± 0.10 mm respectively. The average width (w) was 0.22 ±0.04 mm for fibers and 0.29 ± 0.07 mm for particles. The average aspect ratio (*l/w*) of the milled CCF was 4.27 and 1.3 for fibers and particles. As shown in Figure 4, the wide-area three-dimensional measurement system allows to perform several measurements as the length, diameter and the roughness profile of a single natural fiber. Ra and Rz values obtained as the average value of twenty samples were 0.011 ± 0.004 mm and 0.055 ± 0.021 mm. The natural fibers surface roughness is an important parameter to measure because it plays a significant role in the mechanical interlocking between the fibers and matrix, which is related to biocomposites mechanical properties [18,26,39,40].

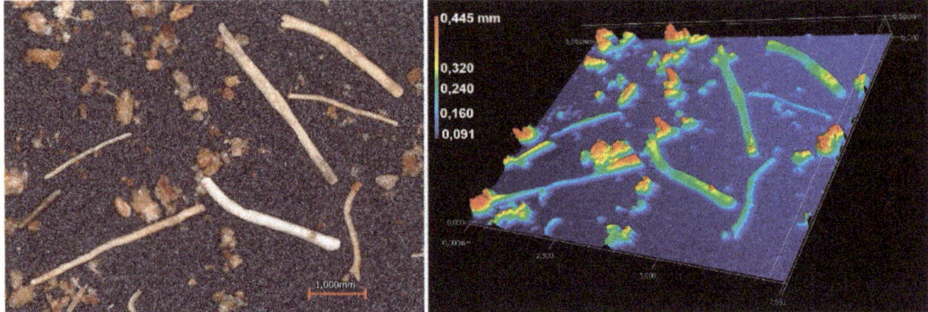

Figure 3. Optical micrograph and 3D height map of milled CCF.

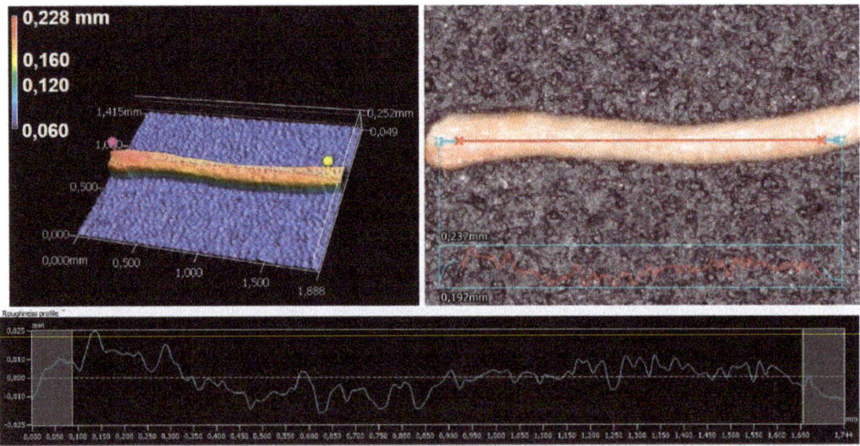

Figure 4. Roughness profile, optical micrograph and 3D height map of a single CCF.

Figure 5 shows the TGA and DTG thermograms of CCF under N_2. Also, the degradation steps, onset temperature (To) and the maximum weight loss rate temperature of the sample (Tmax) are summarized in Table S1 supplementary information. For CCF, the TG curve shows four principal mass loss regions (Figure 5a). These regions are located around 30–100 °C, 100–200 °C, 200–300 °C, and 300–600 °C. The first region is related to the moisture evaporation of the sample with a weight loss of 9.2%. The second region is stable, without weight loss observed related to volatiles or CCF degradation by-products. This region shows that window processing of CCF based biocomposites should be below 200 °C (as indicates in the red circle in Figure 5b) to avoid thermal degradation of the fibers [41]. The third region between 200 °C and 300 °C presents a To of 253 °C and Tmax of 283 °C (Figure 5b), which is related to released elements from the sample like hemicellulose. The last region starts from 323 °C and with a Tmax 336 °C is related to α-cellulose decomposition.

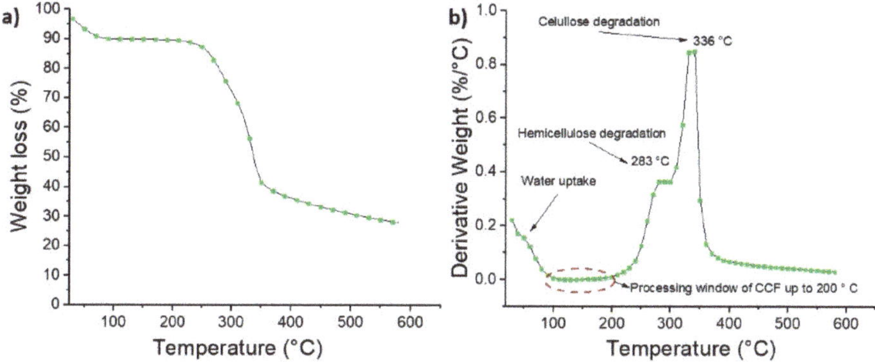

Figure 5. (a) TG and (b) DTG curves of coir coconut fibers at a heating rate of 10 °C/min.

3.2. Biocomposites Characterization

3.2.1. Mechanical Properties

The effect of CCF incorporation on the mechanical properties of the PP-HDPE blend was evaluated. The stress vs. strain graphs obtained from tensile and flexural tests for each material are shown in Figure 6. Also, the mechanical properties calculated from these tests are summarized in Supplementary Table S2.

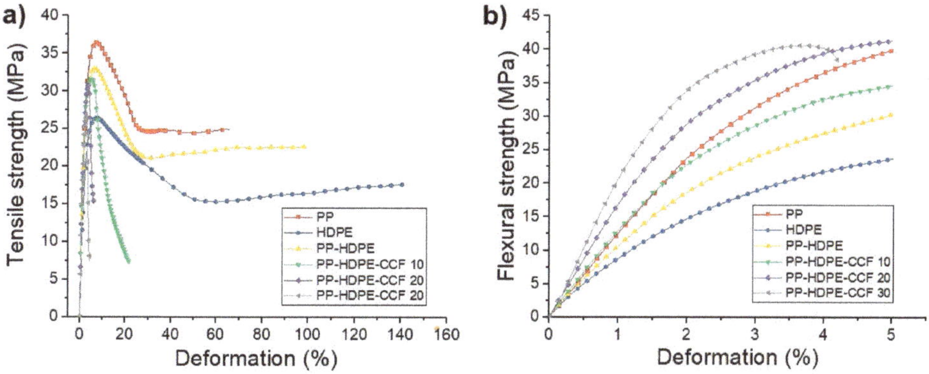

Figure 6. Average tensile stress vs deformation (**a**) and flexural stress vs deformation (**b**) of PP, HDPE, PP-HDPE blend, and their biocomposites.

Regarding neat polymers, it is observed that PP tensile modulus (TM) and tensile strength (TS) values are higher than those of HDPE. However, HDPE deformation at break (ε_b) is higher compared to PP. For PP-HDPE binary blend, TM was not significantly different from PP ($p \geq 0.05$) and its TS and εb values were between neat polymers values. In the case of biocomposites, the results of the test show that CCF addition generates significant increases ($p < 0.05$) in TM values of 16, 35, and 78% compared to the PP-HDPE blend for the biocomposites PP-HDPE-CCF 10, 20 and 30%.

On the other hand, TS values of PP-HDPE- CCF 10, 20 and 30% decrease 6%, 7%, and 20%, respectively, in comparison with PP- HDPE blend ($p < 0.05$). Also, a CCF content increase generates a significant decrease in εb of the PP-HDPE matrix. This decrease in the TS and εb values have already been observed in some biocomposites [16,22,34] and could be related to the weak interfacial bonding between CCF (hydrophilic) and PP-HDPE (hydrophobic) or interface discontinuities that affect the biocomposites deformation capacity (see Section 3.2.4). With CCF loading increase, the weak interfacial area between the polymeric matrix and CCF increases, as a result, TS and εb values decreases.

Flexural test results also show that biocomposites modulus values (FM) increase around 13 and 99% for biocomposites PP-CCF 10 and 30 respectively, compared with PP-HDPE. This stiffening effect is caused by a decrease in the polymeric chain mobility due to CCF and the higher mechanical properties of the CCF compared to the polymeric matrix and could become a decisive property in product applications where the rigidity (related to tensile and flexural modulus) is an essential factor. Also, the FS values of PP-HDPE and biocomposites presents significant differences ($p < 0.05$). FS values of the biocomposites increases between 14% and 35% for PPP-HDPE-CCF 10 and 30 respectively [42]. These differences observed in the strength values of both mechanical tests were also observed on PP-Rice Husk and PP-CCF biocomposites and can be due to a higher interaction natural fiber-polymeric matrix under compression stresses generates during bending [16,34].

The results of the Notched IZOD impact tests (Supplementary Table S2) shows that HDPE has a better impact performance than PP and could be related to its higher deformation and energy absorption capability. As expected, the PP-HDPE blend impact strength was 67% greater than that of neat PP ($p < 0.05$). On the other hand, CCF addition leading to a reduction of impact strength between 44 and 64% for PP-HDPE-CCF 10 and 30% respectively in comparison with PP-HDPE ($p < 0.05$), these results are in the range of those published by other authors for PP-CCF biocomposites [42,43]. This reduction in impact strength in the biocomposites with CCF load could be due to the stiffening effect of the matrix observed earlier in the tensile test and a weak interfacial adhesion between the CCF and polymeric matrix. Also, the increase of CCF generates fibers clusters within the biocomposites that could act as crack initiation sites [44].

3.2.2. Thermal Properties

DSC thermograms for each material are shown in Figure 7. Also, the thermal properties obtained from these thermograms were included in Supplementary Table S3. The degree of crystallinity (χ_c) of each material was estimated from Equation (1):

$$\chi_c = \left(\frac{\Delta H_m}{[\Delta H_m^0 * \left(1 - W_{fiber}\right) * \left(w_{pol}\right)]} \right) * 100 \tag{1}$$

where w_{Fiber} is the CCF fraction, w_{pol} is the fraction of each polymer of the blend, ΔH_m is the normalized melting enthalpy of each sample, and ΔH_{m0} is the specific melting enthalpy of 100% crystalline PP and HDPE. These values are 207 and 293 J/g, respectively [45].

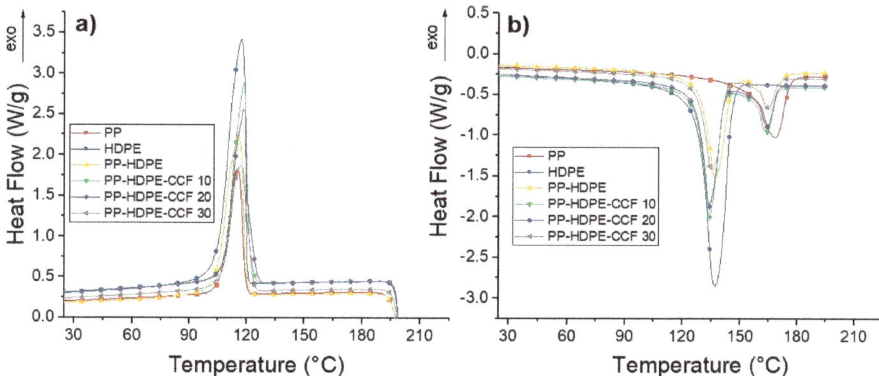

Figure 7. (**a**) Cooling and (**b**) Second heating DSC curves of PP, HDPE, PP-HDPE blend, and their biocomposites.

The cooling cycle of PP, HDPE, and PP-HDPE blend (Figure 7a) shows an exothermic peak located between 115 and 117 °C, which corresponds to PP and HDPE chains crystallization. Regarding biocomposites, it is observed that during cooling CCF addition induces an increase of the PP-HDPE chains crystallization temperature around 2 to 4 °C. This increase indicates that CCF could act as a nucleating agent for polyolefin blends.

The second heating curves (Figure 7b) show single melting endotherms located at 137 and 168 °C for neat PP and HDPE, respectively. These single peaks are associated with the melting temperature (Tm) of PP and HDPE crystals formed during the cooling stage. On the other hand, the PP-HDPE blend show two peaks located at 137 and 166 °C, related to the melting temperatures of each polymer when the blend is formed. This decrease has been observed in PP-PE blends and are related to a lack of miscibility between these polymers [9]. PP-HDPE-CCF biocomposites also exhibit two endothermic peaks around 135 and 165 °C related to the melting of the HDPE and PP phases.

Considering Figure 7b, it is observed that there were small decreases in relation to the Tm of PP-HDPE-CCF biocomposites compared to neat PP-HDPE blend. This thermal behavior has been already observed in several studies regarding biocomposites as PLA-Ramie [46] and PP-NBr-Bagasse fibers [47]. This decrease (albeit small), is due to the incompatibility between non-polar hydrophobic matrices and polar hydrophilic untreated CCF fiber which leads to poor interfacial properties and thus lowering the melting point. Also, the presence of CCF could also disturb the chain arrangement in PP-HDPE blend, thereby decreasing the Tm of the corresponding biocomposites.

TG and DTG thermograms of each material are shown in Figure 8. Also, the thermal parameters obtained are summarized in Supplementary Table S1.

PP and HDPE degradation occur in a one-step process with onset temperatures (To) located at 420 and 464 °C for PP and HDPE, respectively. Also, Tmax values were 457 °C for PP and 485 °C for HDPE. This result shows that HDPE has higher thermal stability than PP. PP-HDPE blend presents a single step degradation with a To located at 413 °C and two Tmax peaks located at 430 and 461 °C. These Tmax peaks are related to PP and HDPE phases of the blend. The residual char after 600 °C was 0.4 and 0.6% for neat polymers and PP-HDPE blend; this low residual char indicates that constituent atoms of the polyolefins (carbon and hydrogen) were volatilized during TGA test.

On the other hand, the thermal degradation of biocomposites takes place in a multi-step process. The first step is related to CCF degradation and presents a To located between 266 and 275 °C, whereas the second step is associated with PP-HDPE thermal degradation. It is observed that CCF addition increases the polymeric matrix thermal stability, to values of PP-HDPE matrix increase between 28 and 40 °C and Tmax increases between 9 and 12 °C as compared to PP-HDPE blend. This increase of the polymeric matrix thermal stability given by CCF addition has already been observed in polyolefins-natural fibers biocomposites. This can be due to the increase of crystallinity and thermal properties of the matrix (as shown in Supplementary Table S3) generated by natural fibers nucleating effect [16,22]. Also, it is observed that the residual char of PP-HDPE-CCF biocomposites increases with the CCF content.

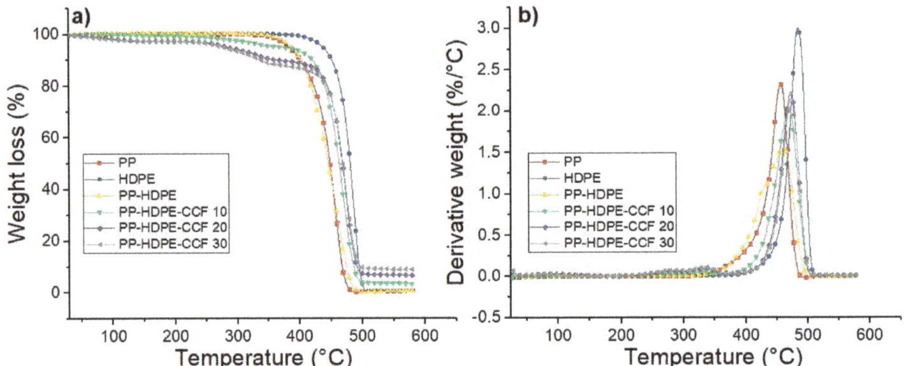

Figure 8. (**a**) TG and (**b**) DTG curves of PP, HDPE, PP-HDPE blend, and their biocomposites.

3.2.3. Dynamic Mechanical Analysis

Figure 9 and Table S4 presents the storage modulus values (E′) of each material vs. temperature. Regarding neat polymer matrices, E′ values of PP are higher than HDPE values in the entire temperature range, whereas, E′ of the PP- HDPE blend values were between neat polyolefins values. Also, E′ values decrease progressively with temperature increase for all materials. This could be due to the softening and the beginning of relaxation processes within the polymer matrix [48]. The CCF addition generates an increase of PP-HDPE blend stiffness, proportionally to the CCF content. This increase in E′ values are related to the stiffening effect given by rigid CCF and is consistent with the results obtained by tensile and flexural tests (Section 3.2.1).

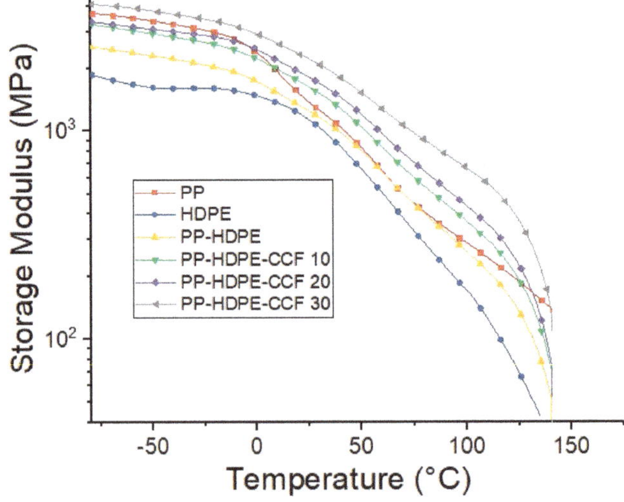

Figure 9. Temperature dependence of storage modulus of PP, HDPE, PP-HDPE blend, and their biocomposites.

However, this stiffening effect is dependent on the glass transition temperature (Tg) of the PP phase. Below Tg (T = 0 °C), E′ increases 28% with 10% wt of CCF and shows a maximum increase of 65% with 30% wt of CCF compared to the PP-HDPE blend. At temperatures above Tg, for example T= 25 °C, this increase was between 31 and 78% for PP-HDPE-CCF 10 and 30% and is greater for temperatures above ambient (T = 80 ° C) where the increase ranges from 35% to 125% with CCF content of 10% and 30% of respectively. This result suggests that at temperatures lower than Tg, the contribution of the fibers to the matrix rigidity is low since the matrix is in a glassy state. As the temperature increases, the drop in the matrix E′ values is compensated by the stiffness of the CCF fibers. In this case the E′ values is controlled by the percentage of fiber and increases with the fiber load in the biocomposite.

On the other hand, loss modulus (E″) vs. temperature of HDPE, PP, PP-HDPE blend, and their biocomposites is shown in Figure 10. E″ plot of neat PP shows a β relaxation around 6 °C related to the glass transition temperature or Tg and a shoulder around 60 °C related to an α relaxation [49,50]. On the other hand, E″ plot of HDPE exhibit a broad α relaxation around 40 °C which is associated with the beginning of the molecular movement of the HDPE crystalline phase. The PP-HDPE blend reveal two peaks related to PP β relaxation and HDPE α relaxation which decline and shift towards high temperatures with 10% wt of CCF.These shifts to higher temperatures are caused by a decrease in the molecular movement of the PP and HDPE chains generated by the presence of the CCF and a dispersed phase on the matrix. With fiber loading increase, the E″ peaks intensity gradually increases and becomes broader. This behavior has been observed in polymer-natural fibers biocomposites [51,52] and reveal that CCF effectively suppress the polymeric chains mobility resulting in a broadening of the Tg range.

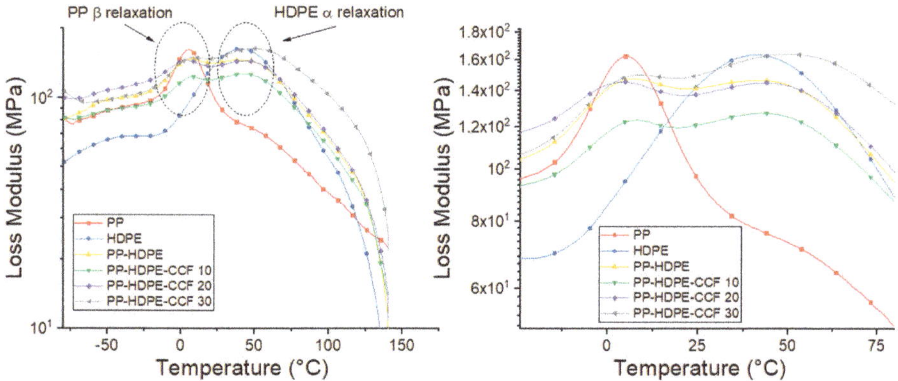

Figure 10. Temperature dependence of loss modulus of PP, HDPE, PP-HDPE blend, and their biocomposites.

Figure 11 shows the variation of Tan δ with temperature. According to Saba et al. [51], Tan δ is the ratio between E″ and E′, and is related with the damping behavior of the polymeric matrix. This graph confirms that neat PP exhibit a β relaxation which corresponds to the glass transition (Tg) and a α relaxation between 60 °C to 75 °C [50], also neat HDPE present the α relaxation observed in E″ graphs (Figure 10). Regarding biocomposites, Tan δ peaks height related to the Tg of the PP phase and the α relaxation of the HDPE phase were observed to gradually decreases and shifts towards higher temperatures with CCF content increase (Supplementary Table S4).This result could be due to the amplified stiffness imparted by the CCF and confirms that fiber addition hinder the molecular movement related to the damping and could be an advantage in some applications where a better performance against mechanical loads is required. Also, the broadness of tan δ peaks, measured as the width at half maximum (FWHM), can be an indicator of the composite homogeneity and the interaction between the matrix and the fibers. In this sense, some author established that larger FWHM values implies heterogeneity and more contact between the phases of a biocomposite [46,53]. Table S4 shows that FWHM values of tan δ peaks is found to be the lowest for PP-HDPE blend and increases gradually with CCF content. This suggests that the heterogeneity of the biocomposite and the interaction between fibers and polymeric matrix increases with CCF content.

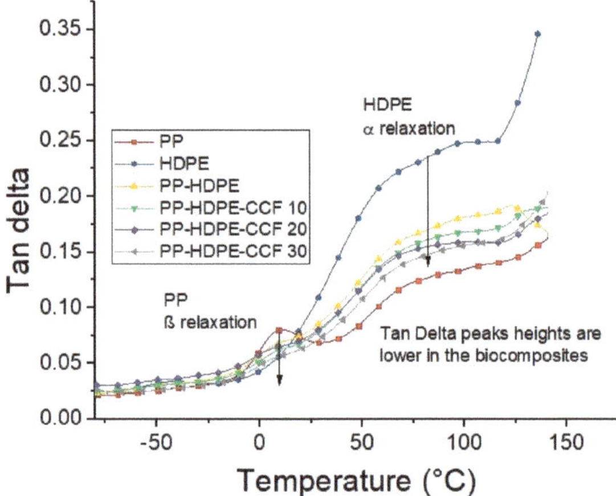

Figure 11. Temperature dependence of tan delta of neat PP, HDPE, PP-HDPE blend, and their biocomposites.

With the aim to further understand the viscoelastic and structural behavior of the studied materials, Cole–Cole diagrams were evaluated. These diagrams are obtained by drawing E″ vs. E′, and can describe the nature of polymeric and composite materials [54,55]. A homogeneous material with a single relaxation time shows a semicircle diagram while a multiphase system with different relaxation times will show two irregularly shaped modified circles [54]. Figure 12 shows the Cole–Cole diagrams for PP, HDPE, PP-HDPE blend, and its biocomposites. The Cole–Cole curves shows that PP and HDPE are homogeneous systems with a concave shape (semicircles), while the PP-HDPE blend is a heterogeneous system which exhibit two semicircles due to two different relaxation mechanisms corresponding to the immisicibility between the polymeric phases. This result is in good agreement with the data obtained from the DSC test (Section 3.2.2). Regarding biocomposites, the Cole–Cole diagram also display two semicircles and a progressive increase in the values of the E′ and E″ with CCF loading [56]. This result show that CCF effectively suppresses the polymeric chains mobility and is an indicative of materials heterogeneity associated with greater differences in relaxation processes of the polymeric matrix when more CCF is incorporated. Also, the Cole–Cole diagram show that among the biocomposites, the one with 30% wt of CCF showed the highest E′ and E″ values. Therefore, it can be inferred that PP-HDPE-30 CCF biocomposite can absorb higher external loads while exhibiting elastic behavior maintaning its characteristics at higher temperatures [57].

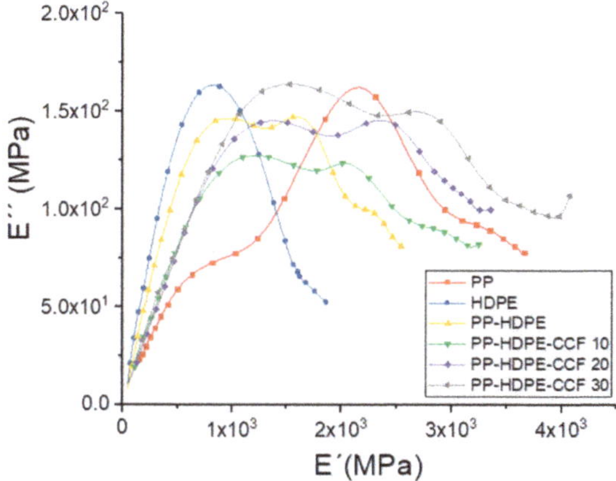

Figure 12. Cole-Cole plots of PP, HDPE, PP-HDPE blend, and their biocomposites.

3.2.4. Morphology

SEM micrographs of neat polymers and PP-HDPE blend are shown in Figure 13a–c, respectively. For PP and HDPE, the fracture surface is smooth, and only one phase is observed in each sample. The PP-HDPE blend micrographs (Figure 13c,d) show a two-phase morphology due to PP and HDPE immiscibility [11]. This phase-separated morphology is consistent with the results obtained in DSC and DMA tests. According to the measured melt flow indexes (MFI values in Section 2.1), PP viscosity is higher in comparison with HDPE viscosity. Thus, PP is not efficiently sheared and separated during melt blending and is the dispersed phase of the blend. Also, HDPE with lower viscosity is the continuous phase [11]. The average diameter of the dispersed phase for the PP-HDPE blend was 0.32 ± 0.06 μm.

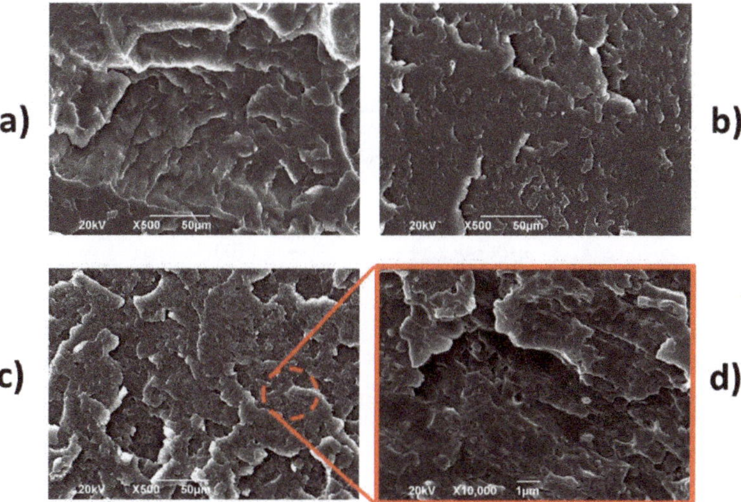

Figure 13. SEM micrographs for (**a**) PP, (**b**) HDPE and PP-HDPE blend (**c,d**).

The biocomposite PP-HDPE-CCF 10 presents a rough fracture surface with a dispersed phase morphology (average diameter of 0.34 ± 0.08 μm) and dispersed CCF (Figure 14a,b). This result shows that 10% of CCF addition did not disturb the dispersed phase formation. Also, the interphase gaps between the CCF and the matrix (yellow circles in Figure 14c) indicate a poor interfacial adhesion between the polymeric matrix and CCF and could be related to the decreasing of tensile and impact properties (Section 3.2.1) [14,16,22]. On the other hand, for biocomposite PP-HDPE-CCF 30 (Figure 14d,e), the dispersed phase presents an oriented and elongated shape. In this case, a higher CCF proportion increases the contact surface between the polymeric phases and could reduce the surface tension between them. This behavior can change the spherical shape of the dispersed phase into the irregular spheroidal shape observed.

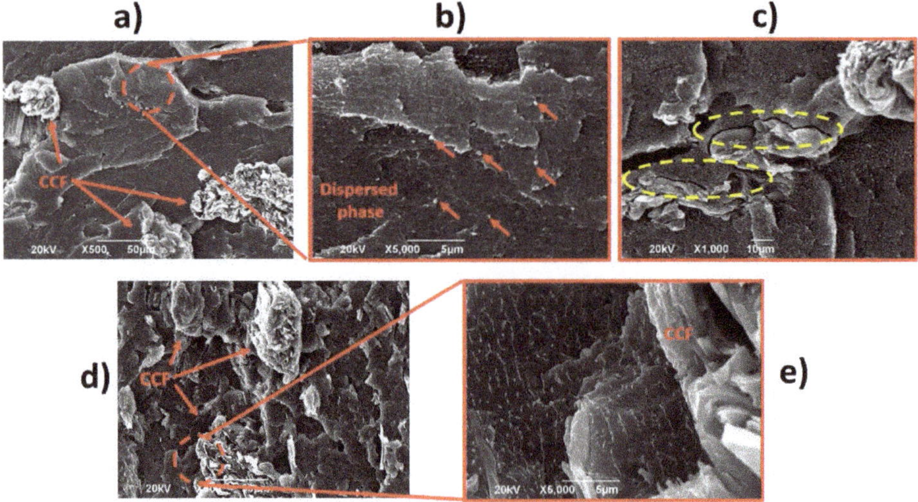

Figure 14. SEM micrographs for PP-HDPE-CCF 10 (**a–c**) and PP-HDPE-CCF 30 (**d,e**).

3.2.5. Linear Shrinkage and 3D Surface Characterization

In this study, the shrinkage was determined as the difference among the linear magnitudes of the mold cavity and that of the injected specimen at room temperature two days after the injection [58]. Also, the following equations were applied to calculate the shrinkage of the molded specimens.

$$Sf(\%) = \frac{L_{f\,specimen} - L_{f\,mold}}{L_{f\,mold}} * 100 \tag{2}$$

$$St(\%) = \frac{L_{t\,specimen} - L_{t\,mold}}{L_{t\,mold}} * 100 \tag{3}$$

where $L_{f\,mold}$ and $L_{t\,mold}$ are the cavity mold dimensions, measured at the flow and transverse direction (as indicated in Figure 15a), $L_{f\,specimen}$ and $L_{t\,specimen}$ are the dimensions of the injected specimen in the two directions (Figure 15b).

Linear shrinkage results of the specimens are shown in Table S5. These values are similar to those observed by Crawford et al., [59] for several polyolefins and show that all specimens present some degree of linear shrinkage during the injection process. However, it is observed that linear shrinkage decreases proportionally with the CCF content. Shrinkage is a frequent defect resulting from the injection molding process that can affect the quality and functionality of the final product. Some studies reported that overall shrinkage is affected by several parameters as the thermodynamic behavior of the injected polymer, the geometry of the injected part, the mold design, and the processing parameters among others [58,60].

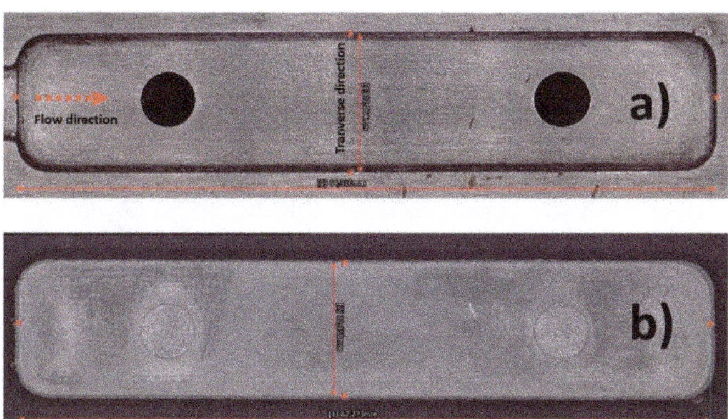

Figure 15. (**a**) mold cavity and (**b**) injected flexural specimen.

On the other hand, the surface characterization performed on the PP-HDPE blend and PP-HDPE-CCF 10, PP-HDPE-CCF 30 flexural specimens is shown in Figure 16. These measures were taken from the edge to the center of the specimen (as indicates in the yellow circle in Figure 16). For the specimen PP-HDPE (Figure 16a), the three-dimensional heights map shows a progressively decrease of the thickness values next to the edge of the specimen (blue and green zones) with a lower height value located at 0.541 mm. This effect is related to the shrinkage of the PP- HDPE matrix during the injection molding. Regarding biocomposites, a decrease in the shrinking with CCF addition was observed. For PP-HDPE-CCF 10 (Figure 16b) the lower height value was 0.250 mm whereas PP-HDPE-CCF 30 specimens present a lower height value located at the edge of 0.193 mm, which is uniform over the studied surface of the specimen (Figure 16c).

This result shows that CCF particles addition enhance the dimensional stability of the PP-HDPE matrix and decrease manufacturing defects as shrinkage in injected specimens and could be an alternative for other additives commonly used to reduce injection molding defects in polyolefins such as talc, calcium carbonate or foaming agents [61]. Also, this behavior is in good agreement with those obtained by several researchers which studied the injection molding of biocomposites with engineering simulation and 3D design software and concluded that natural fibers addition reduces the appearance of processing defects as volumetric shrinkage and warpage in injection molding products [62–64].

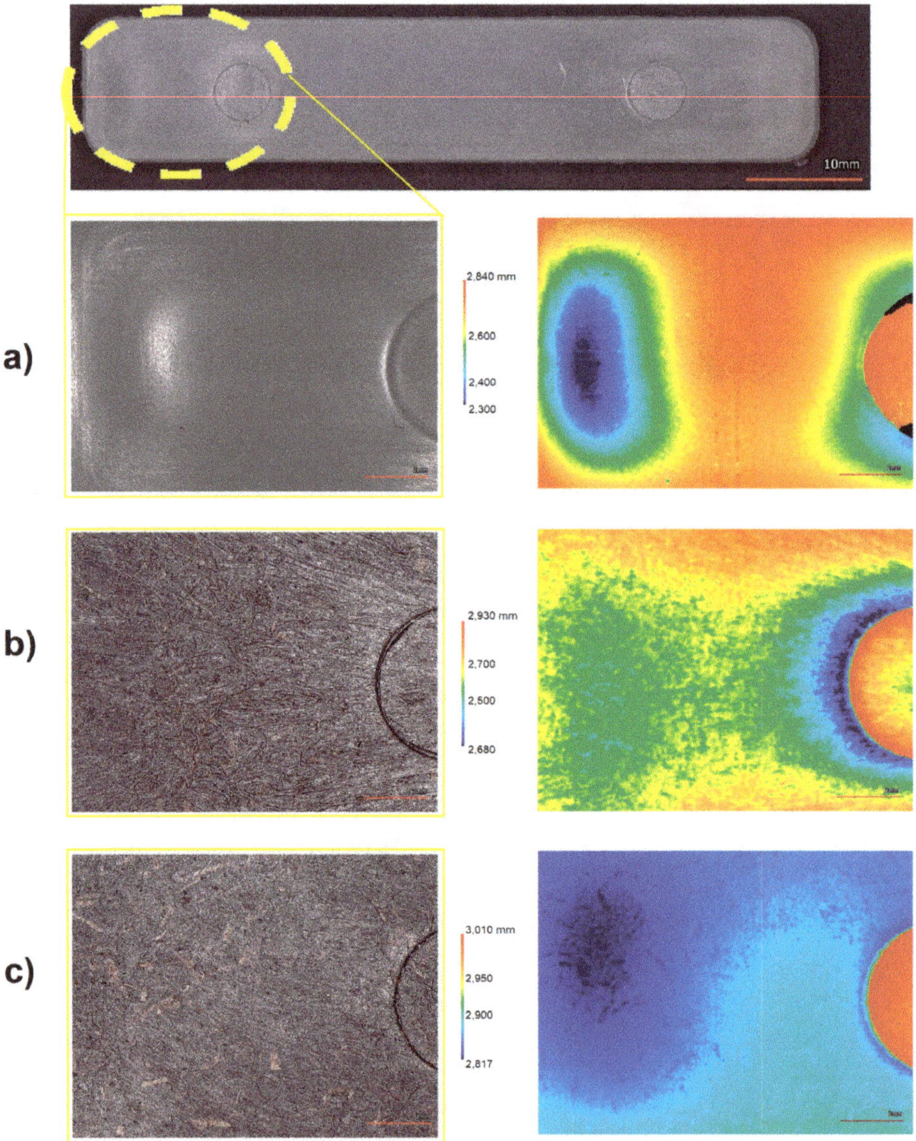

Figure 16. Surface characterization of (**a**) PP-HDPE blend (**b**) PP-HDPE-CCF 10 and (**c**) PP-HDPE-CCF 30 flexural specimens.

4. Conclusions

In this research, PP-HDPE-CCF biocomposites (up to 30% CCF by weight) were processed using extrusion following by injection molding. The objective was to valorize the CCF for their use in polyolefin-natural fiber biocomposites. This CCF is an agro-industrial by-product of the Colombian food industry generated after the separation process of the coconut pulp. The characterization results show that CCF addition generates mechanical properties and thermal stability improvements without affecting the PP-HDPE melting behavior. Also, the dynamic mechanical analysis combined with the three-dimensional microscopy study analysis were sensitive tools for data generation that defines the dynamic mechanical properties, service temperatures and dimensional stability of polymers and biocomposites that support product development, particularly in construction and automotive applications. This study shows that CCF could be an alternative for other additives used to reduce injection molding defects such as talc, calcium carbonate or foaming agents. Finally, PP-HDPE-CCF biocomposites are alternative materials for the design and manufacture of products by injection molding that due to their availability and recyclability potential could generate some economic and environmental benefits in the search for sustainability in the plastics industry facing a circular plastics economy.

Supplementary Materials: The following are available online at http://www.mdpi.com/2073-4360/12/7/1507/s1, Table S1: Thermogravimetric data of the studied materials, Table S2: Mechanical properties of the studied materials, Table S3: Differential scanning calorimetry data of the studied materials, Table S4: DMA results of the studied materials, Table S5: Linear shrinkage of injected specimens at flow (Sf) and transverse (St) directions.

Author Contributions: Conceptualization, M.A.H.-S.; Data curation, M.A.H.-S.; Formal analysis, M.A.H.-S., and J.P.C.-A.; Funding acquisition, M.A.H.-S.; Investigation, M.A.H.-S., J.P.C.-A., S.G.-N. and L.R.-B.; Methodology, M.A.H.-S. and J.P.C.-A.; Supervision, M.A.H.-S.; Writing—original draft, M.A.H.-S. and J.P.C.-A.; Writing—review & editing, M.A.H.-S., J.P.C.-A., S.G.-N. and L.R.-B. All authors have read and agreed to the published version of the manuscript.

Funding: This research received no external funding.

Acknowledgments: The authors acknowledge the Autónoma de Occidente University (Cali- Colombia) and Colombian Ministry of Science for technical and financial support; Research Group of New Solids with Industrial Application-GINSAI (Autónoma de Occidente University) for its support in the thermal characterization; AIMPLAS (Valencia-Spain) for its academic and processing advice in the development of this work; Servicio Nacional de Aprendizaje-SENA (Cali-Colombia) for its technical support through injection molds manufacturing; University del Valle (Cali- Colombia) for SEM microscopy. Also, the authors wish to thank "Super de Alimentos" (Manizales, Colombia) for providing the raw coir coconut fibers (CCF).

Conflicts of Interest: The authors declare no conflict of interest.

References

1. Robeson, L.M. Emerging Technology Involving Polymer Blends. In *Polymer Blends*; Carl Hanser Verlag GmbH & Co. KG: München, Germany, 2007; pp. 415–438.

2. Xavier, S.F.; Xavier, S.F. Properties and performance of polymer blends. In *Polymer Blends Handbook*; Springer: Amsterdam, The Netherlands, 2014; pp. 1031–1201. ISBN 9789400760646.

3. Utracki, L.A.; Wilkie, C.A. *Polymer Blends Handbook*; Springer: Amsterdam, The Netherlands, 2014; ISBN 9789400760646.

4. Mittal, V. *Functional Polymer Blends Synthesis, Properties, and Performances*; CRC Press: Boca Raton, FL, USA, 2012; ISBN 9781138074347.

5. ACOPLASTICOS Plastics in Colombia. Available online: http://www.acoplasticos.org/index.php/mnu-nos/mnu-pyr/pec (accessed on 27 August 2018).

6. Hubo, S.; Delva, L.; Van Damme, N.; Ragaert, K. Blending of recycled mixed polyolefins with recycled polypropylene: Effect on physical and mechanical properties. *AIP Conf. Proc.* **2016**, *1779*, 140006.

7. Aumnate, C.; Rudolph, N.; Sarmadi, M. Recycling of polypropylene/polyethylene blends: Effect of chain structure on the crystallization behaviors. *Polymers* **2019**, *11*, 1456. [CrossRef] [PubMed]

8. Bertin, S.; Robin, J.-J. Study and characterization of virgin and recycled LDPE/PP blends. *Eur. Polym. J.* **2002**, *38*, 2255–2264. [CrossRef]

9. Vranjes, N.; Rek, V. Effect of EPDM on Morphology, Mechanical Properties, Crystallization Behavior and Viscoelastic Properties of iPP+HDPE Blends. *Macromol. Symp.* **2007**, *258*, 90–100. [CrossRef]

10. Rachtanapun, P.; Selke, S.E.M.; Matuana, L.M. Microcellular foam of polymer blends of HDPE/PP and their composites with wood fiber. *J. Appl. Polym. Sci.* **2003**, *88*, 2842–2850. [CrossRef]

11. Lin, J.-H.; Pan, Y.-J.; Liu, C.-F.; Huang, C.-L.; Hsieh, C.-T.; Chen, C.-K.; Lin, Z.-I.; Lou, C.-W. Preparation and Compatibility Evaluation of Polypropylene/High Density Polyethylene Polyblends. *Materials* **2015**, *8*, 8850–8859. [CrossRef]

12. Van Eygen, E.; Laner, D.; Fellner, J. Circular economy of plastic packaging: Current practice and perspectives in Austria. *Waste Manag.* **2018**, *72*, 55–64. [CrossRef]

13. Curtzwiler, G.W.; Schweitzer, M.; Li, Y.; Jiang, S.; Vorst, K.L. Mixed post-consumer recycled polyolefins as a property tuning material for virgin polypropylene. *J. Clean. Prod.* **2019**, *239*. [CrossRef]

14. Dikobe, D.G.; Luyt, A.S. Comparative study of the morphology and properties of PP/LLDPE/wood powder and MAPP/LLDPE/wood powder polymer blend composites. *Express Polym. Lett.* **2010**, *4*, 729–741. [CrossRef]

15. Tanguy, M.; Bourmaud, A.; Beaugrand, J.; Gaudry, T.; Baley, C. Polypropylene reinforcement with flax or jute fibre; Influence of microstructure and constituents properties on the performance of composite. *Compos. Part B Eng.* **2018**, *139*, 64–74. [CrossRef]

16. Hidalgo-Salazar, M.A.; Salinas, E. Mechanical, thermal, viscoelastic performance and product application of PP- rice husk Colombian biocomposites. *Compos. Part B Eng.* **2019**, 107135. [CrossRef]

17. Verma, D.; Gope, P.C. The use of coir/coconut fibers as reinforcements in composites. *Biofiber Reinf. Compos. Mater.* **2015**, 285–319. [CrossRef]

18. Radoor, S.; Karayil, J.; Rangappa, S.M.; Siengchin, S.; Parameswaranpillai, J. A review on the extraction of pineapple, sisal and abaca fibers and their use as reinforcement in polymer matrix. *Express Polym. Lett.* **2020**, *14*, 309–335. [CrossRef]

19. Mochane, M.J.; Mokhena, T.C.; Mokhothu, T.H.; Mtibe, A.; Sadiku, E.R.; Ray, S.S.; Ibrahim, I.D.; Daramola, O.O. Recent progress on natural fiber hybrid composites for advanced applications: A review. *Express Polym. Lett.* **2019**, *13*, 159–198. [CrossRef]

20. Muñoz-Vélez, M.; Hidalgo-Salazar, M.; Mina-Hernández, J.; Muñoz-Vélez, M.F.; Hidalgo-Salazar, M.A.; Mina-Hernández, J.H. Effect of Content and Surface Modification of Fique Fibers on the Properties of a Low-Density Polyethylene (LDPE)-Al/Fique Composite. *Polymers* **2018**, *10*, 1050. [CrossRef]

21. Hidalgo, M.A.; Muñoz, M.F.; Quintana, K.J. Mechanical behavior of polyethylene aluminum composite reinforced with continuous agro fique fibers. *Rev. Lat. Met. Mat.* **2011**, *31*, 187–194.

22. Hidalgo Salazar, M.A.; Muñoz Velez, M.F.; Quintana Cuellar, K.J. Análisis mecánico del compuesto polietileno aluminio reforzado con fibras cortas de fique en disposición bidimensional (Mechanical analysis of polyethylene aluminum composite reinforced with short fique fibers available a in two-dimensional arrangement). *Rev. Latinoam. Metal. y Mater.* **2012**, *32*, 89–95.

23. Siengchin, S. Editorial corner—A personal view Potential use of 'green' composites in automotive applications. *Express Polym. Lett.* **2017**, *11*, 600. [CrossRef]

24. Correa, J.P.; Montalvo-Navarrete, J.M.; Hidalgo-Salazar, M.A. Carbon footprint considerations for biocomposite materials for sustainable products: A review. *J. Clean. Prod.* **2019**, *208*, 785–794. [CrossRef]

25. Väisänen, T.; Das, O.; Tomppo, L. A review on new bio-based constituents for natural fiber-polymer composites. *J. Clean. Prod.* **2017**, *149*, 582–596. [CrossRef]

26. Pickering, K.L.; Efendy, M.G.A.; Le, T.M. A review of recent developments in natural fibre composites and their mechanical performance. *Compos. Part A Appl. Sci. Manuf.* **2016**, *83*, 98–112. [CrossRef]

27. Alvarado, K.; Blanco, A.; Taquechel, A. Fibra de coco: Una alternativa ecológica como sustrato agrícola. *Agric. Org.* **2008**, *3*, 30–31.

28. Beigbeder, J.; Soccalingame, L.; Perrin, D.; Bénézet, J.C.; Bergeret, A. How to manage biocomposites wastes end of life? A life cycle assessment approach (LCA) focused on polypropylene (PP)/wood flour and polylactic acid (PLA)/flax fibres biocomposites. *Waste Manag.* **2019**, *83*, 184–193. [CrossRef] [PubMed]

29. Sapuan, S.M. Materials Selection for Composites: Concurrent Engineering Perspective. In *Composite Materials*; Butterworth-Heinemann: Oxford, UK, 2017; Chapter 6.

30. Mohan Bhasney, S.; Kumar, A.; Katiyar, V. Microcrystalline cellulose, polylactic acid and polypropylene biocomposites and its morphological, mechanical, thermal and rheological properties. *Compos. Part B Eng.* **2020**, *184*, 107717. [CrossRef]

31. Essabir, H.; Raji, M.; Laaziz, S.A.; Rodrique, D.; Bouhfid, R.; Qaiss, A. el kacem Thermo-mechanical performances of polypropylene biocomposites based on untreated, treated and compatibilized spent coffee grounds. *Compos. Part B Eng.* **2018**, *149*, 1–11. [CrossRef]

32. Bledzki, A.K.; Franciszczak, P.; Osman, Z.; Elbadawi, M. Polypropylene biocomposites reinforced with softwood, abaca, jute, and kenaf fibers. *Ind. Crops Prod.* **2015**, *70*, 91–99. [CrossRef]

33. Mir, S.S.; Nafsin, N.; Hasan, M.; Hasan, N.; Hassan, A. Improvement of physico-mechanical properties of coir-polypropylene biocomposites by fiber chemical treatment. *Mater. Des.* **2013**, *52*, 251–257. [CrossRef]

34. Haque, M.; Rahman, R.; Islam, N.; Huque, M.; Hasan, M. Mechanical properties of polypropylene composites reinforced with chemically treated coir and abaca fiber. *J. Reinf. Plast. Compos.* **2010**, *29*, 2253–2261. [CrossRef]

35. Adeniyi, A.G.; Onifade, D.V.; Ighalo, J.O.; Adeoye, A.S. A review of coir fiber reinforced polymer composites. *Compos. Part B Eng.* **2019**, *176*, 107305. [CrossRef]

36. Burgstaller, C. A comparison of processing and performance for lignocellulosic reinforced polypropylene for injection moulding applications. *Compos. Part B Eng.* **2014**, *67*, 192–198. [CrossRef]

37. Pickering, K.L. *Properties and Performance of Natural-Fibre Composites*; Woodhead Publishing Limited-CRC Press: Boca Raton, FL, USA, 2008; ISBN 9781855737396.

38. Alves Fidelis, M.E.; Pereira, T.V.C.; Gomes, O.D.F.M.; De Andrade Silva, F.; Toledo Filho, R.D. The effect of fiber morphology on the tensile strength of natural fibers. *J. Mater. Res. Technol.* **2013**, *2*, 149–157. [CrossRef]

39. Kalagi, G.R.; Patil, R.; Nayak, N. Experimental Study on Mechanical Properties of Natural Fiber Reinforced Polymer Composite Materials for Wind Turbine Blades. In *Materials Today: Proceedings*; Elsevier Ltd.: Amsterdam, The Netherlands, 2018; Volume 5, pp. 2588–2596.

40. Latif, R.; Wakeel, S.; Zaman Khan, N.; Noor Siddiquee, A.; Lal Verma, S.; Akhtar Khan, Z. Surface treatments of plant fibers and their effects on mechanical properties of fiber-reinforced composites: A review. *J. Reinf. Plast. Compos.* **2019**, *38*, 15–30. [CrossRef]

41. Saheb, D.N.; Jog, J.P. Natural fiber polymer composites: A review. *Adv. Polym. Technol.* **1999**, *18*, 351–363. [CrossRef]

42. Sudhakara, P.; Jagadeesh, D.; Wang, Y.; Venkata Prasad, C.; Devi, A.P.K.; Balakrishnan, G.; Kim, B.S.; Song, J.I. Fabrication of Borassus fruit lignocellulose fiber/PP composites and comparison with jute, sisal and coir fibers. *Carbohydr. Polym.* **2013**, *98*, 1002–1010. [CrossRef] [PubMed]

43. Zainudin, E.S.; Yan, L.H.; Haniffah, W.H.; Jawaid, M.; Alothman, O.Y. Effect of coir fiber loading on mechanical and morphological properties of oil palm fibers reinforced polypropylene composites. *Polym. Compos.* **2014**, *35*, 1418–1425. [CrossRef]

44. Sohn, J.S.; Cha, S.W. Effect of chemical modification on mechanical properties of wood-plastic composite injection-molded parts. *Polymers* **2018**, *10*, 1391. [CrossRef]

45. Blaine, R.L. Thermal Applications Note—Polymer Heat of Fusion. Available online: http://www.tainstruments.com/pdf/literature/TN048.pdf (accessed on 15 July 2019).

46. Xu, H.; Liu, C.Y.; Chen, C.; Hsiao, B.S.; Zhong, G.J.; Li, Z.M. Easy alignment and effective nucleation activity of ramie fibers in injection-molded poly(lactic acid) biocomposites. *Biopolymers* **2012**, *97*, 825–839. [CrossRef]

47. Zainal, M.; Santiagoo, R.; Ayob, A.; Ghani, A.A.; Mustafa, W.A.; Othman, N.S. Thermal and mechanical properties of chemical modification on sugarcane bagasse mixed with polypropylene and recycle acrylonitrile butadiene rubber composite. *J. Thermoplast. Compos. Mater.* **2019**, 089270571983207. [CrossRef]

48. Hassaini, L.; Kaci, M.; Touati, N.; Pillin, I.; Kervoelen, A.; Bruzaud, S. Valorization of olive husk flour as a filler for biocomposites based on poly(3-hydroxybutyrate-co-3-hydroxyvalerate): Effects of silane treatment. *Polym. Test.* **2017**, *59*, 430–440. [CrossRef]

49. Guo, C.; Song, Y.; Wang, Q.; Shen, C. Dynamic-mechanical analysis and SEM morphology of wood flour/polypropylene composites. *J. For. Res.* **2006**, *17*, 315–318. [CrossRef]

50. Hidalgo-Salazar, M.; Luna-Vera, F.; Pablo Correa-Aguirre, J. Biocomposites from Colombian Sugarcane Bagasse with Polypropylene: Mechanical, Thermal and Viscoelastic Properties. In *Characterizations of Some Composite Materials*; IntechOpen: London, UK, 2019.

51. Saba, N.; Jawaid, M.; Alothman, O.Y.; Paridah, M.T. A review on dynamic mechanical properties of natural fibre reinforced polymer composites. *Constr. Build. Mater.* **2016**, *106*, 149–159. [CrossRef]

52. Chaitanya, S.; Singh, I.; Song, J. Il Recyclability analysis of PLA/Sisal fiber biocomposites. *Compos. Part B Eng.* **2019**, *173*, 106895. [CrossRef]

53. Manikandan Nair, K.C.; Thomas, S.; Groeninckx, G. Thermal and dynamic mechanical analysis of polystyrene composites reinforced with short sisal fibres. *Compos. Sci. Technol.* **2001**, *61*, 2519–2529. [CrossRef]

54. Bagotia, N.; Sharma, D.K. Systematic study of dynamic mechanical and thermal properties of multiwalled carbon nanotube reinforced polycarbonate/ethylene methyl acrylate nanocomposites. *Polym. Test.* **2019**, *73*, 425–432. [CrossRef]

55. Mohana Krishnudu, D.; Sreeramulu, D.; Venkateshwar Reddy, P. A study of filler content influence on dynamic mechanical and thermal characteristics of coir and luffa cylindrica reinforced hybrid composites. *Constr. Build. Mater.* **2020**, *251*, 119040. [CrossRef]

56. Romanzini, D.; Ornaghi, H.L.; Amico, S.C.; Zattera, A.J. Influence of fiber hybridization on the dynamic mechanical properties of glass/ramie fiber-reinforced polyester composites. *J. Reinf. Plast. Compos.* **2012**, *31*, 1652–1661. [CrossRef]

57. Sathyaseelan, P.; Sellamuthu, P.; Palanimuthu, L. Dynamic mechanical analysis of areca/kenaf fiber reinforced epoxy hybrid composites fabricated in different stacking sequences. *Mater. Today Proc.* **2020**. [CrossRef]

58. Fischer, J.M. *Handbook of Molded Part Shrinkage and Warpage*, 2nd ed.; Elsevier Inc.: Amsterdam, The Netherlands, 2012; ISBN 9781455730575.

59. Crawford, R.J.; Throne, J.L. Mechanical Part Design. In *Rotational Molding Technology*; Elsevier Inc.: Amsterdam, The Netherlands, 2002; pp. 307–365.

60. Masato, D.; Rathore, J.; Sorgato, M.; Carmignato, S.; Lucchetta, G. Analysis of the shrinkage of injection-molded fiber-reinforced thin-wall parts. *Mater. Des.* **2017**, *132*, 496–504. [CrossRef]

61. Tolinski, M. *Additives for Polyolefins: Getting the Most Out of Polypropylene, Polyethylene and TPO*, 2nd ed.; William Andrew: Oxford, UK, 2015; ISBN 9780323371773.

62. Azaman, M.D.; Sapuan, S.M.; Sulaiman, S.; Zainudin, E.S.; Khalina, A. Shrinkages and warpage in the processability of wood-filled polypropylene composite thin-walled parts formed by injection molding. *Mater. Des.* **2013**, *52*, 1018–1026. [CrossRef]

63. Azaman, M.D.; Sapuan, S.M.; Sulaiman, S.; Zainudin, E.S.; Abdan, K. An investigation of the processability of natural fibre reinforced polymer composites on shallow and flat thin-walled parts by injection moulding process. *Mater. Des.* **2013**, *50*, 451–456. [CrossRef]

64. Santos, J.D.; Fajardo, J.I.; Cuji, A.R.; García, J.A.; Garzón, L.E.; López, L.M. Experimental evaluation and simulation of volumetric shrinkage and warpage on polymeric composite reinforced with short natural fibers. *Front. Mech. Eng.* **2015**, *10*, 287–293. [CrossRef]

 *polymers*

Article

The Effects of Reprocessing and Fiber Treatments on the Properties of Polypropylene-Sugarcane Bagasse Biocomposites

Juan P. Correa-Aguirre [1], Fernando Luna-Vera [2], Carolina Caicedo [2], Bairo Vera-Mondragón [2] and Miguel A. Hidalgo-Salazar [1,*]

[1] Research Group for Manufacturing Technologies GITEM, Universidad Autónoma de Occidente, Cali 760030, Colombia; jpcorrea@uao.edu.co

[2] Research Group for Development of Materials and Products GIDEMP, National Center for Technical Assistance to Industry (ASTIN-SENA), Cali 760003, Colombia; fernandolunavera@gmail.com (F.L.-V.); ccaicedo60@misena.edu.co (C.C.); bvera@sena.edu.co (B.V.-M.)

* Correspondence: mahidalgo@uao.edu.co; Tel.: +57-2-3188-000

Received: 16 April 2020; Accepted: 22 June 2020; Published: 27 June 2020

Abstract: This study explores the reprocessing behavior of polypropylene-sugarcane bagasse biocomposites using neat and chemically treated bagasse fibers (20 wt.%). Biocomposites were reprocessed 5 times using the extrusion process followed by injection molding. The mechanical properties indicate that microfibers bagasse fibers addition and chemical treatments generate improvements in the mechanical properties, reaching the highest performance in the third cycle where the flexural modulus and flexural strength increase 57 and 12% in comparison with neat PP. differential scanning calorimetry (DSC) and TGA characterization show that bagasse fibers addition increases the crystallization temperature and thermal stability of the biocomposites 7 and 39 °C respectively, without disturbing the melting process of the PP phase for all extrusion cycles. The rheological test shows that viscosity values of PP and biocomposites decrease progressively with extrusion cycles; however, Cole–Cole plots, dynamic mechanical analysis (DMA), width at half maximum of tan delta peaks and SEM micrographs show that chemical treatments and reprocessing could improve fiber dispersion and fiber–matrix interaction. Based on these results, it can be concluded that recycling potential of polypropylene-sugarcane bagasse biocomposites is huge due to their mechanical, thermal and rheological performance resulting in advantages in terms of sustainability and life cycle impact of these materials.

Keywords: biocomposites; recycling; rheological properties; DMA; injection molding

---

1. Introduction

The reinforcement of polymers with natural fibers such as coir coconut, hemp, sisal, pineapple leaf fibers, sugarcane bagasse, fique and their combinations to create biocomposites has been studied in recent years [1–7]. The term biocomposites refers here to polymeric reinforced composites, where the reinforcing phase and/or the matrix are derived from materials of biological origin. In this sense, several studies have reported the formulation and characterization of biocomposites, which have a status of renewable and sustainable materials since they are composed of natural fibers embedded in non-degradable (i.e., polypropylene, polyethylene, polyamides, etc.) and biodegradable polymeric matrices (starch, polylactic acid, and polyhydroxialkanoates) [8,9].

These materials have the potential to replace traditional plastics in commercial applications such as car parts, toys, furniture, reusable cutlery, among others due to their low cost in comparison with traditional fibers and the enhancement of the polymeric matrices properties induced by natural fibers incorporation. These improvements include weight reduction, better specific properties, dimensional stability, biodegradability, recyclability, decrease in the embodied energy of the products, carbon emissions, and costs due to the polymeric substitution fraction that reduces the amount of plastic material needed to manufacture products [3,10–14].

Sugarcane is one of the most important crops for sugar production around the world. According to the Food and Agriculture Organization located in Rome, Italy (FAO), Colombia is the second-largest producer of sugarcane in South America, with an estimated 220,000 ha planted in 2019 [15], which produces approximately 6 million tons of bagasse by year [16]. This agro-industrial by-product is generated in sugar factories after the cane stem has been crushed and pressed. Sugarcane fiber is mainly composed of cellulose (37 wt.%), hemicellulose (21 wt.%), lignin (22 wt.%) and pectin (10 wt.%) [17]. The availability of this by-product, its low cost and the possibility of valorization are competitive advantages for the development of bagasse fibers based biocomposites at the regional level.

The combination of natural fibers with polymeric matrices generates a problem associated with the incompatibility between the polar and hygroscopic cellulose of the fibers and the non-polar and hydrophobic polymers. Additionally, other components of the natural fiber like hemicellulose, lignin, pectin and waxes generate a smooth surface that hinders the interlocking and the interfacial bonding between the matrix and the reinforcing phase [2,17].

For this reason, several researchers have performed surface treatments over the natural fibers to improve their compatibility with the polymeric matrix. These surface treatments could exhibit physical or chemical nature according to the mechanism applied to improve the interfacial bonding. The most used surface treatment methods included molecular interdiffusion, electrostatic bonding, mechanical interlocking and chemical modification through bleaching, acetylation, alkaline treatments and chemical bonding by coupling agents such as silanes or maleic anhydride [2,17,18].

Anggono et al. [19] studied the incorporation of bagasse (up to 30 wt.%) in a polypropylene (PP) matrix using injection molding processes. They perform alkali treatments on the fibers with calcium hydroxide ($Ca(OH)_2$) and sodium hydroxide (NaOH) and evaluated the effect of those treatments on the mechanical properties of the biocomposites. The results showed that the tensile strength of the biocomposites increases proportionally with bagasse content and chemical modification of these fibers. Additionally, the biocomposites obtained from NaOH treated fibers present the highest mechanical performance results. Carvahlo et al. [20] studied the effect of bagasse content (up to 20 wt.%) and chemical modification (NaOH and acetylation) on the mechanical performance of recycled high-density polyethylene-(r-PE) biocomposites obtained by extrusion. Their results show that chemical modification increased the compatibility between r-PE and bagasse fibers and improves the mechanical properties of the biocomposites. Zainal et al. [21] studied the mechanical, thermal and morphological properties of biocomposites based on a recycled polypropylene-acrylonitrile rubber blend (PP-NBRr) and chemically modified bagasse fibers (up to 30 wt.%) with NaOH and silanes, prepared using melt blending techniques. Their results showed that chemical modification of the fibers enhances the thermal stability and tensile mechanical properties of the biocomposites. They also observed that among chemical treatments, silanization generate better results on the evaluated properties.

The reviewed literature showed that natural fiber-polyolefin based biocomposites could be processed using high-volume manufacturing processes such as extrusion and injection molding reproducibility and production capacity, advantages for the development of products using these materials. However, these manufacturing processes generate some scrap. In the case of injection molding, the overall process generates waste as gates, runners and sprues, which must be ground after the process. Thus, the recycling of wastes generated after products life cycle ending and during processing is an issue to study further and a lucrative option for the growing biocomposites industry that has not yet been fully explored.

Mechanical recycling of polyolefins like PP has been studied due to its ease of processing, property retention and availability [22]. Martín-Alfonso and Franco studied the recycling of PP using multiple extrusion cycles (up to 10 cycles) [23]. Their results showed that thermo-mechanical reprocessing generates a scission of the PP chains, which generates a progressive decrease in thermal stability, melting temperature, viscosity and viscoelastic properties with reprocessing cycles increase.

Regarding biocomposites, Uitterhaegen et al. [13] studied the mechanical behavior of biocomposites based on polyolefins (PP and Bio-PE) and coriander straw (up to 40 wt.%) ground and reprocessed 5 times using injection molding. The authors reported that mechanical properties did not decrease more than 10% through the reprocessing cycles, giving a high recycling potential to these polyolefin-based biocomposites. In another study, Chaitanya et al. [12] explored the recycling of biodegradable biocomposites based on polylactic acid (PLA) and alkaline treated sisal fibers (30 wt.%). The biocomposites were recycled using extrusion (8 cycles) and it was observed that mechanical properties gradually decreased until the third recycling cycle. Beyond these cycles, a significant reduction in properties was observed due to the decrease in PLA molecular weight and fibers attrition. From these results, the authors conclude that PLA-Sisal biocomposites can be recycled up to 3 times to make low to medium strength commercial products.

In the present research, PP-bagasse microfibers (untreated and chemically modified with NaOH and silanes) biocomposites were obtained through extrusion followed by injection molding processes. The mechanical, thermal, rheological and viscoelastic properties were evaluated and compared in order to understand the effect of chemical modification and reprocessing cycles (up to 5 times) on the microfibers dispersion on the biocomposites properties. We consider that the study of the performance of recycled biocomposites is an excellent contribution that supports the novelty of this article, bearing in mind that the interest in the design and manufacture of sustainable and highly recyclable products by injection molding with biocomposites based on natural fibers is increasing around the world.

2. Materials and Methods

2.1. Materials

PP reference 01H41 was sourced from Essentia (Cartagena, Colombia). Untreated sugarcane bagasse fibers were provided by Sucromiles S.A. (Cali, Colombia). In order to perform the chemical modification of these fibers, analytical-grade reagents hexadecyltrimethoxysilane and NaOH were obtained from Sigma-Aldrich (Milwaukee, WI, USA).

2.2. Methodology

2.2.1. Preparation and Chemical Modification of Sugarcane Bagasse

The bagasse fibers were first washed with distilled water and dried at 60 °C for 48 h to remove soil and residues. Then, clean bagasse fibers were grounded with a lab mill and sieved through a 200 μm sieve. The bagasse fibers were separated into three groups: untreated bagasse, aqueous solution of 8% NaOH treated bagasse and aqueous solution of 8% NaOH following by silanized treated bagasse. The chemical surface treatments were performed according to the procedure described in detail in previous research work reported earlier by our group [6].

2.2.2. Processing of Biocomposites

The reprocessing of the biocomposites was simulated using a continuous extrusions methodology. For this technique, the PP and the bagasse fibers were physically mixed in a bag using 20 wt.% of bagasse. This formulation was selected based on experimental results of our group and reviewed literature regarding the microinjection optimization of PP-Bagasse biocomposites [24]. This mixture was fed into the feed zone of a co-rotating twin-screw extruder HAAKE ™ PolyLab™ (Thermo Scientific-Unites States) with 16 mm diameter and 40 D total length, using a temperature gradient between 140 and 170 °C and a screw speed of 70 rpm. These processing parameters were selected from previously reported studies on polyolefin-based biocomposites [6,25]. Then, the extruded material was cooled in water and subsequently pelletized using a mechanical cutter that generated 5 mm long pellets.

These pellets were dried in an air oven at 85 °C for 8 h after each extrusion cycle. The granules of neat PP and biocomposites were extruded 5 times, generating a total of 20 batches of granules that led to the development of 5 biocomposites and 10 reprocessed biocomposites. For this study, the properties of the materials corresponding to processing cycles 1, 3 and 5 were evaluated. The nomenclature of the prepared biocomposites is listed in Table 1.

Table 1. Nomenclature of processed biocomposites.

Processing Cycle Number	Neat PP (wt.%)	Sugar Cane Bagasse (wt.%)	Chemical Treatment	Designation Along the Document
1	100	-	-	PP 1st cycle
	80	20	-	PP-Bag 1st cycle
			NaOH	PP-Bag +alk. 1st cycle
			NaOH + Silanes	PP-Bag +alk. +sil 1st cycle
3	100	-	-	PP 3rd cycle
	80	20	-	PP-Bag 3rd cycle
			NaOH	PP-Bag +alk. 3rd cycle
			NaOH + Silanes	PP-Bag +alk. +sil 3rd cycle
5	100	-	-	PP 5th cycle
	80	20	-	PP-Bag 5th cycle
			NaOH	PP-Bag +alk. 5th cycle
			NaOH + Silanes	PP-Bag +alk.+sil 5th cycle

Finally, a small quantity of the pellets of the different biocomposites was used for the development of injected specimens for flexural and impact tests using a BOY XS microinjection molding machine (BOY Machines Inc., United States) with a temperature gradient between 180 and 185 °C (from the feeding area to the nozzle), a filling pressure of 80 bars, a holding pressure of 60 bars and a mold clamping force of 30 kN. Figure 1 shows the injected PP specimens corresponding to 1st, 3rd and 5th processing cycles and the biocomposites of the 1st processing cycle. It is observed that after the 3rd processing cycle, the neat PP samples show a yellow shade that indicates thermal degradation of the matrix during reprocessing.

2.2.3. Mechanical Properties

The mechanical properties in terms of flexural and impact performance of the materials were determined following the ASTM D790-17 and D256-10 standards, respectively. Three-point bending tests were performed using an INSTRON 3366 universal testing machine, while impact tests were performed using a 2.5 Joules impact tester. Flexural tests were carried out up to 5% deformation using specimens with a rectangular cross-section and 3.2 mm of thickness. The crosshead speed was 1.36 mm/min and the distance between the support span was 50 mm, while the impact tests were carried out on notched specimens (IZOD). The results were taken as the average of 5 samples and were subjected to an analysis of variance (ANOVA). Post hoc comparison was performed to determine the individual means, which are significantly different from a set of means of each reprocessing group using Tukey's test at a 5% probability level.

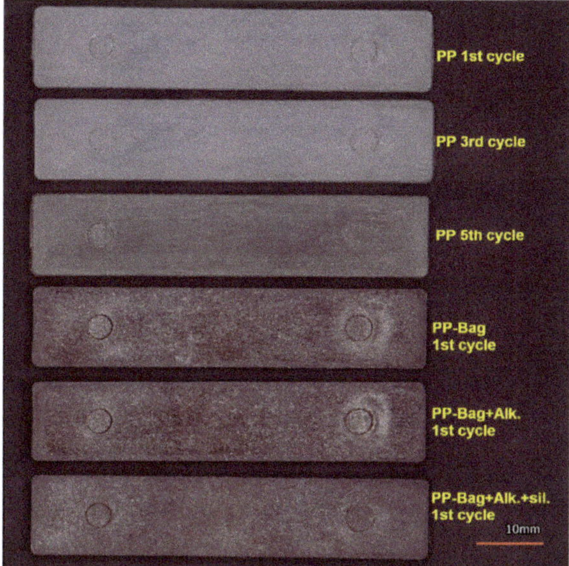

Figure 1. Injected specimens of PP and their bagasse fiber biocomposites.

2.2.4. Thermal Measurements

The thermal stability of the materials was evaluated by thermogravimetric analysis (TGA), measuring the weight loss (%) as a function of temperature using a TA Q500 thermogravimetric analyzer (Texas Instruments, Dallas, TX, USA). These tests were carried out from 25 to 600 °C at a heating rate of 10 °C/min in a nitrogen atmosphere to determine the onset degradation temperature (T_o) and the temperature at the maximum degradation rate (T_{max}). In order to explore the effect of reprocessing cycles and chemical modification of the fibers on the thermal properties of the material, differential scanning calorimetry (DSC) tests were performed. These tests were carried out at a heating–cooling rate of 10 °C/min in a nitrogen atmosphere in several steps: First, the samples were subjected to heating cycle from 20 to 190 °C to erase the thermal history related to processing events, following by a cooling cycle from 190 to 0 °C to determine crystallization temperature (T_c). Finally, a second heating cycle was performed from 0 to 200 °C to determine the melting temperature of the PP phase. Additionally, the degree of crystallinity (χ_c) of each material was calculated from Equation (1) [26]:

$$\chi_c = \left(\frac{\Delta H_m}{\Delta H_m^0 \times W} \times 100 \right) \tag{1}$$

where W represents the PP fraction by weight, ΔH_m is the normalized melting enthalpy of PP of each sample, and ΔH_m^0 (207 J/g) is the melting enthalpy of 100% crystalline PP [27].

2.2.5. Rheological Measurements

The rheological behavior was determined by a rotational rheometer DHR-2 (Texas Instruments, Dallas, TX, USA) equipped with a cone-plate configuration with a diameter of 25 mm and an angle of 5.7°. For this geometry, the cone was truncated to avoid contact between the cone and the plate, and to prevent damage to either with a calibrated distance of 145 μm at the center of the cone. The rheological measurements were performed at 195 °C, the shear rate between 0.1 and 10 s^{-1} and a strain of 1%. Storage modulus (G′), loss modulus (G″) and complex viscosity (η*) were measured.

2.2.6. Dynamic Mechanical Analysis (DMA)

The thermo-mechanical properties of the materials were evaluated using a dynamic mechanical analysis (DMA) RSA-G2 (Texas Instruments, Dallas, TX, USA) with a three-point bending clamp. The equipment was set up as follows: frequency of 1 Hz, 0.01% of strain, temperature range from −50 to 120 °C and a heating rate of 3 °C/min. Storage modulus (E′), loss modulus (E″) and tan δ (loss factor) were measured.

2.2.7. Scanning Electronic Microscopy (SEM)

SEM of the different samples was carried out on the cryogenic fracture surfaces of non-tested injected specimens, operating at a voltage of 10 kV. The samples were previously sputter-coated with gold to increase their electric conductivity. Magnifications of 200× of the fracture surfaces were taken.

3. Results and Discussion

3.1. Mechanical Properties

The influence of bagasse fibers addition and reprocessing cycles on the PP flexural properties were evaluated. With ever increasing demand for high quality and reliable materials and products, flexural tests have become an important tool in both the manufacturing process and research fields to define the material ability to resist deformation under load [28]. Some recent studies have been carried out to study the effects of reprocessing on the flexural and tensile properties of PP reinforced with natural fibers [13,24]. Their results show some differences between flexural and tensile properties, the latter being lower than the former.

During a tensile test, the entire sample is under tensile stress and the rupture begins through the propagation of the largest defect within the specimen. On the other hand, during a flexural test, the maximum stress occurs at the upper and lower surfaces of the specimen where the shear stress is minimum. If the largest defect in the sample is not located in these sections, its influence on the failure mechanism and, therefore, on the flexural strength of the material will be minimal [13]. Therefore, the tensile strength values will be lower as compared to flexural strength values. Despite these differences, in these studies, it was observed that the results of both characterization techniques follow a similar trend with reprocessing cycles. Therefore, the flexural test has been validated as a valuable tool for the mechanical characterization of biocomposites. Figure 2 presents the three-dimensional colormap surface of the flexural modulus and flexural strength of the materials, which are summarized in the Table 2.

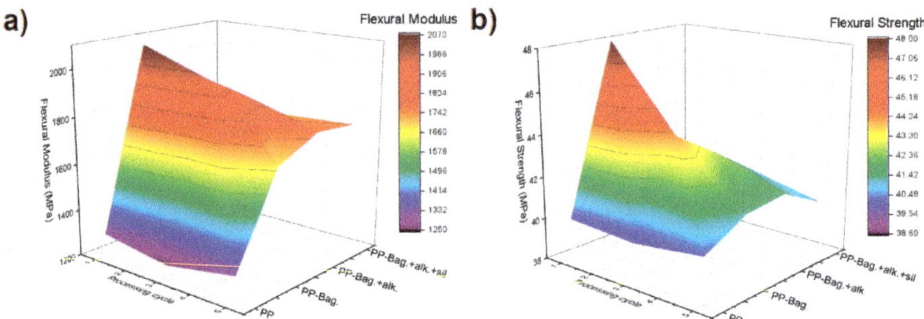

Figure 2. Three-dimensional colormap surface of flexural modulus (**a**) and flexural strength (**b**) of PP and PP-Bagasse biocomposites.

The 3D colormap surface indicates that successive reprocessing cycles did not affect the flexural behavior of the PP. However, bagasse fibers addition and chemical treatments performed on these fibers generate improvements in the flexural properties of the PP matrix. Additionally, it is observed that flexural behavior of biocomposites are dependent on reprocessing cycles of the materials, reaching maximum values around the third cycle. With subsequent reprocessing, a decrease in flexural properties was observed.

Table 2. Flexural properties of PP and PP-Bagasse biocomposites.

Sample	Flexural Modulus (MPa)			Flexural Strength (MPa)		
	Processing Cycle Number					
	1	3	5	1	3	5
PP	1296 ± 70a	1251 ± 54a	1305 ± 36a	40.0 ± 0.7a	39.8 ± 0.9a	40.2 ± 0.4a
PP-Bag.	2069 ± 30b	1969 ± 48b	1673 ± 100b	48.0 ± 1.1b	44.1 ± 0.7b	41.3 ± 1.3a
PP-Bag.+alk.	1847 ± 114c	1853 ± 68c	1761 ± 78b	43.3 ± 0.5c	42.7 ± 0.3c	41.5 ± 1.2a
PP-Bag. +alk.+sil.	1505 ± 94d	1729 ± 66c	1742 ± 116b	38.6 ± 1.9a	41.2 ± 0.4d	40.4 ± 0.9a

(a–d) Different letters in the same column indicate significative differences from a set of means of each reprocessing group ($p < 0.05$).

The first processing cycle shows the flexural modulus (FM) of biocomposites PP-Bag. and PP-Bag. + Alk, increased by 60% and 42% compared to neat PP. Additionally, flexural strength values (FS) increased by 20% and 8%, respectively. For biocomposite PP-Bag. +alk. +sil. FM value increased by 16%; however, no significant differences in the FS value were observed ($p \geq 0.05$). Cerqueira et al. investigated the effect of untreated bagasse addition on the flexural properties of PP and found that FM and FS values increased by 32% and 35% respectively [29]. These improvements in flexural properties due to the addition of natural fibers have been observed in long [30] and short fibers [31,32]. However, it is essential to remark that our study demonstrated that this effect was also generated with the addition of microfibers.

For the third processing cycle, all FM and FS values of the biocomposites show significant differences compared to the FM value of the PP matrix. These increments were 57%, 48% and 38% for PP-Bag., PP-Bag. + Alk. and PP-Bag. +alk. +sil respectively. Additionally, FS values increased by 11%, 7% and 4% respectively. It is interesting to show that FM and FS values of the sample PP-Bag. +alk. +sil. increased by 15% and 7% in comparison with the first processing cycle values. This could be related to a better dispersion state of the silanized bagasse fibers within the PP matrix, the higher thermal stability of chemically modified fibers [6] and a better interaction fiber–matrix generated by the reprocessing cycles.

For the last reprocessing cycle, FM values of the biocomposites show significant differences in comparison with the PP matrix ($p < 0.05$); however, no significant differences were observed among the biocomposites. In the same way, the FS values of the samples were statistically equivalent. These results show that reprocessing could improve fiber dispersion and improve fiber–matrix interaction under compression stresses developed in the biocomposites during bending. However, these improvements seem to achieve a maximum point that in our study corresponded to the third cycle.

Similar behavior was reported by Chaitanya et al., [12], who studied the recyclability of polylactic acid-sisal biocomposites. They found that reprocessing generates a severe reduction in mechanical and viscoelastic properties due to fiber and matrix degradation; therefore, they concluded that recycling of PLA/Sisal biocomposites beyond third reprocessing cycle is not recommended. Figure 3 shows the effect of the addition of bagasse fibers and the reprocessing cycles on the impact strength values.

For the first processing cycle, no significant differences were observed in the impact strength values of the PP matrix and the biocomposites PP-Bag. and PP-Bag. + Alk. However, for the PP-Bag. + Alk. +sil. biocomposite the impact strength increased by around 40% compared to neat PP. This result shows that bagasse fibers treated by silanes agents had improved the capacity of the polymeric matrix to absorb energy. From the revised literature, it can be observed that several factors governed the impact behavior of natural fiber-reinforced biocomposites, for example, chemical treatment applied on the natural fibers, type of natural fiber, interfacial bonding, the composition of the biocomposite and the toughness of the polymeric matrix. In the case of silanes treatments, it was found that this process may have different effects on the impact properties of biocomposite PE-Hemp and PE-Sisal [17].

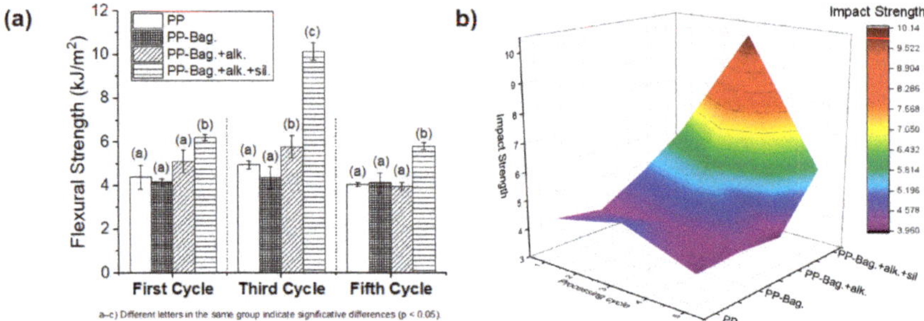

Figure 3. (**a**) Impact properties of PP and PP-Bagasse biocomposites and (**b**) 3D colormap surface of the impact properties.

Biocomposites reprocessing causes interesting changes in the impact properties studied. According to the 3D colormap surface (Figure 3b), the third reprocessing cycle generates a significant increase in the impact values of PP-Bag. + Alk. and PP-Bag. + Alk. +sil. biocomposites. These increases were between 17 and 103% in comparison with reprocessed PP. For the fifth processing cycle, the impact values have similar behavior to that observed in the first cycle. The impact value of biocomposite PP-Bag. + Alk. +sil. increased by 43% compared to the PP. These results are evidence that reprocessing improves the dispersion state of the silanized fibers, fiber–matrix interaction and promotes PP energy absorption. However, as observed in the flexural test these improvements reach their highest point around the third reprocessing cycle.

3.2. Thermal Properties

DSC curves for neat PP and their biocomposites with bagasse fibers at the 1st, 3rd and 5th processing cycles are shown in Figures 4–6. The numerical values of the thermal events of the samples are shown in Table 3. These thermograms do not show any indication of bagasse fibers because the peaks attributed to the different reactions or mechanisms involved in pyrolyzing of the bagasse appears at temperatures higher than those selected for our DSC tests (above 290 °C) [33,34]. However, the fibers effect on the crystallization and melting behavior of the PP phase are discussed in this work.

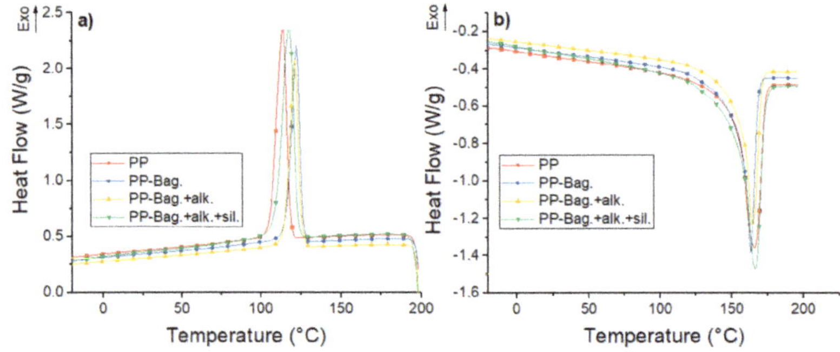

Figure 4. (**a**) Cooling and (**b**) second heating DSC curves for first processing cycle PP and PP-Bagasse biocomposites.

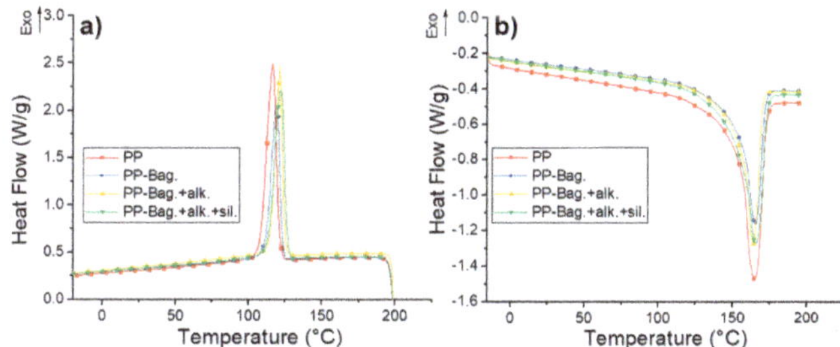

Figure 5. (**a**) Cooling and (**b**) second heating DSC curves for third processing cycle PP and PP-Bagasse biocomposites.

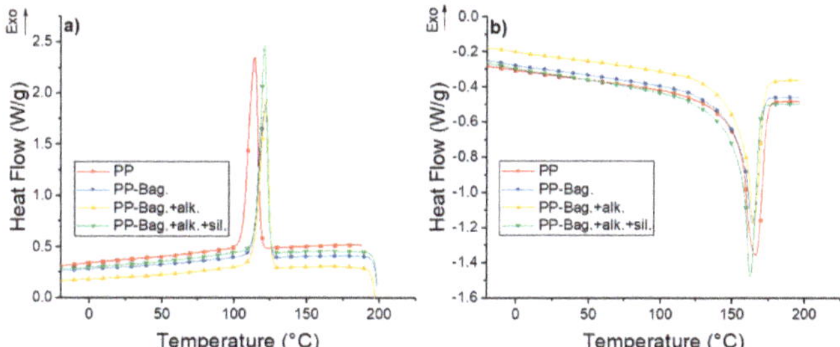

Figure 6. (**a**) Cooling and (**b**) second heating DSC curves for fifth processing cycle PP and PP-Bagasse biocomposites.

Table 3. Thermal properties on cooling and second heating differential scanning calorimetry (DSC) scans of the samples.

Processing Cycle Number	Sample	Cooling		Second Heating	
		T_c * (°C)	T_m * (°C)	ΔH_m (J/g)	X_c (%)
1	PP	114	166	91	44
	PP-Bag.	122	163	75	45
	PP-Bag. +alk.	121	165	74	44
	PP-Bag. +alk. +sil.	118	166	99	59
3	PP	116	165	101	49
	PP-Bag.	119	165	80	49
	PP-Bag. +alk.	121	163	84	50
	PP-Bag. +alk. +sil.	122	165	100	60
5	PP	114	166	95	46
	PP-Bag.	121	164	75	46
	PP-Bag. +alk.	121	164	78	47
	PP-Bag. +alk. +sil.	121	162	83	50

* T_c and T_m were taken at the maximum peak of crystallization and melting peaks.

Cooling thermograms of PP show exothermic peaks located between 114 and 116 °C. These peaks corresponded to the crystallization during the cooling of the PP chains. These crystallization peaks are also observed in PP-Bagasse biocomposites, however, these peaks are located at temperatures between 3 and 7 °C higher compared to the PP in the different extrusion cycles. This shows that bagasse fibers could act as nucleation points that allow the crystallization of PP chains at higher temperatures.

The second heating runs of PP and PP-Bagasse biocomposites present endothermic peaks between 163 and 166 °C related to the melting of the PP matrix. This indicates that bagasse fibers addition did not interfere with the melting process of the PP matrix. In this study, the maximum processing temperature was 185 °C, which is higher than PP melting temperature. This was done with the aim to ensure completely melting of PP crystals and improving the processing of the material without causing degradation to the bagasse fibers. In this aspect, some authors reported that biocomposites processing must be performed below 200 °C to avoid natural fibers degradation [2,17,35].

Additionally, melting enthalpy (ΔH_m) and crystallinity degree (χ_c) values of the biocomposites changed with the number of extrusion cycles and with the bagasse fiber type. For all extrusion cycles, a decrease in the ΔH_m values of PP-Bag and PP-Bag.-alk biocomposites were observed. However, the χ_c of the PP matrix remained similar when the ΔH_m was corrected, considering the weight fraction of bagasse (Equation (1)). This behavior was also observed in other natural fiber-polyolefin biocomposites [5,36]. On the other hand, for biocomposites with silanized bagasse fibers (extrusion cycles 1 and 3), an increase in the χ_c values of around 15% was observed in comparison with the χ_c of the PP matrix. This χ_c increase was slightly for cycle 5; however, it is concluded that the silanized process improved the nucleating effect of the bagasse fibers in the PP. Therefore, the mechanical strength improvement observed in the PP-Bagasse biocomposites could be related to the reinforcement effect of bagasse fibers in the PP and the crystallinity changes of the thermoplastic matrix. Similar results were reported by Zainal et al. [21] on polypropylene-acrylonitrile butadiene rubber-modified bagasse biocomposites. They reported that the chemical treatment of bagasse fibers using silanes increases the nucleation density and the crystallinity degree (%) of the polymeric matrix.

Thermogravimetry (TG) and Derivative Thermogravimetry (DTG) thermograms of PP and PP-Bagasse biocomposites at the 1st, 3rd and 5th processing cycles are shown in Figures 7–9. Additionally, the main thermal parameters obtained from these curves are summarized in Table 4.

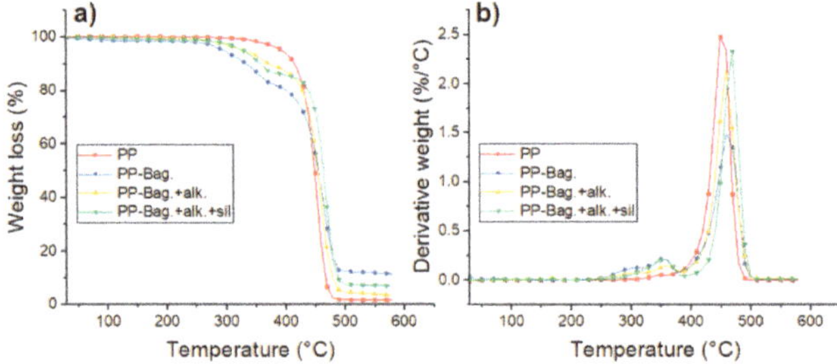

Figure 7. (**a**) TG and (**b**) DTG curves for first processing cycle PP and PP-Bagasse biocomposites.

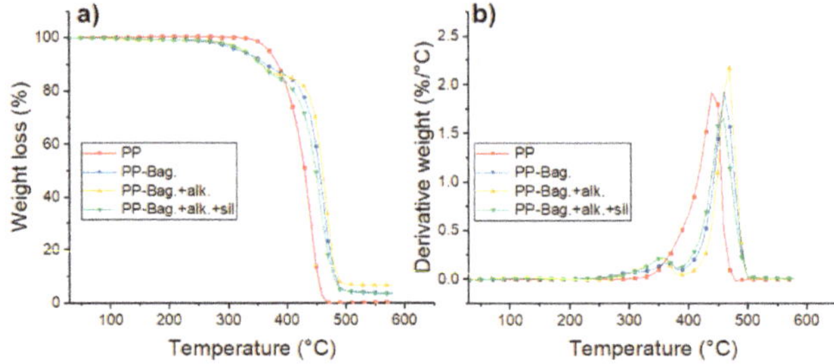

Figure 8. (**a**) TG and (**b**) DTG curves for third processing cycle PP and PP-Bagasse biocomposites.

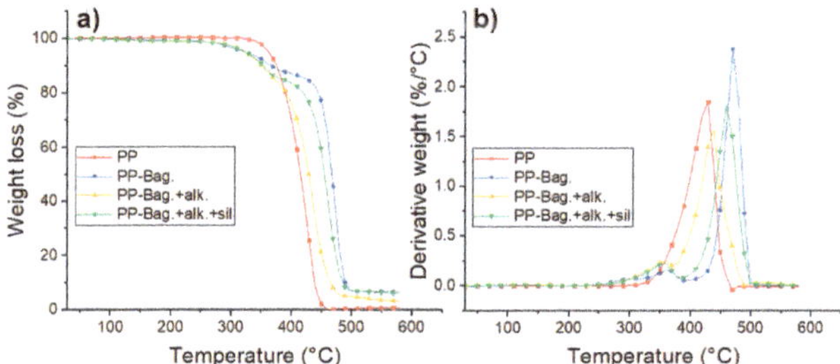

Figure 9. (**a**) TG and (**b**) DTG curves for fifth processing cycle PP and PP-Bagasse biocomposites.

Table 4. Thermal degradation data of PP and PP-Bagasse biocomposites.

Processing Cycle	Sample	Degradation Stage	T_O (°C)	T_{max} (°C)
First cycle	PP	1	408	455
	PP-Bag.	1	271	354
		2	423	462
	PP-Bag. +alk.	1	314	363
		2	438	461
	PP-Bag. +alk. +sil.	1	317	360
		2	447	468
Third cycle	PP	1	373	445
	PP-Bag.	1	271	363
		2	431	463
	PP-Bag. +alk.	1	309	358
		2	431	468
	PP-Bag. +alk. +sil.	1	310	358
		2	428	457
Fifth cycle	PP	1	371	431
	PP-Bag.	1	287	366
		2	430	472
	PP-Bag. +alk.	1	318	360
		2	417	440
	PP-Bag. +alk. +sil.	1	322	360
		2	434	461

The thermal degradation of PP matrices occurs in a single step process. For the first cycle, a T_o of 408 °C and a T_{max} of 455 °C were observed. For reprocessing cycles 3 and 5, To values decreased by 35 and 37 °C while T_{max} decreased by 10 and 24 °C as compared with PP at the first extrusion cycle. This lowering in PP thermal stability with melt reprocessing has been already observed in other studies and could be related to the chain scission mechanism of PP during multiple extrusions [23,37]. Da costa et al. cited that scission of the PP chains during reprocessing generates small and defective molecules, a broader distribution of molecular weights and reduction in the onset degradation temperature of the polymer [38].

For biocomposites, degradation occurs in a two-step process. The first step is related to the decomposition of the bagasse fibers within the biocomposite, while the second step corresponds to the thermal degradation of the PP matrix [6]. The first degradation step show that thermal stability of the chemically treated bagasse fibers is higher than the exhibited by untreated fibers. According to a previous research work published recently by our group the performed chemical treatments could help to extract low thermal stability components of the bagasse fibers like hemicellulose, lignin, pectin and waxes [6]. With only cellulose, the bagasse fibers gain some thermal stability. Additionally, the silanes presence increases thermal stability of the bagasse fibers within the biocomposite, mostly due to the formation of refractory siloxane networks between the fibers and PP after silanization as indicated by literature [39].

The second degradation stage shows that bagasse addition increases the thermal stability of the PP phase. For the first reprocessing cycle, To increased between 15 and 39 °C. Additionally, Tmax increased between 13 and 6 °C in comparison to neat PP as shown in Table 4. This increment in the thermal stability of the biocomposites has been observed in several studies and could be related to the increase of the crystallinity with bagasse addition observed by DSC [40,41]. This behavior is more evident during reprocessing cycles 3 and 5 due to the observed decrease of the thermal stability of neat PP during melt reprocessing. This result shows that bagasse addition improves the thermal stability of the polymer matrix when reprocessing cycles, such as mechanical recycling, are carried out.

After 500 °C, the residue of the samples remains. These residues were composed mainly of ashes and had a weight of 12% for PP-Bag., 8% for PP-Bag. +alk. +sil. and 4% for PP-Bag. +alk. This difference could be related to lignin present on the untreated bagasse, which generates a large number of solid residues after the pyrolysis of the fiber [33].

3.3. Rheological Properties

The influence of bagasse fibers addition and reprocessing cycles on the PP storage modulus (G′) and loss modulus (G″) modulus vs. a frequency is presented in Figure 10. According to Osswald and Rudolph, G′ is a measure for the stored energy and is related to the rigidity and relative entanglement of polymeric chains. On the other hand, G″ is a measure for the lost energy dissipated, for example, as heat or used on the relative movement among polymeric chains [42].

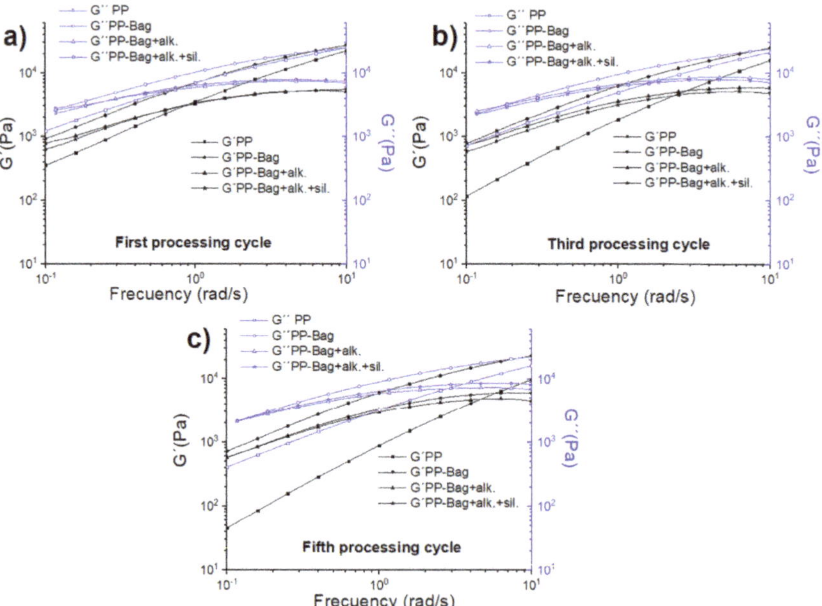

Figure 10. Storage and loss modulus as a frequency function of PP and PP-Bagasse biocomposites: (**a**) First, (**b**) Third and (**c**) Fifth processing cycles.

The results show that G″ is higher than G′ for neat PP and all biocomposites, indicating that these materials present a dominant liquid viscoelastic behavior in the studied frequency range. This behavior has previously been observed in another polyolefin-natural fiber biocomposites with fiber percentages up to 20% by weight [43,44]. Additionally, it is observed that G′ and G″ values of neat PP and biocomposites decrease with successive reprocessing cycles. This decrease has been previously observed and could be related to changes in the length and entanglements of PP polymeric chains caused by multiple extrusion processes [23]. Therefore, the elastic behavior of the biocomposites would be lower with successive reprocessing cycles.

Besides this, a decrease in G′ and G″ values are observed in biocomposites with chemically modified bagasse fibers. These chemical treatments modify the surface of the bagasse fibers; in general terms, these modifications reduce the particle agglomeration, improving the slip or flow between them inside the biocomposite. This is reflected in lower G′ and G″ values. It is interesting to note that the biocomposites with silanized bagasse fibers present the lowest G′ and G″ values for all reprocessing cycles. This could be due to short silane chains that could act as a lubricant at the PP-bagasse fibers interface and could reduce the internal stress generated by fibers agglomeration.

Figure 11 provides the complex viscosity vs. frequency of the neat PP and biocomposites. All materials show shear-thinning behavior, as had been previously observed in PP and PP-natural fiber biocomposites [43]. This behavior is related to the viscoelastic nature of the polymeric matrix and the interaction with the bagasse fibers. At the frequencies studied, the polymeric chains do not have enough recovery time due to the contact between the fibers leading to the non-Newtonian rheological characteristics observed.

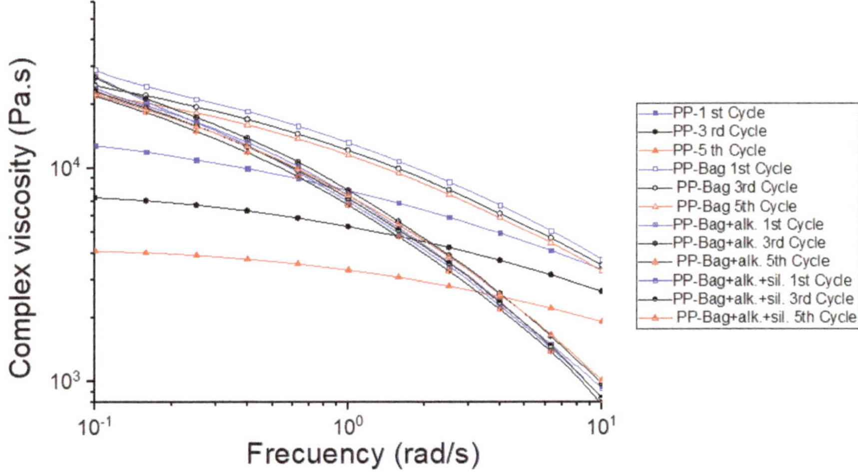

Figure 11. Complex viscosity as a frequency function of PP and PP-Bagasse biocomposites.

The results also show that PP-Bag. biocomposites present the higher viscosity values in the entire frequency range studied. For PP-Bag., the relative movement and disentanglement of bagasse fibers are impeded due to agglomeration of fibers, which hinders polymer chains flow and increasing the viscosity values. For biocomposites obtained from chemically modified bagasse fibers, the viscosity values are lower in comparison to PP-Bag. biocomposites. As mentioned above, the untreated fibers can agglomerate due to adhesive forces between fibers. The performed alkaline treatment aimed to extract the lignin, which is a hydrophobic layer that covers the bagasse fibers. This treatment exposed the cellulose of the bagasse and improved their dispersion within the polymeric matrix, thus decreased the particle–particle interactions, allowing the polymer chains to flow and decreased the viscosity. Furthermore, the silanes treatment produced a functionalized surface with covalent Si-O bonds, which hindered the agglomeration of the fibers and acted as a lubricant, which improved the fibers flowing within the polymeric matrix, causing a decrease in the viscosity values and eased the processability of the biocomposites by conventional plastic transformation processes such as extrusion or injection molding.

Regarding the reprocessing cycles, it is observed that the viscosity values of PP and biocomposites decreased progressively with extrusion cycles. This decrease of the viscosity values was also reported on several PP reprocessing studies and can be related to a decrease of the PP matrix molecular weight due to polymeric chain scission during the multiple extrusion reprocessing steps [23,37,38].

With the aim of further investigate the effect of reprocessing and bagasse fibers addition on the rheological and structural behavior of the materials, Cole–Cole diagrams were used (Figure 12). In this diagram, the imaginary viscosity component (η'') is represented as a function of the real component of the viscosity (η'). The graph should be like a semicircle if the system describes a single relaxation. In heterogeneous melts containing agglomerated fibers, the semicircle shape of the Cole–Cole graph will be modified, the elastic component of the viscosity, and the relaxation time increases [42].

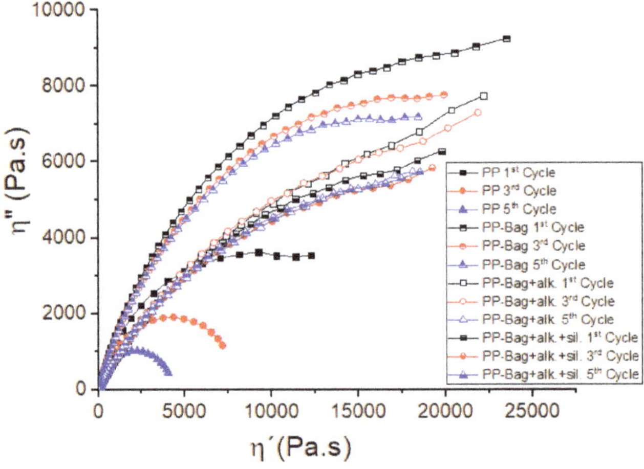

Figure 12. Cole–Cole plots of PP and PP-Bagasse biocomposites.

The Cole–Cole plots of neat PP revealed a semicircle related to single relaxation time. On the other hand, untreated bagasse fibers addition generated an increase in the elastic behavior and relaxation time of the structure, visualized in viscosity components values increments. This behavior indicates the presence of agglomerated fibers and decreased progressively with extrusion cycles. Finally, it is observed that chemical modification of bagasse fibers clearly generated a decrease in the elastic component of the viscosity (n") and shorter relaxation times of the materials. These results could indicate that continuous extrusion processes and chemical modification generate a better dispersion of bagasse fibers within the polymeric matrix.

3.4. Dynamic Mechanical Analysis

According to Saba et al., [45], the storage modulus (E') is related to the ability of a material to store energy during a dynamic test and determine the stiffness of the sample. Additionally, E' is essential for the evaluation of the mechanical properties from the molecular basis because it is sensitive to structural changes within the polymeric matrix such as molecular weight and the interfacial bond between the fibers and the matrix.

Figure 13 shows the E' as a function of temperature for the first processing cycle. It can be observed that E' values of PP increase after the bagasse fibers incorporation. At room temperature, the E' value of neat PP (1481 MPa) increased to 18% with the addition of untreated bagasse fibers. This stiffness increase could be attributed to a decrease in the PP chains mobility generated by the rigid bagasse fibers and indicates that PP-Bag biocomposite had a higher capacity to store energy in comparison with neat PP. This behavior was even more significant after the incorporation of chemically modified bagasse. E′ value increased by 32 and 52% for PP-Bag. +alk. +sil. and PP-Bag. +alk. in comparison with neat PP, respectively. These results show that chemical modifications induced a better adhesion on the interface between bagasse fibers and PP matrix and increased the material capacity to absorb energy. This result was in good agreement with the data obtained from the impact test (Section 3.1).

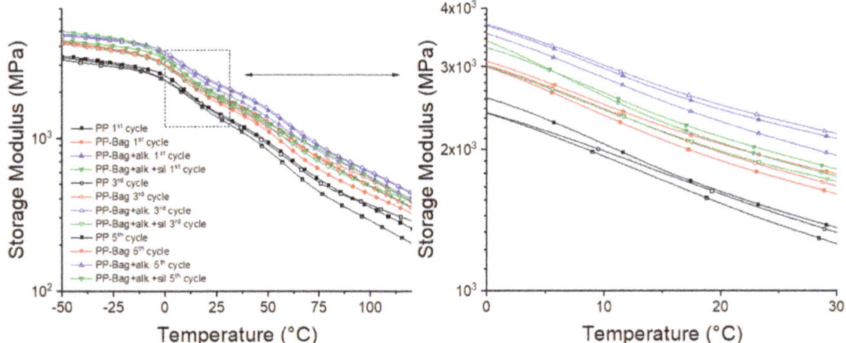

Figure 13. Temperature dependence of the storage modulus of PP and their biocomposites.

Regarding reprocessing cycles, it is observed that E′ values of biocomposites increased progressively with extrusion cycles in comparison with neat PP. For the third reprocessing cycle, E′ values increased up to 27 and 60% for PP-Bag and PP-Bag. +alk., whereas, for the fifth reprocessing cycle, this increment lay between 35 and 57% (again for PP-Bag and PP-Bag. +alk.). This processing event has been already observed during PLA-Sisal biocomposites recycling and indicates that thermo-mechanical reprocessing generate an improvement in the interfacial bonding between the bagasse fibers and the PP matrix [12].

Similar to the observed behavior in E′ graphs, the E″ peaks of PP-bag biocomposites were higher as compared to neat PP (Figure 14). These increases indicate a reduction in the mobility of PP chains due to bagasse fibers. These increases of E″ peaks were higher upon chemically modified bagasse incorporation for all reprocessing cycles and could be related to the enhanced adhesion at the interphase given by the chemical treatments, which suppressed the molecular mobility of the polymeric matrix. This trend has been observed for several polymer-chemical treated fibers biocomposites [6,46,47].

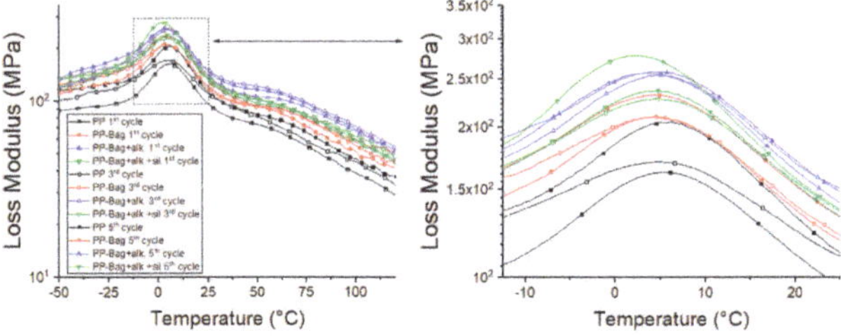

Figure 14. Temperature dependence of loss modulus of PP and their biocomposites.

Tan δ is defined as the ratio between the loss and storage modulus (Tan δ = E″/E′) and is related to the damping properties of the polymeric matrix [45]. The variation of tan δ with temperature is represented in Figure 15. According to this graph, PP shows two main relaxation peaks in the evaluated temperature range, a β relaxation, located around 7–9 °C, which corresponds to the glass transition (Tg) and an α relaxation between 60 and 75 °C [6].

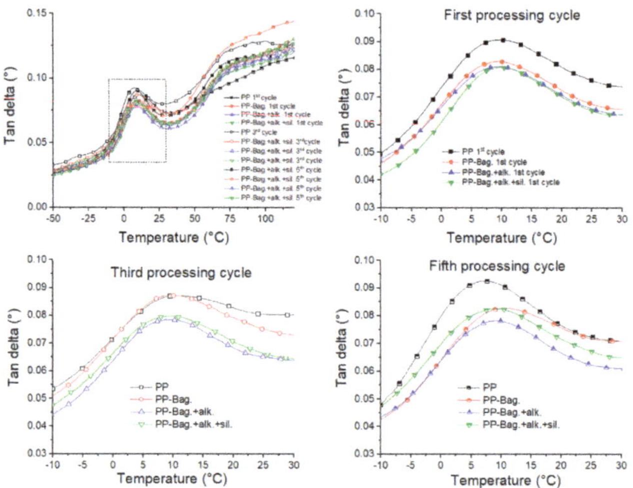

Figure 15. Temperature dependence of tan delta of neat PP and their biocomposites.

Table 5 shows that after untreated bagasse fibers addition into neat PP, Tan δ peaks height decreased for all reprocessing cycles. This is because bagasse fibers hinder the mobility of the polymer chains; also, these fibers support higher stresses fields, which generate less deformation of them at the interface, which causes more energy dissipation. Additionally, it is observed that bagasse fibers addition did not generate significant changes in PP relaxations values (Table 5). For biocomposites obtained from chemically modified bagasse fibers, the Tan δ peaks exhibited even lower magnitude when compared with untreated bagasse biocomposites for all reprocessing cycles. These results, albeit small, show that biocomposites with improved interfacial bonding between natural fibers and PP matrix, given by the chemical modification, will tend to dissipate less energy, showing the lower magnitude of the Tan δ peak in comparison with biocomposites with a weakly bonded interface and can be decisive in product applications that require better mechanical behavior under flexural loads.

To further understand the mobility of the chain segments and the state of dispersion, the full width at half maximum (FWHM) of tan δ peaks was evaluated. Some studies on biocomposites viscoelastic properties have shown that FWHM is a measurement of tan δ curve broadness and could be useful to evaluate the reduction of the molecular mobility during a relaxation like the glass transition [48,49]. According to Manikandan et al., a higher FWHM value implies more interaction and contact between the phases of the composite, which can be associated with low dispersion and a heterogeneous amorphous phase [49]. Among biocomposites, those based on untreated bagasse fibers present the higher FWHM values for all reprocessing cycles. This could be attributed to the suppressed effect of the poor dispersed fibers on the PP matrix molecular mobility. It is also observed that FWHM values decreased for all biocomposites with reprocessing, and it was found to be the lowest for chemically treated fiber-based biocomposites. This result shows that the successive reprocessing improved the homogeneity of the matrix and confirmed that chemical treatments improved the adhesion at the interface, which suppressed the molecular movement of the polymer matrix. This means that biocomposites obtained from chemically modified fibers acted more elastic and confirmed that under a load, these biocomposites had more potential to store energy instead of dissipating it.

Table 5. Dynamic mechanical analysis (DMA) results of the studied materials.

Processing Cycle Number	Sample	Tg (°C) *	E′ (MPa) at 25 °C	Full Width at Half Maximum (FWHM) of Tan δ Peaks	Tan δ Peaks Height
1	PP	7.9	1481	24.4	0.091
	PP-Bag.	8.2	1741	23.3	0.083
	PP-Bag. +alk.	8.1	2261	21.9	0.081
	PP-Bag. +alk. +sil.	8.6	1962	22.2	0.081
3	PP	8.4	1453	24.9	0.087
	PP-Bag.	8.3	1810	22.5	0.087
	PP-Bag. +alk.	7.3	2319	20.7	0.078
	PP-Bag. +alk. +sil.	7.9	1835	20.5	0.080
5	PP	7.0	1383	23.4	0.093
	PP-Bag.	8.9	1905	22.7	0.083
	PP-Bag. +alk.	8.1	2111	21.1	0.080
	PP-Bag. +alk. +sil.	8.2	1899	21.8	0.078

* Tg values of the PP phase were taken at the maximum peak of the tan delta curves.

3.5. Morphology

Figure 16 shows the cryogenic fracture surface of neat PP (Figure 16a) and PP-bag. Biocomposites at the first (Figure 16b,c) and fifth reprocessing cycles (Figure 16d,e). Neat PP presents an irregular surface fracture, which is caused by the inherent semi-crystalline structure of this polymer whereas untreated bagasse incorporation generates a rough fracture surface.

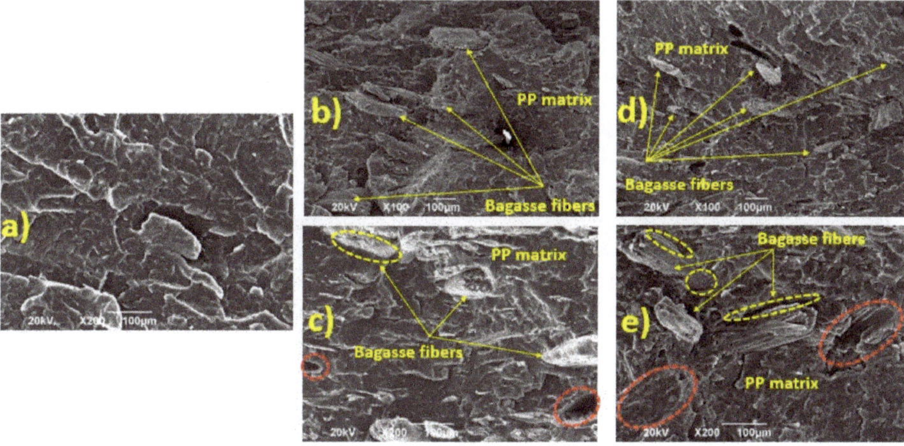

Figure 16. SEM micrograph of the fracture surface of (**a**) neat PP, (**b,c**) PP-Bag. 1st cycle and (**d,e**) PP-Bag. 5th cycle.

Regarding the first cycle, the lower magnification allows one to observe bundles of bagasse fibers within the PP matrix with an average length of 213 ± 24 micrometers (Figure 16b). On the other hand, higher magnification (Figure 16c) show some voids resulting from the fibers pulled out from the matrix (as shown in red circles) and well-delineated gaps at the interface between the bagasse fibers and the PP matrix (yellow circles). These gaps are evidence of the low interfacial adhesion between the PP and the untreated bagasse fibers. This microstructural behavior at the interface was observed in different polymer matrix biocomposites and was related to the low chemical affinity between the inherent hydrophobic polyolefin matrix and the hydrophilic natural fiber.

After five reprocessing cycles, a better distribution of the fibers and a decrease in their length to 165 ± 17 micrometers was observed (Figure 16c). This result shows that during the five extrusion and injection cycles the fibers get crushed to some extent and their length was reduced by 20% and could be related to the decrease observed in the mechanical performance observed in PP-bag. biocomposites considering all reprocessing cycles (Table 2).

Several studies have shown that reprocessing of biocomposites generates a decrease of 60% in the fibers length. The fibers length decreasing is higher than that observed in our study, however the initial size of the fibers used must be considered. In these studies, the initial size of the fibers was 1–0.8 mm, while in our study, the initial average size was around 0.2 mm. However, the authors report that biocomposites retain its mechanical performance throughout cycles better than fiberglass reinforced composites due to the inherent flexibility of the natural fibers and the ability to resist external mechanical forces.

Reprocessing cycles will inevitably impact the mechanical properties of biocomposites; however, our study shows that PP-Bag. biocomposites' mechanical performance was maintained up to three reprocessing cycles without the addition of virgin material. Based on this result, it could be concluded that recycling potential of biocomposites was huge due to their mechanical performance retention resulting in advantages in terms of sustainability and life cycle impact of these materials.

Figure 17 shows the cryogenic fracture surface corresponding to the biocomposites of the fifth reprocessing cycle along with the optical micrograph of the surface of each sample at the same lighting level. For Figure 17b,c, chemical treatments reduced the gaps between the bagasse fibers and the PP matrix and improved the interface of the biocomposites. Additionally, the optical micrographs of chemically modified fibers based biocomposites show a decrease of the dark areas related to particles agglomeration on the surface of the sample. This behavior is observed in better detail in the sample with silanized bagasse and could be due to the lubricant effect given by silanes, which improved the dispersion of the fibers. This result was also consistent with the data obtained by rheology and DMA measurements and confirmed that chemical treatments generated a bonding effect at the PP and natural fibers interface and exposed the cellulose of the bagasse improving their dispersion within the polymeric matrix.

The scanning electronic microscopy gives valuable information about biocomposites morphological characterization. Several authors have made conclusions studying the fiber content (composition), the chemical treatment of natural fibers, and biocomposites reprocessing using SEM [12,20,21,44]. This technique, together with the performed rheological and dynamic mechanical characterization, could provide valuable evidence of the dispersion state of the fibers within the biocomposites. In our case, the Cole–Cole plots and FWHM of tan δ peaks gave evidence for concluding that multiple extrusion cycles could decrease the particle agglomeration and generated a better dispersion of the fibers within the matrix.

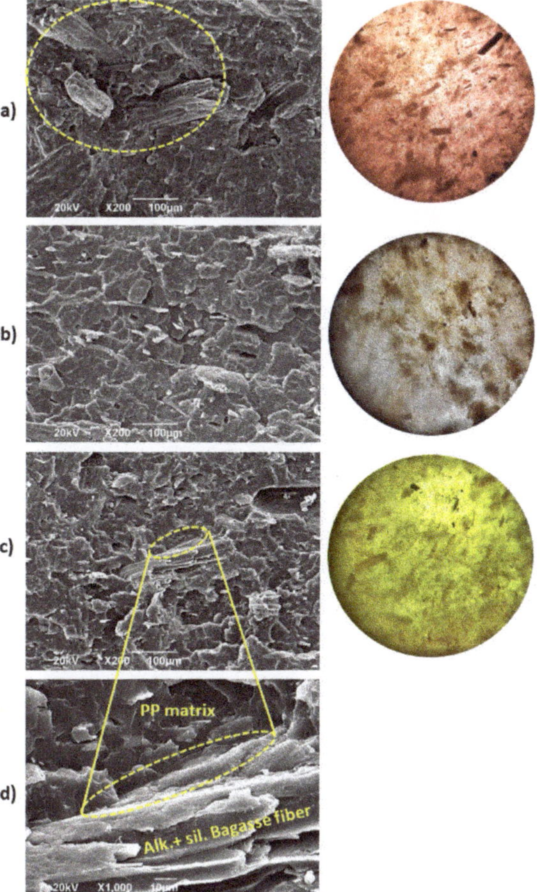

Figure 17. SEM (**left**) and optical micrographs (**right**) for 5th processing cycle: (**a**) PP.Bag., (**b**) PP-Bag + Alk. and PP-Bag + Alk. +sil. (**c,d**).

4. Conclusions

In this research, PP-bagasse biocomposites were prepared by incorporating 20% by weight of bagasse fibers treated by alkaline treatment with NaOH and silanization after the alkaline treatment.

These biocomposites were reprocessed 5 times using the extrusion process followed by injection molding after each reprocessing cycle in order to evaluate the effects of reprocessing and chemical treatments on the morphology, mechanical, thermal, as well as viscoelastic properties of these materials. Doing so, the following conclusions could be obtained from the present research:

- The mechanical properties indicate that reprocessing and chemical treatments performed to bagasse microfibers could improve fiber dispersion and fiber–matrix interaction under compression stresses developed in the biocomposites during bending, and promoted PP energy absorption. These mechanical improvements achieved a maximum point that, in our study, corresponded to the third cycle.
- Thermal characterization revealed that bagasse fibers addition increased the crystallization temperature and the thermal stability of the PP phase for all extrusion cycles without disturbing

the melting process of the PP matrix. Additionally, silanized fibers based biocomposites presented the highest thermal stability for all processing cycles.

- The rheological test shows that the viscosity values of PP and biocomposites decreased progressively with extrusion cycles. Additionally, biocomposites obtained from chemically modified bagasse particles presented lower viscosity values in comparison with neat bagasse based biocomposites. However, Cole–Cole plots indicates that continuous extrusion processes and chemical modification generated a better dispersion of bagasse fibers within the polymeric matrix.

- DMA results included a complete analysis of the height and broadness of tan δ peaks and show that reprocessing and chemical modifications induced a better adhesion on the interface between bagasse fibers and PP matrix and increased the PP capacity to absorb energy.

- SEM micrographs show that during reprocessing the bagasse fibers got crushed to some extent and length was reduced around 20%. However it is important to remark that despite this decrease in fiber length, PP-Bag. biocomposites' mechanical performance was maintained up to three reprocessing cycles without the addition of virgin material. Additionally, chemical treatments generated a bonding effect at the PP and natural fibers interface and exposed the cellulose of the bagasse, improving their dispersion within the polymeric matrix.

- Based on these findings, it could be concluded that bagasse fibers show an interesting potential for biocomposites production with a high potential of application in the design and manufacture of sustainable and highly recyclable products by injection molding. This could generate some economic and environmental benefits in the search for sustainability in the plastics industry.

Author Contributions: Conceptualization, F.L.-V. and M.A.H.-S.; Formal analysis, J.P.C.-A., F.L.-V., C.C. and M.A.H.-S.; Funding acquisition, B.V.-M. and M.A.H.-S.; Investigation, J.P.C.-A., F.L.-V., C.C. and M.A.H.-S.; Methodology, J.P.C.-A., F.L.-V., C.C. and M.A.H.-S.; Project administration, B.V.-M. and M.A.H.-S.; Resources, B.V.-M. and M.A.H.-S.; Supervision, B.V.-M.; Writing—original draft, J.P.C.-A., F.L.-V. and M.A.H.-S.; Writing—review and editing, J.P.C.-A., F.L.-V., C.C., B.V.-M. and M.A.H.-S. All authors have read and agreed to the published version of the manuscript.

Funding: This research received no external funding.

Acknowledgments: The authors acknowledge the Autónoma de Occidente University (Cali-Colombia) for technical and financial support; Research Group of New Solids with Industrial Application-GINSAI (Autónoma de Occidente University) for its support in the thermal characterization; National Center for Technical Assistance to Industry-ASTIN SENA (Cali-Colombia) for its technical support through injection molds manufacturing and rheological measurements; University del Valle (Cali-Colombia) for SEM microscopy. Also, the authors wish to thank to the company "Sucromiles" (Cali-Colombia), for providing the raw sugarcane bagasse.

Conflicts of Interest: The authors declare no conflict of interest.

References

1. Mochane, M.J.; Mokhena, T.C.; Mokhothu, T.H.; Mtibe, A.; Sadiku, E.R.; Ray, S.S.; Ibrahim, I.D.; Daramola, O.O. Recent progress on natural fiber hybrid composites for advanced applications: A review. *Express Polym. Lett.* **2019**, *13*, 159–198. [CrossRef]

2. Pickering, K.L.; Efendy, M.G.A.; Le, T.M. A review of recent developments in natural fibre composites and their mechanical performance. *Compos. Part A Appl. Sci. Manuf.* **2016**, *83*, 98–112. [CrossRef]

3. Hidalgo-Salazar, M.A.; Correa-Aguirre, J.P.; Montalvo-Navarrete, J.M.; Lopez-Rodriguez, D.F.R.-G.A. Recycled Polypropylene-Coffee Husk and Coir Coconut Biocomposites: Morphological, Mechanical, Thermal and Environmental Studies. In *Thermosoftening Plastics*; IntechOpen: London, UK, 2020.

4. KC, B.; Faruk, O.; Agnelli, J.A.M.; Leao, A.L.; Tjong, J.; Sain, M. Sisal-glass fiber hybrid biocomposite: Optimization of injection molding parameters using Taguchi method for reducing shrinkage. *Compos. Part A Appl. Sci. Manuf.* **2016**, *83*, 152–159. [CrossRef]

5. Hidalgo-Salazar, M.A.; Correa, J.P. Mechanical and thermal properties of biocomposites from nonwoven industrial Fique fiber mats with Epoxy Resin and Linear Low Density Polyethylene. *Results Phys.* **2018**, *8*, 461–467. [CrossRef]

6. Hidalgo-Salazar, M.; Luna-Vera, F.; Pablo Correa-Aguirre, J. Biocomposites from Colombian Sugarcane Bagasse with Polypropylene: Mechanical, Thermal and Viscoelastic Properties. In *Characterizations of Some Composite Materials*; IntechOpen: London, UK, 2019.

7. Radoor, S.; Karayil, J.; Rangappa, S.M.; Siengchin, S.; Parameswaranpillai, J. A review on the extraction of pineapple, sisal and abaca fibers and their use as reinforcement in polymer matrix. *Express Polym. Lett.* **2020**, *14*, 309–335. [CrossRef]

8. Beigbeder, J.; Soccalingame, L.; Perrin, D.; Bénézet, J.C.; Bergeret, A. How to manage biocomposites wastes end of life? A life cycle assessment approach (LCA) focused on polypropylene (PP)/wood flour and polylactic acid (PLA)/flax fibres biocomposites. *Waste Manag.* **2019**, *83*, 184–193. [CrossRef] [PubMed]

9. Sapuan, S.M. Chapter 6 - Materials Selection for Composites: Concurrent Engineering Perspective. In *Composite Materials*; Butterworth-Heinemann: Oxford, UK, 2017.

10. Väisänen, T.; Das, O.; Tomppo, L. A review on new bio-based constituents for natural fiber-polymer composites. *J. Clean. Prod.* **2017**, *149*, 582–596. [CrossRef]

11. Correa, J.P.; Montalvo-Navarrete, J.M.; Hidalgo-Salazar, M.A. Carbon footprint considerations for biocomposite materials for sustainable products: A review. *J. Clean. Prod.* **2018**. [CrossRef]

12. Chaitanya, S.; Singh, I.; Song, J. Il Recyclability analysis of PLA/Sisal fiber biocomposites. *Compos. Part B Eng.* **2019**, *173*, 106895. [CrossRef]

13. Uitterhaegen, E.; Parinet, J.; Labonne, L.; Mérian, T.; Ballas, S.; Véronèse, T.; Merah, O.; Talou, T.; Stevens, C.V.; Chabert, F.; et al. Performance, durability and recycling of thermoplastic biocomposites reinforced with coriander straw. *Compos. Part A Appl. Sci. Manuf.* **2018**, *113*, 254–263. [CrossRef]

14. Hidalgo Salazar, M.A.; Muñoz Velez, M.F.; Quintana Cuellar, K.J. Análisis mecánico del compuesto polietileno aluminio reforzado con fibras cortas de fique en disposición bidimensional (Mechanical analysis of polyethylene aluminum composite reinforced with short fique fibers available a in two-dimensional arrangement). *Rev. Latinoam. Metal. Mater.* **2011**, *32*, 89–95.

15. FAO Food Outlook: Sugar chapter. Available online: http://www.fao.org/fileadmin/templates/est/COMM_MARKETS_MONITORING/Sugar/Documents/sugar_assessment_food_outlook_may_2019.pdf (accessed on 3 April 2020).

16. Carlos Cueva-Orjuela, J.; Hormaza-Anaguano, A.; Merino-Restrepo, A. Sugarcane bagasse and its potential use for the textile effluent treatment. *Rev. DYNA* **2017**, *84*, 291–297. [CrossRef]

17. Latif, R.; Wakeel, S.; Zaman Khan, N.; Noor Siddiquee, A.; Lal Verma, S.; Akhtar Khan, Z. Surface treatments of plant fibers and their effects on mechanical properties of fiber-reinforced composites: A review. *J. Reinf. Plast. Compos.* **2019**, *38*, 15–30. [CrossRef]

18. Muñoz-Vélez, M.; Hidalgo-Salazar, M.; Mina-Hernández, J.; Muñoz-Vélez, M.F.; Hidalgo-Salazar, M.A.; Mina-Hernández, J.H. Effect of Content and Surface Modification of Fique Fibers on the Properties of a Low-Density Polyethylene (LDPE)-Al/Fique Composite. *Polymers* **2018**, *10*, 1050. [CrossRef]

19. Anggono, J.; Sugondo, S.; Sewucipto, S.; Purwaningsih, H.; Henrico, S. The use of sugarcane bagasse in PP matrix composites: A comparative study of bagasse treatment using calcium hydroxide and sodium hydroxide on composite strength. *AIP Conf. Proc.* **2017**, *1788*, 030055.

20. de Carvalho Neto, A.G.V.; Ganzerli, T.A.; Cardozo, A.L.; Fávaro, S.L.; Pereira, A.G.B.; Girotto, E.M.; Radovanovic, E. Development of composites based on recycled polyethylene/sugarcane bagasse fibers. *Polym. Compos.* **2014**, *35*, 768–774. [CrossRef]

21. Zainal, M.; Santiagoo, R.; Ayob, A.; Ghani, A.A.; Mustafa, W.A.; Othman, N.S. Thermal and mechanical properties of chemical modification on sugarcane bagasse mixed with polypropylene and recycle acrylonitrile butadiene rubber composite. *J. Thermoplast. Compos. Mater.* **2019**, 089270571983207. [CrossRef]

22. Achilias, D.S.; Antonakou, E.; Roupakias, C.; Megalokonomos, P.; Lappas, A. Recycling techniques of polyolefins from plastic wastes. *Glob. Nest J.* **2008**, *10*, 114–122. [CrossRef]

23. Martín-Alfonso, J.E.; Franco, J.M. Influence of polymer reprocessing cycles on the microstructure and rheological behavior of polypropylene/mineral oil oleogels. *Polym. Test.* **2015**, *45*, 12–19. [CrossRef]

24. Lila, M.K.; Singhal, A.; Banwait, S.S.; Singh, I. A recyclability study of bagasse fiber reinforced polypropylene composites. *Polym. Degrad. Stab.* **2018**, *152*, 272–279. [CrossRef]

25. Hidalgo-Salazar, M.A.; Salinas, E. Mechanical, thermal, viscoelastic performance and product application of PP- rice husk Colombian biocomposites. *Compos. Part B Eng.* **2019**, *176*, 107135. [CrossRef]

26. Mazian, B.; Bergeret, A.; Benezet, J.C.; Malhautier, L. Impact of field retting and accelerated retting performed in a lab-scale pilot unit on the properties of hemp fibres/polypropylene biocomposites. *Ind. Crops Prod.* **2020**, *143*, 111912. [CrossRef]

27. Blaine, R.L. Thermal Applications Note - Polymer Heat of Fusion. Available online: http://www.tainstruments.com/pdf/literature/TN048.pdf (accessed on 15 July 2019).

28. Hodgkinson, J.M. Testing the strength and stiffness of polymer matrix composites. In *Failure Mechanisms in Polymer Matrix Composites*; Elsevier: Amsterdam, The Netherlands, 2012; pp. 129–182.

29. Cerqueira, E.F.; Baptista, C.A.R.P.; Mulinari, D.R. Mechanical behaviour of polypropylene reinforced sugarcane bagasse fibers composites. *Procedia Eng.* **2011**, *10*, 2046–2051. [CrossRef]

30. Hidalgo, M.A.; Muñoz, M.F.; Quintana, K.J. Mechanical behavior of polyethylene aluminum composite reinforced with continuous agro fique fibers. *Rev. Lat. Met. Mat.* **2011**, *31*, 187–194.

31. Hidalgo, M.; Muñoz, M.; Quintana, K. Mechanical analysis of polyethylene aluminum composite reinforced with short fique fibers available a in two-dimensional arrangement. *Rev. Latinoam. Metal. Mater.* **2012**, *32*, 89–95.

32. Bourmaud, A.; Baley, C. Investigations on the recycling of hemp and sisal fibre reinforced polypropylene composites. *Polym. Degrad. Stab.* **2007**, *92*, 1034–1045. [CrossRef]

33. Yang, H.; Yan, R.; Chen, H.; Lee, D.H.; Zheng, C. Characteristics of hemicellulose, cellulose and lignin pyrolysis. *Fuel* **2007**, *86*, 1781–1788. [CrossRef]

34. Fernandes Pereira, P.H.; Cornelis, H.; Voorwald, J.; Odila, M.; Cioffi, H.; Mulinari, D.R.; Da Luz, S.M.; Lucia, M.; Pinto, C.; Silva, D. Sugarcane Bagasse Pulping and Bleaching: Thermal and Chemical Characterization. *BioRes* **2011**, *6*, 2471–2482.

35. Ramamoorthy, S.K.; Skrifvars, M.; Persson, A. A Review of Natural Fibers Used in Biocomposites: Plant, Animal and Regenerated Cellulose Fibers. *Polym. Rev.* **2015**, *55*, 107–162. [CrossRef]

36. Hidalgo-Salazar, M.A.; Munõz, M.F.; Mina, J.H. Influence of Incorporation of Natural Fibers on the Physical, Mechanical, and Thermal Properties of Composites LDPE-Al Reinforced with Fique Fibers. *Int. J. Polym. Sci.* **2015**. [CrossRef]

37. Caicedo-Cano, C.; Crespo-Delgado Lina, M.; De La Cruz-Rodríguez Hever, Á.-J.N.A. Thermo-mechanical properties of Polypropylene: Effects during reprocessing. *Ing. Investig. Technol.* **2017**, *18*, 245–252. [CrossRef]

38. Da Costa, H.M.; Ramos, V.D.; Rocha, M.C.G. Rheological properties of polypropylene during multiple extrusion. *Polym. Test.* **2005**, *24*, 86–93. [CrossRef]

39. Kabir, M.M.; Wang, H.; Lau, K.T.; Cardona, F. Chemical treatments on plant-based natural fibre reinforced polymer composites: An overview. *Compos. Part B Eng.* **2012**, *43*, 2883–2892. [CrossRef]

40. Luz, S.M.; Gonçalves, A.R.; Del'arco, A.P.; Ferrão, P.M.C. Composites from Brazilian natural fibers with polypropylene: Mechanical and thermal properties. *Compos. Interfaces* **2008**, *15*, 841–850. [CrossRef]

41. Motaung, T.E.; Linganiso, L.Z.; John, M.; Anandjiwala, R.D.; Motaung, T.E.; Linganiso, L.Z.; John, M.; Anandjiwala, R.D. The Effect of Silane Treated Sugar Cane Bagasse on Mechanical, Thermal and Crystallization Studies of Recycled Polypropylene. *Mater. Sci. Appl.* **2015**, *6*, 724–733. [CrossRef]

42. Osswald, T.; Rudolph, N. *Polymer Rheology. Fundamentals and Applications*; Hanser Publications: Munich, Germany, 2015; ISBN 9781569905173.

43. Ares, A.; Bouza, R.; Pardo, S.G.; Abad, M.J.; Barral, L. Rheological, mechanical and thermal behaviour of wood polymer composites based on recycled polypropylene. *J. Polym. Environ.* **2010**, *18*, 318–325. [CrossRef]

44. Escócio, V.A.; Pacheco, E.B.A.V.; Silva, A.L.N.; da Cavalcante, A.d.P.; Visconte, L.L.Y. Rheological Behavior of Renewable Polyethylene (HDPE) Composites and Sponge Gourd (*Luffa cylindrica*) Residue. *Int. J. Polym. Sci.* **2015**, *2015*, 714352. [CrossRef]

45. Saba, N.; Jawaid, M.; Alothman, O.Y.; Paridah, M.T. A review on dynamic mechanical properties of natural fibre reinforced polymer composites. *Constr. Build. Mater.* **2016**, *106*, 149–159. [CrossRef]

46. Shinoj, S.; Visvanathan, R.; Panigrahi, S.; Varadharaju, N. Dynamic mechanical properties of oil palm fibre (OPF)-linear low density polyethylene (LLDPE) biocomposites and study of fibre-matrix interactions. *Biosyst. Eng.* **2011**, *109*, 99–107. [CrossRef]

47. Mohanty, S.; Verma, S.K.; Nayak, S.K. Dynamic mechanical and thermal properties of MAPE treated jute/HDPE composites. *Compos. Sci. Technol.* **2006**, *66*, 538–547. [CrossRef]

48. Xu, H.; Liu, C.Y.; Chen, C.; Hsiao, B.S.; Zhong, G.J.; Li, Z.M. Easy alignment and effective nucleation activity of ramie fibers in injection-molded poly(lactic acid) biocomposites. *Biopolymers* **2012**, *97*, 825–839. [CrossRef] [PubMed]

49. Manikandan Nair, K.; Thomas, S.; Groeninckx, G. Thermal and dynamic mechanical analysis of polystyrene composites reinforced with short sisal fibres. *Compos. Sci. Technol.* **2001**, *61*, 2519–2529. [CrossRef]

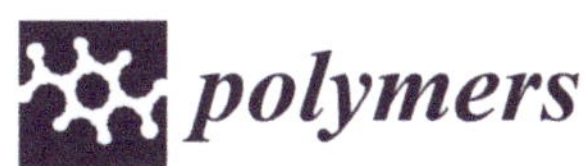 

Article

Characterization, Biocompatibility, and Optimization of Electrospun SF/PCL/CS Composite Nanofibers

Hua-Wei Chen * and Min-Feng Lin

Department of Chemical and Materials Engineering, National Ilan University, Yilan 26047, Taiwan; superlemi@gmail.com
* Correspondence: hwchen@niu.edu.tw

Received: 9 March 2020; Accepted: 25 June 2020; Published: 27 June 2020

Abstract: In this study, composite nanofibers (SF/PCL/CS) for the application of dressings were prepared with silk fibroin (SF), polycaprolactone (PCL), and chitosan (CS) by electrospinning techniques, and the effect of the fiber diameter was investigated using the three-stage Taguchi experimental design method (L9). Nanofibrous scaffolds were characterized by the combined techniques of scanning electron microscopy (SEM) and transmission electron microscopy (TEM), a cytotoxicity test, proliferation tests, the antimicrobial activity, and the equilibrium water content. A signal-to-noise ratio (S/N) analysis indicated that the contribution followed the order of SF to PCL > flow rate > applied voltage > CS addition, possibly owing to the viscosity and formation of the beaded fiber. The optimum combination for obtaining the smallest fiber diameter (170 nm) with a smooth and uniform distribution was determined to be a ratio of SF to PCL of 1:2, a flow rate of 0.3 mL/hr, and an applied voltage of 25 kV at a needle tip-to-collector distance of 15 cm (position). The viability of these mouse fibroblast L929 cell cultures exceeded 50% within 24 hours, therefore SF/PCL/CS could be considered non-toxic according to the standards. The results proposed that the hydrophilic structure of SF/PCL/CS not only revealed a highly interconnected porous construction but also that it could help cells promote the exchange of nutrients and oxygen. The SF/PCL/CS scaffold showed a high interconnectivity between pores and porosity and water uptake abilities able to provide good conditions for cell infiltration and proliferation. The results from this study suggested that SF/PCL/CS could be suitable for skin tissue engineering.

Keywords: chitosan; composite nanofibers; electrospinning; silk fibroin; polycaprolactone; Taguchi

---

1. Introduction

In recent decades, nanofiber membranes have been widely used in various fields and attracted more attention due to their unique properties, such as large specific surface areas, high porosity, interconnected pores, and high functionality. Electrospun composite nanofibers possess great potentialities in biomedical applications, such as tissue engineering [2–4], wound healing [5–7], and drug delivery [8–10], as well as other applications such as magnetism [11], photonics [12], filtration [13], composites [14], shape memory [15], and lithium batteries [16]. The diameter of electrospun nanofibers affects many important properties, such as the melting point, tensile modulus, hardness, drug delivery, biological factors, and cell growth of nonwoven fabrics [17,18]. Research [19] has demonstrated that fiber diameter plays a key role in cell adhesion, proliferation, and cell migration on the scaffold.

Silk fibroin (SF), which is extracted mainly from silkworms, has various properties, including good biocompatibility, biodegradability, morphological flexibility, mechanical properties, a low inflammatory response, non-toxicity, and non-carcinogenicity, and it can promote cell adhesion, migration, and the proliferation of cell ligands [8,20,21]. However, the β-sheet secondary structure of pure SF seems to impede the electrospinning process, and the mechanical properties of neat SF electrospun fibers were poor [20]. Herein, polycaprolactone (PCL), the most commonly used synthetic polymer, was chosen to be blended with SF. PCL is widely used in tissue engineering and drug delivery applications due to its good mechanical properties and biodegradability. PCL has limitations to its biological activity, hydrophobicity, and bacterial degradation, therefore PCL cannot provide the adhesion environment required for cells [2]. PCL has renowned mechanical properties and does not have ionizable side groups in its structures, such as –COOH and –NH$_2$, which occur on the natural polymer chitosan (CS) and on several anionic polysaccharides and proteins, respectively. The CS is an amino polysaccharide derived from chitin that has excellent biological properties, such as biocompatibility, biodegradability, hydrophilicity, non-toxicity, and antithrombotic and antimicrobial activities. Chitosan that possesses positively charged groups (amine groups) is likely to interact with negatively charged cell membranes via electrostatic interaction. Furthermore, the antimicrobial activity of chitosan effectively increases the permeability of negatively charged cell membranes to disrupt and release intracellular compounds [22]. Therefore, investigations of SF and its association with other components (PCL and CS) are carried out to give better mechanical properties to the smooth nanofibers for the tissue engineering. The Taguchi method is an effective method to find the influence of different factors on the target results, thereby improving the manufacturability, reliability, and quality of a product and reducing the number of experiments and calculation time [23,24]

Thus, SF/PCL/CS composite nanofiber scaffolds for the application of dressings and tissue engineering were fabricated with PCL polymer as a precursor using electrospinning techniques in this study. Further, the effect of chitosan additions on the nanofiber diameter was also investigated via the analysis of the antimicrobial activity and equilibrium water content. The optimum combination of parameters obtained from the ratio of silk fibroin to polycaprolactone, chitosan additions, flow rate, and applied voltage in response to minimizing diameter size and its variation for SF/PCL/CS composite nanofibers was determined by means of the Taguchi DoE method. Herein, the biocompatible properties were evaluated with cytotoxicity tests and proliferation tests so as to determine the optimal SF/PCL/CS scaffolds.

2. Experimental Method

2.1. Preparation of Regenerated Silk Fibroin (SF)

All the materials, solvents, and reagents were purchased from commercial suppliers and used as received. Cocoons were received from Paolun farm (Taiwan). Polycaprolactone (PCL, Mw = 80,000) was purchased from Sigma-Aldrich (St. Louis, MO). Polyethylene oxide (PEO, Mw = 60,000–100,000), chitosan (CS, Mw = 10,000–30,000), and formic acid were purchased from Acros Organics. The preparation of the silk fibroin (SF) used in this study made reference to previous research [1], with some modification. The cocoons were boiled in a 0.5% (w/v) Na$_2$CO$_3$ aqueous solution at a temperature of 100 °C for one hour. The silk fibers were rinsed with distilled water for 30 min to remove the Na$_2$CO$_3$ aqueous solution and then rinsed with deionized water for 30 min to remove the sericin. After being dried to a constant weight in an oven at 80 °C, the degummed silk fibers were dissolved in 40% aqueous CaCl$_2$ at 100 °C. The SF solution was replaced with a dialyzed membrane (molecular weight cut-off (MWCO) = 12,000–14,000 Da) for three days to eliminate small molecular impurities and calcium chloride. Lastly, the SF solution was lyophilized in a freeze dryer and stored at room temperature.

2.2. Preparation of the Electrospinning Solutions and Electrospinning

The SF and PCL were dissolved in formic acid to obtain 10 wt.% concentrations. Different ratios of SF/PCL (5.00%:5.00%, 3.33%:6.67%, and 2.50%:7.50%) were dissolved in formic acid and stirred at room temperature for two hours. Subsequently, 1 wt.% PEO as a thickener was added to the solution. Finally, from 0.5 to 1 wt.% CS was added to the electrospinning solutions.

The SF/PCL/CS composite nanofibers were obtained from the electrospinning of the prepared suspensions through a FES-COS electrospinning apparatus (Falco Co, Taipei, Taiwan). Briefly, the suspensions were drawn into a 10 mL syringe with a 21-gauge needle. The electrospinning was performed under ambient conditions (a temperature of 24.5 to 27.5 °C and a relative humidity of 45% to 50%). The following optimized electrospinning parameters were kept constant throughout the experiments: 100 rpm roller collector to collect fibers, 15 cm TCD (tip to collector distance), 15 kV–25 kV applied voltage, and 0.2 mL/h–0.4 mL/h feeding rate. The SF/PCL/CS homogeneous solutions were electrospun on a rotating cylindrical drum covered with an aluminum layer during the process. Finally, the collected membranes were taken from the surface of the collector and conserved in a sealed container for further experiments.

2.3. Taguchi DOE Parameter Setting

Numerous references [19–24] state that the ratio of silk fibroin to polycaprolactone, the chitosan content, the flow rate, and the applied voltage have significant effects on the average diameter and uniformity of fibers; thus, these four concentration and electrospinning parameters were selected for this experiment (Table 1).

Table 1. Actors and levels used in the experiment.

	Ratio of Silk Fibroin and Polycaprolactone	Chitosan Content (wt.%)	Flow Rate (mL/h)	Voltage (kV)
Level 1	1:1 (5.00%:5.00%)	0.50	0.2	15
Level 2	1:2 (3.33%:6.67%)	0.75	0.3	20
Level 3	1:3 (2.50%:7.50%)	1.00	0.4	25

The full factorial experiment of 81 (3^4) trials could be completed in just 27 runs due to the slope collector; however, that would entail a large number of tests, which would be significant in both experimental cost and time. As a result, the Taguchi design of experiments (DoE) layouts were more applicable when compared to a traditional full-factorial counterpart because they reduced the number of tests to a practical level. The L9 DoE orthogonal array was selected with the assumption of no factorial interactions, resulting in nine trials, as illustrated in Table 2.

Table 2. Experimental results of the fiber fineness of the composite nanofibers planned by the $L_9(3^4)$ orthogonal table.

$L_9(3^4)$	Ratio (SF: PCL)	Chitosan Addition (%)	Flow Rate (mL/h)	Voltage (kV)	S/N	Means of Diameter (nm)	Porosity (%)	WVTR (g m^{-2}·24 h)
1	1:1	0.50	0.2	15	12.65	232.55±60.28	88.01±4.32	4417.29±87.27
2	1:1	0.75	0.3	20	12.71	229.09±60.66	83.10±4.21	4362.97±91.67
3	1:1	1.00	0.4	25	13.61	208.39±55.81	91.96±5.16	4641.38±19.21
4	1:2	0.50	0.3	25	15.37	170.00±55.76	92.05±3.70	4768.71±85.04
5	1:2	0.75	0.4	15	11.92	253.42±65.37	91.43±5.50	4636.29±25.46
6	1:2	1.00	0.2	20	13.11	220.67±62.13	85.37±4.04	4432.89±45.84
7	1:3	0.50	0.4	20	10.86	285.71±78.08	92.60±4.10	4581.97±92.58
8	1:3	0.75	0.2	25	11.47	266.75±76.71	84.74±5.70	4330.71±5.09
9	1:3	1.00	0.3	15	12.68	231.94±63.64	82.03±1.45	4405.41±74.86

In the "larger the better" characteristic, the formula for calculating the ratio of S/N as the best parameter for calculating the factors was calculated by the following Equation (1):

$$S/N = -10 \times \log\left(\frac{1}{n} \sum_{i=1}^{n} \frac{1}{y_i^2} \right) \tag{1}$$

where n and y denote the number of measurements and observed data, respectively.

2.4. Characterization of Nanofiber Scaffolds

The morphology of the nanofiber scaffold was detected using scanning electron microscopy (SEM; TS 5136MM, TESCAN, Czech Republic). The average fiber diameter was determined by measuring 100 fibers selected randomly from each sample. Chemical analysis was performed using a Fourier transform infrared spectrometer (FTIR; Spectrum 100, Perkin Elmer, USA) with a scan range of 4000 to 450 cm^{-1} and an accumulation of 16 scans.

According to standard method of Japanese Industrial Standards (JIS) 10099A, the water vapor transmission rate (WVTR) is a measure of the passage of water vapor through a substance. In addition to measurements of the permeability of the vapor barriers, the porosity of the SF/PCL/CS was also examined by the study [25]. For the antibacterial assay, the inhibitory effects of chitosan on bacterial growth were detected by the plate well diffusion method [26] via the formation of a zone of inhibition. To attain this figure, the sequential dilution was necessary (six for *E. coli* and for *S. aureus*) according to the simultaneous counting of plate colonies (CFU). The procedure used in this analysis followed the agar diffusion method according to the previous literature [27], in which small circular cavities were punctured in the culture medium for each chitosan concentration.

The equilibrium water content (EWC) was measured by the conventional gravimetric method. The pre-weighed dry samples were immersed in deionized water, and the excess surface water was blotted out with absorbent paper. The swelling procedure was repeated until there was no further weight increase. The EWC was calculated as the weight increase with respect to the weight of the swollen samples within 24 h using the following Equation (2):

$$EWC = \frac{W_{\text{wet}} - W_{\text{dry}}}{W_{\text{wet}}} \times 100\% \tag{2}$$

where W_{wet} and W_{dry} denote the weights of the swollen and dry samples, respectively.

A cytotoxicity assay is a test for analyzing the cytotoxic effects of materials and medical devices on living organisms [28]. The following cell culture-based tests, as recommended by ISO 10993-5, used a direct contact test. In all the tests (blank, negative control, positive control, and sample), the incubation time of the mouse fibroblast L929 cell cultures was 24 h. Cell culture is the process by which cells from human tissue are grown in an incubator under controlled conditions in order to provide sufficient material for testing. After solubilization, the solutions were transferred to fresh flat bottom 96-well plates, and the absorbance (570 nm) of each well was measured by spectrophotometry. The background absorbance at 650 nm was subtracted from the readings at 570 nm to obtain the final optical density (OD). The following Equation (3) was used to calculate the reduction in the culture viability of the cells exposed to a tested sample (i.e., SF/PCL/CS under optimal conditions) in comparison to the cell culture viability of group b:

$$\text{Viability } (\%) = \frac{OD_{570e}}{OD_{570b}} \times 100\% \tag{3}$$

where OD_{570e} is the average OD of the respective groups that were in contact with different lots of the product and OD_{570b} is the average OD of all the wells of group b. All the values were final ODs after the subtraction of background absorbance.

3. Results and Discussion

3.1. Optimum Combination of Factors for the Application of Dressings

3.1.1. Taguchi Method to Optimize Dressings

Uniform fiber diameters and smaller diameters in the scaffold for the application of dressings can provide a higher surface area and interconnected holes to promote the exchange of nutrients and oxygen and enhance the proliferation ability of cells. In the formation of nanofibers, the ratio of silk fibroin to polycaprolactone, the chitosan content, the flow rate, and the voltage are crucial factors affecting the nanofiber diameter. In Taguchi designed experiments, the higher values of the signal-to-noise ratio (S/N) identify control factor settings that minimize the effects of the noise factors. Using the Equation (1) of the smaller the better method (i.e., the smallest diameter of the nanofibers was selected based on maximum S/N ratio), the S/N ratio of the fiber fineness (the diameter of the nanofibers) in the SEM micrograph of the nanofiber could be calculated, and the results are shown in Table 2. According to the results from the L_9 (3^4) sample, the average S/N ratio of the four factor levels was calculated to perform the next analysis, as listed in Table 3. The influence of the four factors on the fiber diameter followed the order of the ratio of silk fibroin and polycaprolactone (Δ = 1.80) > flow rate (Δ = 1.46) > applied voltage (Δ = 1.26) > chitosan addition (Δ = 1.10). This finding suggested that the ratio of silk fibroin and polycaprolactone was the most significant factor for achieving a small electrospun nanofiber diameter for the application of dressings. Table 3 also shows the contribution of the four parameters to the influence of the SF/PCL/CS composite nanofiber diameter. The ratio of the silk fibroin and polycaprolactone was an important factor affecting the diameter of the nanofibers due to the highest contribution percentage (32.0%). Furthermore, the results indicated that the ratio of silk fibroin and polycaprolactone affected the viscosity of the electrospinning solution to produce a stable Taylor cone. The results were in accordance with the Taguchi experimental S/N design.

Table 3. Smaller the better of the fiber fineness signal-to-noise ratio (S/N) analysis.

	Ratio of Silk Fibroin and Polycaprolactone	Chitosan Content	Flow Rate	Voltage
1	12.99	12.96	12.41	12.42
2	13.47	12.03	13.59	12.23
3	11.67	13.13	12.13	13.48
Δ	1.80	1.10	1.46	1.26
Factor influence order	1	4	2	3
Contribution (%)	32.0	19.1	26.2	22.7

The average value of the confirmation experiments was 170.00±55.76 nm under the optimal parameters of a silk fibroin to polycaprolactone ratio of 1:2, a chitosan addition of 0.5%, a rate of advancement of 0.3 mL/h, and an operating voltage of 25 kV (Table 2), which provided the highest average value compared with the nine groups of quality data. As shown in Figures 1 and 2, the results indicated that the optimal parameters inferred by the Taguchi method had a smaller fiber diameter, were smoother, and had a low distribution. Conclusively, the comparison of the optimal parameters inferred by the Taguchi method with the data results of the orthogonal table proved that the inferred optimal parameters were appropriate.

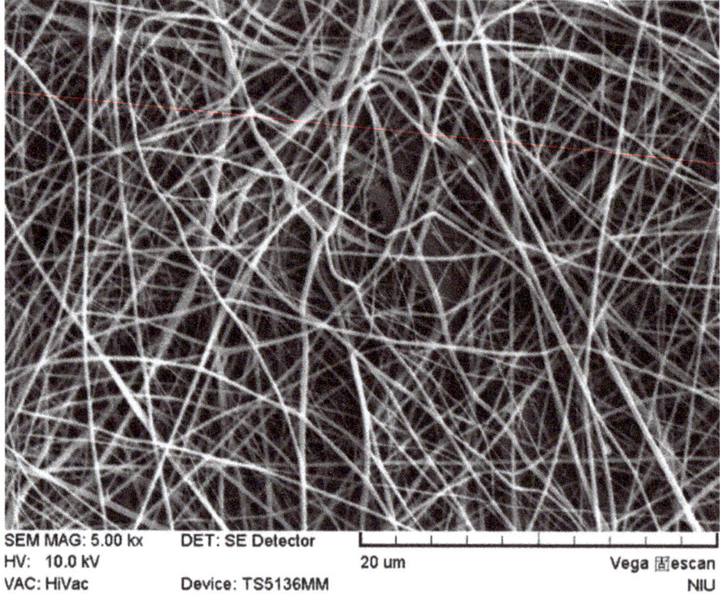

Figure 1. SEM photomicrograph of the silk fibroin (SF), polycaprolactone (PCL), and chitosan (CS) (SF/PCL/CS) nanofibers under optimal conditions.

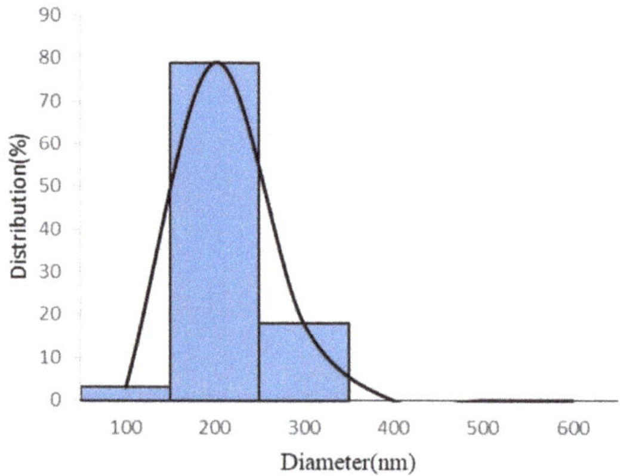

Figure 2. Fiber diameter distribution of the SF/PCL/CS nanofibers under optimal conditions.

3.1.2. Porosity and Water Uptake Abilities

The adequate pore size and interconnected pores of a scaffold provides a sufficient opportunity for cell migration and proliferation. The ability of a dressing to control water loss can be determined by the water vapor transmission rate (WVTR). The ability of a scaffold to preserve water is also important in order to evaluate its property for tissue engineering. Table 2 describes the WVTR of the SF/PCL/CS nanofibers. The porosity of the nine samples (L_9 (3^4)) of the SF/PCL/CS nanofibers in this study was more than 80% and was significantly higher than the pure PCL scaffolds, which had a 70% porosity, as reported by [29]. The highly porous SF/PCL/CS nanofibers could provide an appropriate environment for initial cell growth (by their structural stability), accelerated degradation (by their large surface area), and the sustained delivery of bioactive molecules (by their high porosity).

In this study, SF/PCL/CS nanofibers with graded WVTR were prepared by changing the porosity of the membrane (Table 2). The corresponding average WVTR of the samples was in the range of between 4330.71 and 4768.71 g m^{-2}·24 h (extremely high permeability, L_9 (3^4)). An extremely high WVTR may lead to the dehydration of a wound, whereas an unacceptably low WVTR may cause the accumulation of wound exudates. Hence, a dressing with a suitable WVTR is required to provide a moist environment that can establish the best environment for natural healing. Thus, a dressing prepared by optimal condition with a WVTR of approximately 4768.71 g/m^2·24 h could also maintain the optimal moisture content for the proliferation and function of cells and fibroblasts. According to SEM and the appropriate porosity of the SF/PCL/CS nanofiber, it was considered to have great potential for skin tissue engineering due to its interconnected pore network and suitable WVTR.

3.1.3. Cytotoxicity Tests

Cytotoxicity assays are necessary for the assessment and characterization of the potentially toxic and harmful effects of a biomaterial's compounds [28]. They are a feasible and reliable in vitro technique used for the biocompatibility evaluation of materials. Table 4 shows phase contrast images of the cultures in the experiment after 24 h, including blank, negative control (polyethylene, PE), positive control (dimethylsulfoxide, DMSO), and SF/PCL/CS under optimal conditions. In cultures exposed to blank, negative control, positive control, and SF/PCL/CS, the viability was 100, 100, 13, and 57% at 24 h of continued growth in the culture, respectively. A tested product (SF/PCL/CS) has non-cytotoxic potential for application to tissue engineering when the cell culture viability increases to >50% in comparison to the positive control (dimethylsulfoxide, DMSO), which was set at a 13% viability. The study also proposed the similar results that the tested product could be considered non-toxic as the viability of these cultures exceeded 50% [30]. Furthermore, the viability of 57% for the tested product (SF/PCL/CS) in this study was higher than that of 50.2% for the CNTs-doped PLGA nanofibers [31]. The research [31] show that the PLGA and the HNTs- or CNTs-doped PLGA nanofibers display appreciable MTT formazan dye sorption, corresponding to a 35.6–50.2% deviation from the real cell viability assay data. From Figure 3, the DMSO treatment substantially altered the morphology and attachment of cells in concurrence with a significant reduction in the cell viability. The results showed that the electrospun scaffolds (SF/PCL/CS) could support the attachment and the proliferation of mouse fibroblast L929 cells. In addition, the cells cultured on the scaffolds exhibited normal cell shapes. The obtained results confirmed the potential for use of the electrospun fiber as scaffolds for skin tissue engineering.

Table 4. Viability of cell L 929 according to the MTT test method.

Test Item	Absorbance (%)	Viability (%)
Blank	0.504 ± 0.011	100
Negative control	0.502 ± 0.005	100
Positive control	0.064 ± 0.002	13
Sample (SF/PCL/CS)	0.288 ± 0.020	57

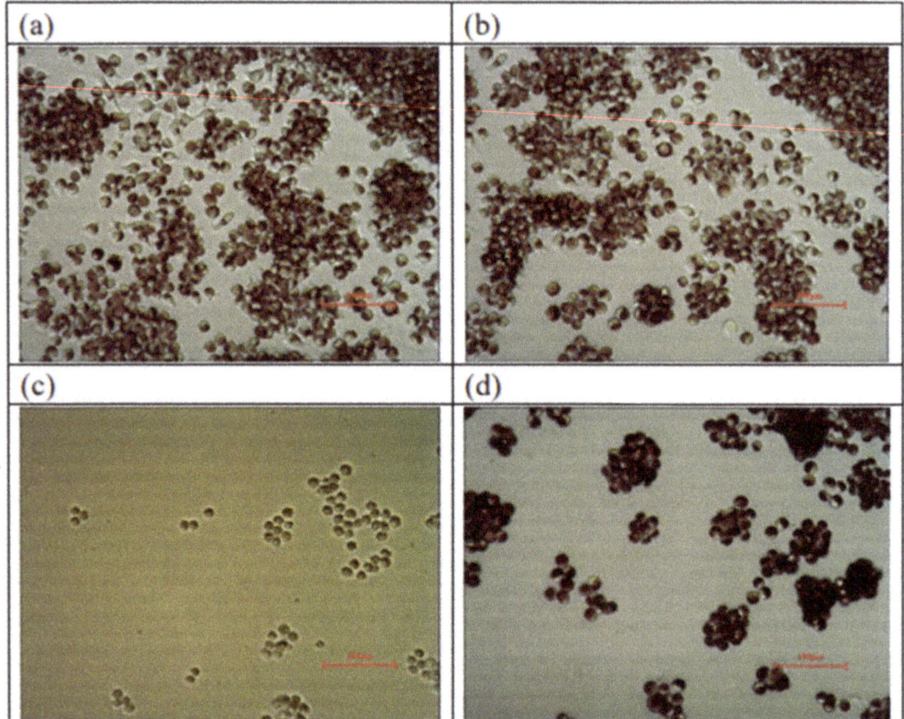

Figure 3. Photomicrograph of cell L 929 by direct contact method within 24 hours: (**a**) blank; (**b**) negative control (polyethylene, PE); (**c**) positive control (dimethylsulfoxide, DMSO); and (**d**) the sample (SF/PCL/CS under optimal conditions).

3.2. Effect of the Ratio of Silk Fibroin to Polycaprolactone on Fiber Diameter

In the electrospinning process, the solution concentration is considered to be the most important parameter affecting the fiber morphology [31,32]. Figure 4 shows the effect of the ratio of silk fibroin to polycaprolactone on the fiber diameter. At a silk fibroin to polycaprolactone ratio of 1.0, the entanglement between the polymer chains formed obvious beads because the low viscosity of the solution did not provide a stable jet. As the ratio of silk fibroin to polycaprolactone decreased from 0.5 to 0.25, the average fiber diameter increased from 208.27±68.27 nm to 664.23±131.54 nm in this study. The main reason was that the increase in viscosity hindered the bending stability of the jet to produce a coarser fiber [33,34].

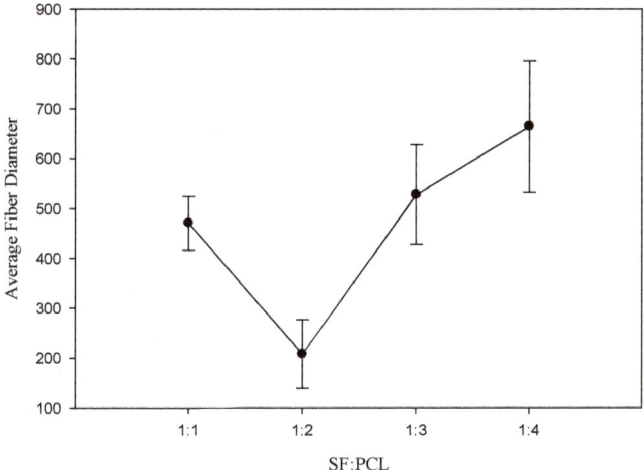

Figure 4. Effect of the ratio of silk fibroin to polycaprolactone on the fiber diameter.

3.3. Effect of Chitosan Addition on Fiber Diameter, Antimicrobial Activity, and Equilibrium Water Content

Figure 5 depicts the evaluation results of the fiber diameter of the electrospinning scaffold with different chitosan additions. The average fiber diameter decreased from 523.23±92.60 mm to 181.45±41.57 nm with the increased chitosan addition from 0.25% to 1.00% because of the charge density. The chitosan addition enhanced the higher charge density of the jet to produce the thinner fibers due to the increase in conductivity [21,35]. Compared to the chitosan addition of 1.00%, the average fiber diameter of the electrospinning scaffold was increased to 308.90±74.98 nm with a chitosan addition of 1.50%, because the increase in viscosity caused the bending instability of the jet and the accelerating solidification of the polymer jet [34,36].

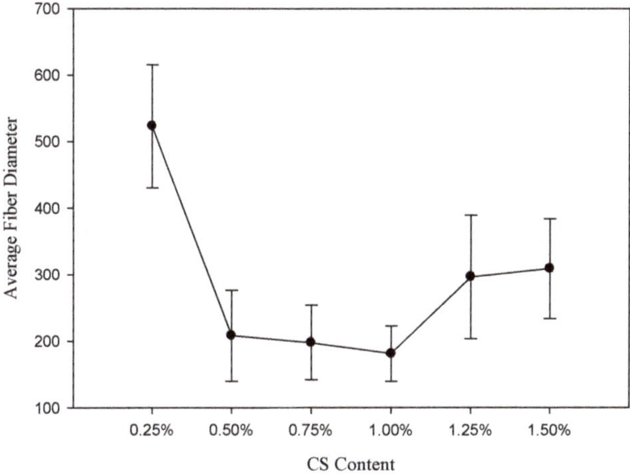

Figure 5. Effect of the chitosan addition on the fiber diameter.

The results from Table 5 revealed the mean diameter of inhibition zone for *E. coli* with the chitosan amounts of 0.25%, 0.5%, 0.75%, and 1.00% were 65, 63, 64, and 63 mm, respectively, proving the strong antibacterial property. Moreover, the samples with the chitosan amounts of 0.25%, 0.5%, 0.75%, and 1.00% exhibited the inhibition zone diameters for *S. aureus* of 58, 55, 51, and 55 mm, respectively. As illustrated in Figures 6 and 7, the activity intensity could be visually determined by agar well diffusion assay testing via assessing the local inhibition. The results of the experiments conducted using different chitosan additions had an inhibitory effect on the mean diameter of the inhibition zone for two types of bacteria (*E. coli* and *S. aureus*). The results exhibited better inhibitory effects against gram-positive bacterium *S. aureus* compared to the gram-negative bacterium *E. coli.*, which was in agreement with the results attained in previously published works [37,38]. The results in this study proposed that unmodified chitosan generally acts stronger on gram-negative strains than on gram-positive strains, owing to the electrostatic interaction between positively charged $R-N(CH_3)_3^+$ sites and negatively charged microbial cell membranes.

Table 5. Effect of chitosan addition on the mean diameter of the inhibition zone of *E. coli* and *S. aureus* at different chitosan amounts.

Name of the Sample	Conc. (%)	Mean Diameter of Inhibition Zone (mm) Test Organisms	
		Staphylococcus Aureus	**Escherichia Coli**
Chitosan	0.25	58 ± 4	65 ± 0
	0.50	55 ± 1	63 ± 4
	0.75	51 ± 1	64 ± 0
	1.00	55 ± 1	63 ± 2

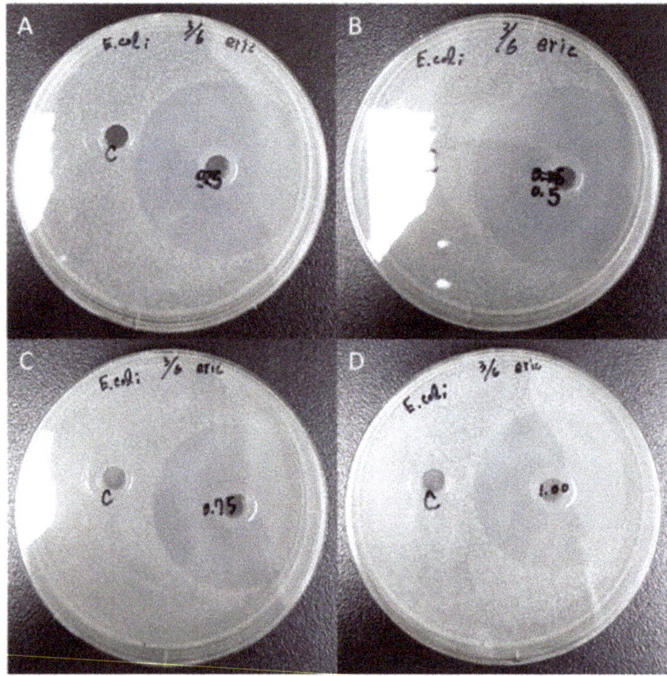

Figure 6. Effect of the chitosan addition on the antimicrobial activity of E. coli at different chitosan amounts: (A) 0.25; (B) 0.50; (C) 0.75; (D) 1.00 wt%.

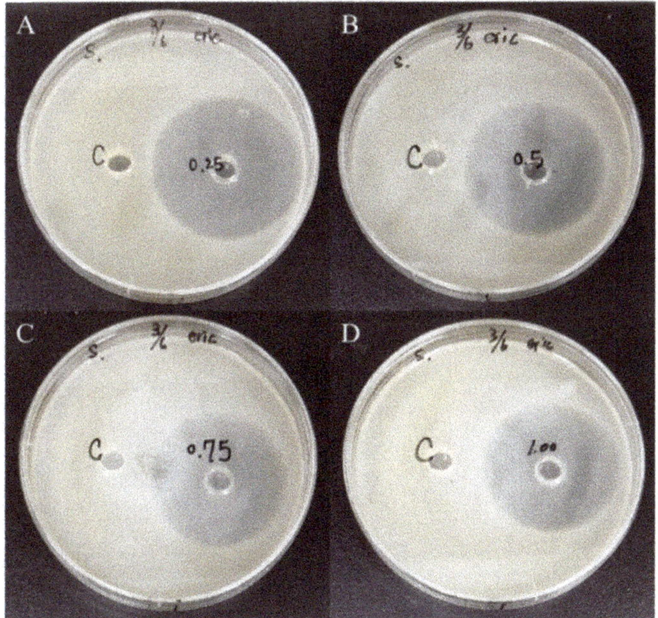

Figure 7. Effect of chitosan addition on the antimicrobial activity of *S. aureus* at different chitosan amounts: (**A**) 0.25; (**B**) 0.50; (**C**) 0.75; (**D**) 1.00 wt%.

The hydrophilicity of nanofibers (SF/PCL/CS) with different chitosan additions (0.25%, 0.50%, 0.75%, and 1.00%) was measured by gravimetric analysis and designated by the percentage equilibrium water content (EWC), as shown in Figure 8. The EWC was increased with the increasing chitosan addition due to the functional groups of –OH and –NH$_2$. The equilibrium water content (EWC) of all the nanofibers (chitosan additions of 0.25%, 0.50%, 0.75%, and 1.00%) was in the range of 500% to 950% and was significantly higher than the YY0148-2006 standard for medical dressings [39] and the hydrogels for tissue engineering [40]. The copolymerization with the zwitterionic comonomer leads hydrogels with a high equilibrium water content (EWC) of up to 700% while maintaining mechanical robustness [40].

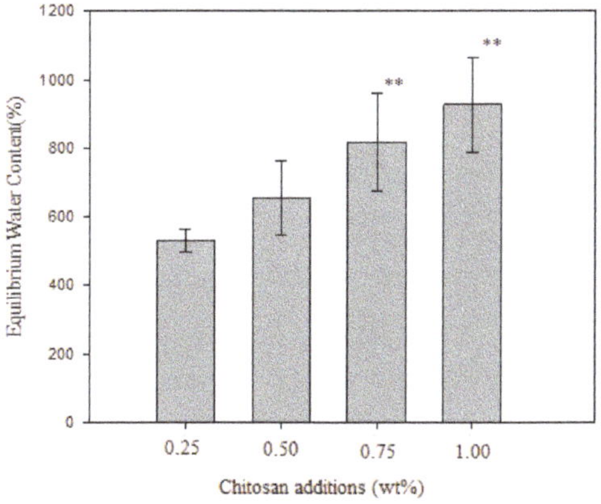

Figure 8. Effect of chitosan addition on the equilibrium water content.

3.4. Effect of Flow Rate on Fiber Diameter

In the electrospinning process, the flow rate is considered to be the most important parameter affecting the fiber uniformity [31,41]. The effect of the flow rate on the fiber diameter is shown in Figure 9a. As shown in Figure 9b, uniform beadless electrospun nanofibers were prepared via a critical flow rate of 0.3 mL/h for the polymeric solution. The shape of the Taylor cone at the tip of the capillary would not be maintained if the flow of the solution through the capillary was insufficient, and the insufficient intermolecular surface tension would be unable to resist the Coulomb force to maintain a stable jet [41]. However, at a higher flow rate, the solution would be wasted without differentiating adequately into the fibers. The short evaporation time of the solvent could result in the formation of beads and increase the average fiber diameter [32]. The morphology of the SF/PCL/CS nanofibers appeared to be inhomogeneous under a higher flow rate, as depicted in Figure 9c. The results could be explained by the inharmony between the fiber formation speed and the solution feed rate.

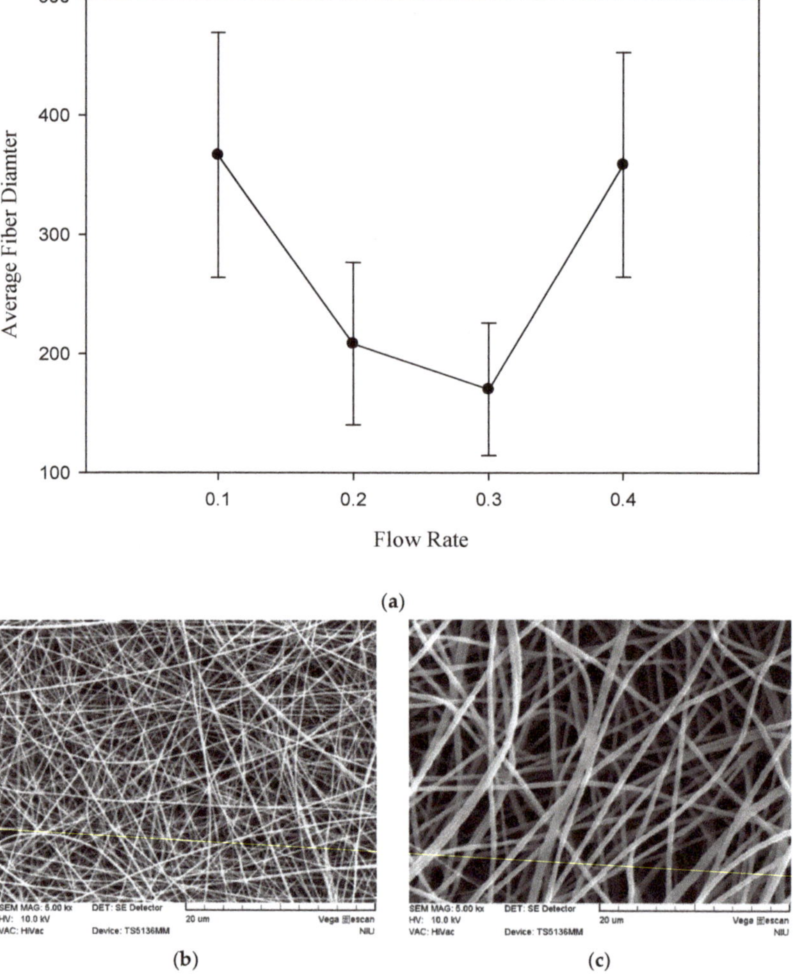

Figure 9. (a) Effect of flow rate on the fiber diameter of the SF/PCL/CS nanofibers, (b) SEM micrograph at 0.3 mL/h, (c) SEM micrograph at 0.4 mL/h.

3.5. Effect of Applied Voltage on Fiber Diameter

In general, the operating applied voltage must exceed the minimum threshold applied voltage before the Taylor cone appears to form ultrafine nanofibers [23]. As shown in Figure 10, the results revealed that the average fiber diameter significantly decreased from 440.69 ± 105.47 nm to 208.27 ± 55.76 nm as the applied voltage increased from 15 to 25 kV. This result could be explained by the force imbalance between the repulsive Columbic force and the contracting viscoelastic force. With the increasing voltage, the intrinsic equilibrium became difficult to restore when the fibers were collected on the metal rod accompanied by the charge transfer.

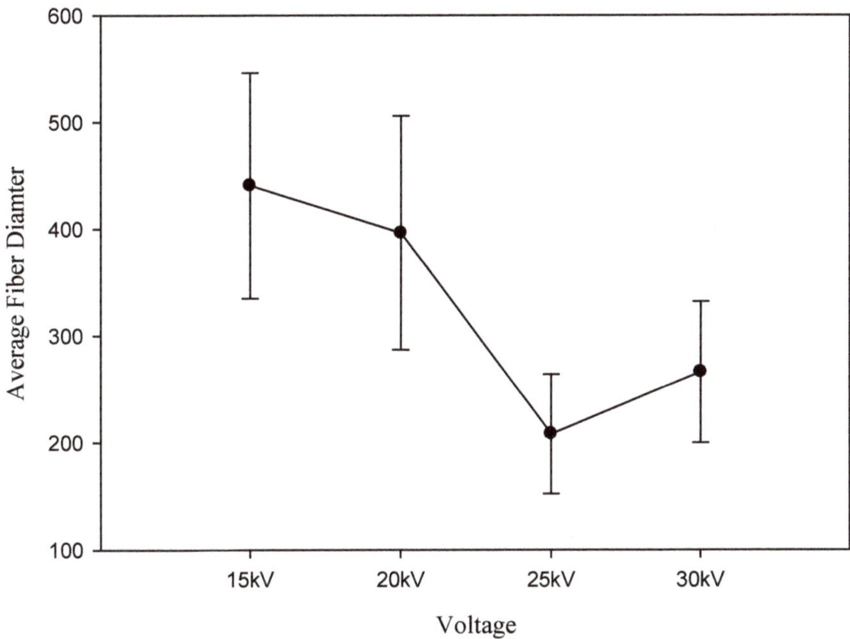

Figure 10. Effect of applied voltage on fiber diameter.

4. Conclusion

SF, PCL, and CS were blended in different concentrations and compositions and were evaluated for their fiber diameter to examine the optimum values for nanofibers. The S/N analysis via the Taguchi experimental design showed that the ratio of SF to PCL was the most influential parameter on the fiber diameter. A smooth and uniform distributed SF/PCL/CS nanofiber with a fiber diameter of 170.0 nm was synthesized at the optimal parameters of a 1:2 ratio of silk fibroin to polycaprolactone, a 0.5% chitosan addition, a 0.3 mL/h rate of advancement, and an operating voltage of 25 kV. The SF/PCL/CS (under optimal conditions) with a WVTR of approximately 4768.71 g/m^2·24 h could maintain the optimal moisture content for the proliferation and function of cells and fibroblasts. The EWC of the SF/PCL/CS nanofiber was increased from 500% to 950% by increasing the chitosan additions from 0.25% to 1.00% and was significantly higher than the YY0148-2006 standard for medical dressings. The porosity of the nine samples (L$_9$ (3^4)) of SF/PCL/CS in this study was more than 80% and was significantly higher than that of the pure PCL scaffolds. According to SEM and the appropriate porosity of the SF/PCL/CS nanofiber, it was considered to have great potential for skin tissue engineering due to its interconnected pore network and suitable WVTR.

Polymers **2020**, *12*, 1439

Author Contributions: Project administration, M.-F.L.; Supervision, H.-W.C. All authors have read and agreed to the published version of the manuscript.

Funding: This research was supported in part by grant MOST 107-2221-E-197-001-MY2 from the Ministry of Science and Technology, Republic of China.

Conflicts of Interest: The authors declare no conflict of interest.

References

1. Ayutsede, J.; Gandhi, M.; Sukigara, S.; Micklus, M.; Chen, H.-E.; Ko, F. Regeneration of Bombyx mori silk by electrospinning. Part 3: Characterization of electrospun nonwoven mat. *Polymer* **2005**, *46*, 1625–1634. [CrossRef]

2. Yao, Y.; Wang, J.; Cui, Y.; Xu, R.; Wang, Z.; Zhang, J.; Wang, K.; Li, Y.; Zhao, Q.; Kong, D. Effect of sustained heparin release from PCL/chitosan hybrid small-diameter vascular grafts on anti-thrombogenic property and endothelialization. *Acta Biomater.* **2014**, *10*, 2739–2749. [CrossRef]

3. Dias, J.R.; Baptista-Silva, S.; Sousa, A.; Oliveira, A.L.; Bártolo, P.J.; Granja, P.L. Biomechanical performance of hybrid electrospun structures for skin regeneration. *Mater. Sci. Eng. C* **2018**, *93*, 816–827. [CrossRef]

4. Avsar, G.; Agirbasli, D.; Agirbasli, M.A.; Gunduz, O.; Oner, E.T. Levan based fibrous scaffolds electrospun via co-axial and single-needle techniques for tissue engineering applications. *Carbohydr. Polym.* **2018**, *193*, 316–325. [CrossRef] [PubMed]

5. Surucu, S.; Sasmazel, H.T. Development of core-shell coaxially electrospun composite PCL/chitosan scaffolds. *Int. J. Biol. Macromol.* **2016**, *92*, 321–328. [CrossRef] [PubMed]

6. Pakravan, M.; Heuzey, M.-C.; Ajji, A. A fundamental study of chitosan/PEO electrospinning. *Polymer* **2011**, *52*, 4813–4824. [CrossRef]

7. Narayanan, G.; Gupta, B.S.; Tonelli, A.E. Enhanced mechanical properties of poly(ε-caprolactone) nanofibers produced by the addition of non-stoichiometric inclusion complexes of poly(ε-caprolactone) and α-cyclodextrin. *Polymer* **2015**, *73*, 321–330. [CrossRef]

8. Yang, X.; Fan, L.; Ma, L.; Wang, Y.; Lin, S.; Yu, F.; Pan, X.; Luo, G.; Zhang, D.; Wang, H. Green electrospun Manuka honey/silk fibroin fibrous matrices as potential wound dressing. *Mater. Des.* **2017**, *119*, 76–84. [CrossRef]

9. Lin, L.; Zhu, Y.; Li, C.; Liu, L.; Surendhiran, D.; Cui, H. Antibacterial activity of PEO nanofibers incorporating polysaccharide from dandelion and its derivative. *Carbohydr. Polym.* **2018**, *198*, 225–232. [CrossRef] [PubMed]

10. Ahmed, R.; Tariq, M.; Ali, I.; Asghar, R.; Khanam, P.N.; Augustine, R.; Hasan, A. Novel electrospun chitosan/polyvinyl alcohol/zinc oxide nanofibrous mats with antibacterial and antioxidant properties for diabetic wound healing. *Int. J. Biol. Macromol.* **2018**, *120*, 385–393. [CrossRef]

11. Zhang, D.; Karki, A.B.; Rutman, D.; Young, D.P.; Wang, A.; Cocke, D.; Ho, T.H.; Guo, Z. Electrospun polyacrylonitrile nanocomposite fibers reinforced with Fe_3O_4 nanoparticles: Fabrication and property analysis. *Polymer* **2009**, *50*, 4189–4198. [CrossRef]

12. Zhu, J.; Wei, S.; Patil, R.; Rutman, D.; Kucknoor, A.S.; Wang, A.; Guo, Z. Ionic liquid assisted electrospinning of quantum dots/elastomer composite nanofibers. *Polymer* **2011**, *52*, 1954–1962. [CrossRef]

13. Aslan, T.; Arslan, S.; Eyvaz, M.; Güçlü, S.; Yüksel, E.; Koyuncu, İ. A novel nanofiber microfiltration membrane: Fabrication and characterization of tubular electrospun nanofiber (TuEN) membrane. *J. Membr. Sci.* **2016**, *520*, 616–629. [CrossRef]

14. Jiang, L.; Tu, H.; Lu, Y.; Wu, Y.; Tian, J.; Shi, X.; Wang, Q.; Zhan, Y.; Huang, Z.; Deng, H. Spherical and rodlike inorganic nanoparticle regulated the orientation of carbon nanotubes in polymer nanofibers. *Chem. Phys. Lett.* **2016**, *650*, 82–87. [CrossRef]

15. Yao, Y.; Wei, H.; Wang, J.; Lu, H.; Leng, J.; Hui, D. Fabrication of hybrid membrane of electrospun polycaprolactone and polyethylene oxide with shape memory property. *Compos. B. Eng.* **2015**, *83*, 264–269. [CrossRef]

16. Rao, M.; Geng, X.; Liao, Y.; Hu, S.; Li, W. Preparation and performance of gel polymer electrolyte based on electrospun polymer membrane and ionic liquid for lithium ion battery. *J. Membr. Sci.* **2012**, *399–400*, 37–42. [CrossRef]

17. Narayanan, G.; Tekbudak, M.Y.; Caydamli, Y.; Dong, J.; Krause, W.E. Accuracy of electrospun fiber diameters: The importance of sampling and person-to-person variation. *Polym Test.* **2017**, *61*, 240–248. [CrossRef]

18. Yu, W.-J.; Xu, S.-M.; Zhang, L.; Fu, Q. Morphology and mechanical properties of immiscible polyethylene/polyamide12 blends prepared by high shear processing. *Chinese J. Polym. Sci.* **2017**, *35*, 1132–1142. [CrossRef]

19. Park, Y.R.; Ju, H.W.; Lee, J.M.; Kim, D.-K.; Lee, O.J.; Moon, B.M.; Park, H.J.; Jeong, J.Y.; Yeon, Y.K.; Park, C.H. Three-dimensional electrospun silk-fibroin nanofiber for skin tissue engineering. *Int. J. Biol. Macromol.* **2016**, *93*, 1567–1574. [CrossRef]

20. Aznar-Cervantes, S.D.; Daniel, V.C.; Luis, M.O.; Cenis, J.L.; Abel, L.P.A. Influence of the protocol used for fibroin extraction on the mechanical properties and fiber sizes of electrospun silk mats. *Mater. Sci. Eng. C* **2013**, *33*, 1945–1950. [CrossRef]

21. Calamak, S.; Aksoy, E.A.; Ertas, N.; Erdogdu, C.; Sagıroglu, M.; Ulubayram, K. Ag/silk fibroin nanofibers: Effect of fibroin morphology on Ag+ release and antibacterial activity. *Eur. Polym. J.* **2015**, *67*, 99–112. [CrossRef]

22. Vaz, J.M.; Taketa, T.B.; Hernandez-Montelongo, J.; Chevallier, P.; Cotta, M.A.; Mantovani, D.; Beppu, M.M. Antibacterial properties of chitosan-based coatings are affected by spacer-length and molecular weight. *Appl. Surf.* **2018**, *445*, 478–487. [CrossRef]

23. Albetran, H.; Dong, Y.; Low, I.M. Characterization and optimization of electrospun TiO_2/PVP nanofibers using Taguchi design of experiment method. *J. Asian Ceram. Soc.* **2015**, *3*, 292–300. [CrossRef]

24. Senthil, T.; Anandhan, S. Electrospinning of non-woven poly(styrene-co-acrylonitrile) nanofibrous webs for corrosive chemical filtration: Process evaluation and optimization by Taguchi and multiple regression analyses. *J. Electrostat.* **2015**, *73*, 43–55. [CrossRef]

25. Ryan, M.P.; Rea, M.C.; Hill, C.; Ross, R.P. An application in cheddar cheese manufacture for a strain of Lactococcus lactis producing a novel broad-spectrum bacteriocin, lacticin 3147. *Appl. Environ. Microbiol.* **1996**, *62*, 612–619. [CrossRef]

26. Duffy, G.; Whiting, R.C.; Sheridan, J.J. The effect of a competitive microflora, pH and temperature on the growth kinetics of *Escherichia coli* O157:H7. *Food Microbiol.* **1999**, *16*, 299–307. [CrossRef]

27. Rosengren, A.; Faxius, L.; Tanaka, N.; Watanabe, M.; Bjursten, L.M. Comparison of implantation and cytotoxicity testing for initially toxic biomaterials. *J. Biomed. Mater. Res. A* **2005**, *75*, 115–122. [CrossRef]

28. Izquierdo, R.; Garcia-Giralt, N.; Rodriguez, M.T.; Cáceres, E.; García, S.J.; Ribelles, J.L.G.; Joan, M.M.; Monllau, C.; Suay, J. Biodegradable PCL scaffolds with an interconnected spherical network for tissue engineering. *J. Biomed. Mater. Res. A* **2008**, *85*, 25–35. [CrossRef]

29. Mehrabani, M.G.; Karimian, R.; Rakhshaei, R.; Pakdel, F.; Eslami, H.; Fakhrzadeh, V.; Rahimi, M.; Salehi, R.; Kafil, H.S. Chitin/silk fibroin/TiO_2 bio-nanocomposite as a biocompatible wound dressing bandage with strong antimicrobial activity. *Int. J. Biol. Macromol.* **2018**, *116*, 966–976. [CrossRef]

30. Zhou, J.; Cao, C.; Ma, X. A novel three-dimensional tubular scaffold prepared from silk fibroin by electrospinning. *Int. J. Biol. Macromol.* **2009**, *45*, 504–510. [CrossRef]

31. Qi, R.; Shen, M.; Cao, X.; Guo, R.; Tian, X.; Yu, J.; Shi, X. Exploring the dark side of MTT viability assay of cells cultured onto electrospun PLGA-based composite nanofibrous scaffolding materials. *Analyst* **2011**, *14*, e1800403. [CrossRef] [PubMed]

32. Sutasinpromprae, J.; Jitjaicham, S.; Nithitanakul, M.; Meechaisue, C.; Supaphol, P. Preparation and characterization of ultrafine electrospun polyacrylonitrile fibers and their subsequent pyrolysis to carbon fibers. *Polym. Int.* **2006**, *55*, 825–833. [CrossRef]

33. Jacobs, V.; Anandjiwala, R.D.; Maaza, M. The influence of electrospinning parameters on the structural morphology and diameter of electrospun nanofibers. *J. Appl. Polym. Sci.* **2010**, *115*, 3130–3136. [CrossRef]

34. Jia, Y.-T.; Gong, J.; Gu, X.-H.; Kim, H.-Y.; Dong, J.; Shen, X.-Y. Fabrication and characterization of poly (vinyl alcohol)/chitosan blend nanofibers produced by electrospinning method. *Carbohydr. Polym.* **2007**, *67*, 403–409. [CrossRef]

35. Vrieze, S.D.; Westbroek, P.; Camp, T.V.; Clerck, K.D. Solvent system for steady state electrospinning of polyamide 6.6. *J. Appl. Polym. Sci.* **2010**, *115*, 837–842. [CrossRef]

36. Da Silva, L.P.; De Britto, D.; Seleghim, M.H.R.; Assis, O.B.G. In vitro activity of water-soluble quaternary chitosan chloride salt against E coli. *World J. Microbiol. Biotechnol.* **2010**, *26*, 2089–2092. [CrossRef]

37. No, H.K.; Park, N.Y.; Lee, S.H.; Meyers, S.P. Antibacterial activity of chitosans and chitosan oligomers with different molecular weights. *Int. J. Food Microbiol.* **2002**, *75*, 65–72. [CrossRef]

38. Li, S.; Li, L.; Guo, C.; Qin, H.; Yu, X. A promising wound dressing material with excellent cytocompatibility and proangiogenesis action for wound healing: Strontium loaded Silk fibroin/Sodium alginate (SF/SA) blend films. *Int. J. Biol. Macromol.* **2017**, *104*, 969–978. [CrossRef]
39. Someswararao, M.V.; Dubey, R.S.; Subbarao, P.S.V.; Singh, S. Electrospinning process parameters dependent investigation of TiO_2 nanofibers. *Results Phys.* **2018**, *11*, 223–231. [CrossRef]
40. Kostina, N.Y.; Blanquer, S.; Pop-Georgievski, O.; Rahimi, K.; Dittrich, B.; Höcherl, A.; Michálek, J.; Grijpma, D.W.; Rodriguez-Emmenegger, C. Zwitterionic functionalizable scaffolds with gyroid pore Architecture for tissue engineering. *Macromol. Biosci.* **2019**, *19*, e1800403. [CrossRef]
41. Jin, H.-J.; Fridrikh, S.V.; Rutledge, G.C.; Kaplan, D.L. Electrospinning Bombyx mori silk with poly(ethylene oxide). *Biomacromolecules* **2002**, *3*, 1233–1239. [CrossRef] [PubMed]

Article

Green Synthesis of Metal-Organic Framework Bacterial Cellulose Nanocomposites for Separation Applications

Radwa M. Ashour [1,2,*,†], Ahmed F. Abdel-Magied [1,2,*,†], Qiong Wu [3], Richard T. Olsson [3,*] and Kerstin Forsberg [1,*]

[1] Department of Chemical Engineering, KTH Royal Institute of Technology, 100 44 Stockholm, Sweden
[2] Nuclear Materials Authority, P.O. Box 530, ElMaadi, Cairo 11381, Egypt
[3] Department of Fibre and Polymer Technology, KTH Royal Institute of Technology, 100 44 Stockholm, Sweden; qiongwu@kth.se
* Correspondence: radwa@kth.se (R.M.A.); fawzy@kth.se (A.F.A.-M.); rols@kth.se (R.T.O.); kerstino@kth.se (K.F.)
† These authors contributed equally to this work.

Received: 30 March 2020; Accepted: 9 May 2020; Published: 13 May 2020

Abstract: Metal organic frameworks (MOFs) are porous crystalline materials that can be designed to act as selective adsorbents. Due to their high porosity they can possess very high adsorption capacities. However, overcoming the brittleness of these crystalline materials is a challenge for many industrial applications. In order to make use of MOFs for large-scale liquid phase separation processes they can be immobilized on solid supports. For this purpose, nanocellulose can be considered as a promising supporting material due to its high flexibility and biocompatibility. In this study a novel flexible nanocellulose MOF composite material was synthesised in aqueous media by a novel and straightforward in situ one-pot green method. The material consisted of MOF particles of the type MIL-100(Fe) (from Material Institute de Lavoisier, containing Fe(III) 1,3,5-benzenetricarboxylate) immobilized onto bacterial cellulose (BC) nanofibers. The novel nanocomposite material was applied to efficiently separate arsenic and Rhodamine B from aqueous solution, achieving adsorption capacities of 4.81, and 2.77 mg g^{-1}, respectively. The adsorption process could be well modelled by the nonlinear pseudo-second-order fitting.

Keywords: bacterial cellulose; metal organic framework; nanocomposite; adsorption

---

1. Introduction

Since the advent of the industrial revolution, dumping of large amounts of industrial waste including dyes and toxic metal ions has contributed to the serious issue of water pollution [1]. Consequently, various techniques for the removal of organic dyes and toxic metal ions from aqueous solutions, including adsorption [2–4], chemical precipitation [5], ion exchange [6], and membrane separation [7] have been evaluated. Amongst these, adsorption is proven to be an effective and convenient method for water purification due to the ease of operation and the low cost. However, for nanoadsorbents, complicated and tedious high-speed centrifugation or separation of the adsorbent using filtration is required, hindering the extensive application of such adsorbents. Therefore, development of novel materials for water treatment is of great interest.

Metal-organic frameworks (MOFs) are porous hybrid materials composed of metal ions bridged by polydentate organic ligands. Since the synthesis of MOF-5 reported by Yaghi and co-workers (1999) [8], MOFs have received significant attention due to their high crystallinity, large surface areas, thermal stability and unique porosity. MOF applications are numerous, including catalysis [9–11], drug delivery [12], gas storage [13–15], chemical sensing [16], and separation [17–19]. The properties of the MOF structures can be easily tuned by selecting different metal ions and bridging organic polydentate ligands, and the design and preparation of new MOFs into various structures is still of great interest [20]. However, the handling and processing of MOFs are also challenging due to the crystalline nature of the MOFs, making them brittle and fragile as inorganic materials [21,22]. To tackle this problem, efforts have been made to entrap MOFs onto various substrates, thus realizing a combination of their advantages, while circumventing, and to different extents addressing the MOF's shortcomings. Various MOF composites have been successfully prepared via direct deposition of MOF particles on solid substrate surfaces [23]. However, although such synthetic protocols produce a new types of functional materials, their applications are limited by their constrained morphologies and synthetic protocols. Usually, substrate surface modifications are needed to increase the MOF loading and conditions must be applied that in many cases involve environmentally hazardous reagents [24].

Membrane adsorption can be applied to separate metal ions in larger scale. For this purpose, a column is used that is packed with a short stack of porous membranes with a large diameter, to avoid high pressure drops. In membrane adsorption fibres can act as an adsorbing medium. One advantage of this technology is that process flow rates can be orders of magnitudes higher than for packed beds, without elevated pressure. The challenge is to achieve a comparable adsorption capacity to that of packed beds. However, by using functionalised nanofibers the adsorption capacities can even exceed packed bed resins due to the very high surface area of the fibres. Refined low-cost biopolymers such as nanocellulose can here be considered as a promising supporting material due to its very interesting properties, such as high chemical purity and crystallinity, flexibility and biocompatibility [25]. Furthermore, nanocellulose is the most abundant and renewable green biopolymer and a sustainable raw material. Bacterial cellulose (BC) fibrils are one of the most frequently reported nanocellulose types. It is produced by cultivating *Acetobacter xylinum* in the presence of sugar [26]. BC nanofibrillar cellulose shows great promise to be used as a substrate since it can be generated in high yield from its natural source containing >70 wt% crystalline cellulose. It also shows high aspect ratio due to its fibrillar length reaching up to several micrometers while at the same time providing average fiber diameters of 20–100 nm [27]. To date, only a few studies have reported integration of MOFs onto nanocellulose. Matsumoto and co-workers reported the successful growth of MOFs (up to 44% loading) at carboxyl groups on crystalline TEMPO-oxidized cellulose nanofibers and prepared densely packed films coated on a filter paper, which demonstrated high gas separation selectivity [28]. Zhu and co-workers have developed a strategy for combining MOFs and cellulose nanocrystals into a highly functional aerogel for separation applications, where the MOF loading can be easily tailored by changing the initial ratio of the components [29]. Recently, Au-Duong and co-workers successfully demonstrated that a flexible nanocomposite pellicle combining imidazolate framework-8 (ZIF-8) and wet BC could be simply synthesized when polydopamine surface coating on cellulose nanofibers is applied in advance [30].

Herein, we report a new, straightforward, environmentally friendly in situ green strategy to integrate MOF Material Institute de Lavoisier-100(Fe), Fe-BTC (alternatively known as MIL-100(Fe); BTC = 1,3,5-benzenetricarboxylate) [31] into BC in aqueous media. This method makes it possible to synthesize cellulose-based metal organic framework nanocomposite material without using any chemical modifier. The synthetic approach provides a general strategy to prepare such nanocomposites for various applications. The synthesized MIL-100(Fe)@BC nanocomposite not only maintained high micro/ mesoporosities of the MIL-100(Fe) particles, but also combined flexibility and shapeability of the BC crystal support. This combination produces a shapeable, low-cost, chemically inert and scalable product that can be used in various applications, for example, water purification. The synthesized hybrid MIL-100(Fe)@BC nanocomposite was evaluated in water purification as an efficient adsorbent for the removal of arsenic As(III), and Rhodamine B.

2. Materials and Methods

Materials: Iron(III) chloride hexahydrate (≥98%), 1,3,5-benzenetricarboxylate (98%) were purchased from VWR (Stockholm, Sweden). Rhodamine B (≥95%) was purchased from *Sigma-Aldrich* (Stockholm, Sweden). Arsenic pure single-element standard (1000 mg L^{-1}) was purchased from PerkinElmer (Stockholm, Sweden). All chemicals were used as received without further purification. Deionized water (resistivity: 18.2 MΩ/cm) was used to prepare aqueous solutions. The bacterial cellulose nanofibrils were extracted from a uniformly grown bacterial cellulose pellicle, using 1 L aqueous sulphuric acid solution containing 30 vol% of reagent grade acid (Merck KGaA, Stockholm, Sweden) under stirring at 60 ± 0.5 °C (300 rpm). The total time for the extraction was 7 h, i.e., until no visible cellulose pieces were apparent and the solution had acquired an even beige color. The supernatant was decanted after a first isothermal centrifugation at 10,000 rpm for 10 min and replaced with fresh MilliQ water. The procedure was repeated twice until a neutral pH was obtained. The yield of the extraction was ca. 60 wt% based on the dry weight of the bacterial cellulose. The surface area of the bacterial cellulose nanofibrils was approximated to ca. 159 m^2 g^{-1} from size distributions and counting of a minimum of 500 fibrils deposited on TEM grids. The approximated surface area was in relatively good agreement with the surface area of 189 m^2 g^{-1} reported by Roman et al. [32] while larger than the experimental value of 103 m^2 g^{-1} for dry bacterial cellulose fibrils by Olsson et al. [26].

In situ one-pot synthesis of MIL-100(Fe)@BC: Iron(III) chloride hexahydrate (164 mg) was added to 50 mL of well-sonicated BC (2.0 wt% dry contents) and the reaction mixture was refluxed at 100 °C under mechanical stirring (350 rpm). After 30 min, different concentrations of 1,3,5-benzenetricarboxylate were dissolved in 4 mL of deionized water and added to the reaction mixture, which was kept for an additional 30 min at 100 °C. The final suspension was centrifuged and washed 3 times with acetone followed by deionized water. The product was placed in a plastic tube that was directly frozen by immersing the tube in liquid N$_2$ followed by freeze-drying to obtain the final hybrid nanocomposite MIL-100(Fe)@BC, that was activated at 100 °C for 12 h before being used as an adsorbent.

Adsorption: The adsorption experiments were carried out at room temperature. Adsorption studies were carried out by soaking MIL-100(Fe)@BC (0.18 g) in aqueous solution containing different contaminants (20 mg L^{-1} of As(III) (50 mL) and 10 mg L^{-1} of Rhodamine B (40 mL)) for a certain amount of time. After adsorption, the MIL-100(Fe)@BC nanocomposite was separated from the aqueous solution, and the dye concentration was analyzed by UV-vis based on a calibration curve prepared from solutions with known contaminant concentration at the maximum wavelengths (554 nm). The concentration of As(III) in the supernatant was determined by inductively coupled plasma atomic emission spectroscopy (ICP-OES, Waltham, MA, USA). The adsorption experiments were performed in duplicate/triplicate and the average values of total adsorption are reported. The adsorption capacity (q_t, mg g^{-1}) and the removal % at time t were determined using Equations (1) and (2).

$$q_t = \frac{(C_0 - C_t)V}{m} \tag{1}$$

$$Removal\ (\%) = \frac{C_0 - C_t}{C_0} \times 100 \tag{2}$$

where C_0 and C_t (mg L^{-1}) are the initial concentration and concentration at time t in aqueous solution, respectively, V (L) is the volume of the aqueous phase, and m (g) is the mass of the MIL-100(Fe)@BC nanocomposite. The kinetics of the adsorption process was investigated by fitting the nonlinear forms of pseudo-first order and pseudo-second order models to the data using Equations (3) and (4).

$$q_t = q_e\left(1 - e^{-k_1 t}\right) \tag{3}$$

$$q_t = \frac{k_2 q_e^2 t}{1 + k_2 q_e t} \tag{4}$$

where k_1 (min^{-1}) and k_2 (g mg^{-1} min) are the pseudo-first-order and pseudo-second-order rate constants of adsorption, q_t and q_e (mg g^{-1}) are the adsorption capacity at a given time t and at equilibrium, respectively.

Characterization: Scanning electron microscopy (SEM) measurements were conducted on a Hitachi S-4800 microscope (Tokyo, Japan) using accelerating voltages of 2 kV to 10 kV. High resolution transmission electron microscopy (HRTEM) was performed at 90 K using a JEOL-2011 (Tokyo, Japan) at 200 kV. Powder X-ray diffraction was conducted on a Bruker D8 advance X-ray diffractometer (Madison, WI, USA). The concentration of the contaminated water was determined using an UV-vis (ultraviolet-visible) spectrophotometer (DR 3900, Hach, Stockholm, Sweden). The total concentrations of arsenic ions were determined by ICP-OES (Thermo Fisher iCAP 7400, Waltham, MA, USA). The pH value of the solution was measured by a pH-meter (ORION Star A211, Thermo ScientificTM, Waltham, MA, USA). Thermogravimetric analysis (TGA) was performed under a nitrogen atmosphere from 25 °C to 700 °C with a heating rate of 5 °C min^{-1} using a thermogravimetric analyzer (TGA/SDTA 851e, Mettler Toledo, Mississauga, ON, Canada).

3. Results and Discussion

The hybrid MIL-100(Fe)@BC nanocomposites were synthesized by mixing BC and Fe(III) under refluxing conditions for 30 min followed by the addition of BTC at different molar ratios of BTC/Fe(III), as shown in Scheme 1. When Fe(III) ions are added to the BC solution in the preparation of the MIL-100(Fe)@BC nanocomposite, complexation interaction will take place between the Fe(III) ions and the BC crystals hydroxyl groups, and the Fe(III) ions concentration will increase in the vicinity of the fibrillar BC crystals surface. Once the ligand precursor BTC was added to the reaction mixture, the binding Fe(III) ions would participate to form MIL-100(Fe) crystals, that grew gradually on the BC network that became partially immobilized as the inorganic condensations occurred. At a BTC/Fe(III) molar ratio of 25, aggregation of MIL-100(Fe) crystals with particles of average size > 400 nm appeared on the surface of the BC and the nanofibers could barely be observed after freeze-drying (Figure S1). However, as the BTC/Fe(III) molar ratio increased to 120, MIL-100(Fe) particles with smaller sizes were uniformly observed on the surface of the BC. These were stable enough to persist during the freeze-drying and form the hybrid nanocomposite MIL-100(Fe)@BC (Figure 1a,b).

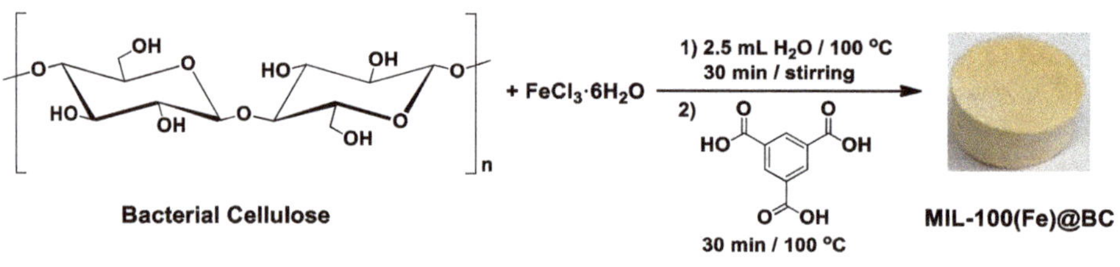

Bacterial Cellulose + FeCl$_3$·6H$_2$O → 1) 2.5 mL H$_2$O / 100 °C, 30 min / stirring; 2) 30 min / 100 °C → MIL-100(Fe)@BC

Scheme 1. General procedure for preparation of the MIL-100(Fe)@BC nanocomposite.

An increased ligand/metal ion molar ratio resulted in the formation of smaller MOF particle sizes on the surface of BC. The MOF size reduction upon increasing the ligand/metal ion molar ratio has also been observed by other researchers [29]. Such behaviour can be explained by the high supersaturation driving force at higher ligand/metal ion ratios, leading to high nucleation in the early stage of MOF synthesis and the formation of small MOF crystals [33]. The resultant hybrid MIL-100(Fe)@BC nanocomposite was flexible, ultralight and mechanically robust enough to be easily processed and further evaluated without any loss or damage of the MOFs structural integrity.

The powder X-ray diffraction patterns for BC and 100(Fe)@BC nanocomposite confirmed the formation of MIL-100(Fe) within the BC crystal network (Figure S2). Brunauer-Emmett-Teller (BET) analysis using N$_2$ sorption was also employed to determine the surface area of the prepared

MIL-100(Fe)@BC nanocomposite (Figure 1c). The BET surface area of the MIL-100(Fe)@BC nanocomposite is 47.13 ± 0.15 m^2/g.

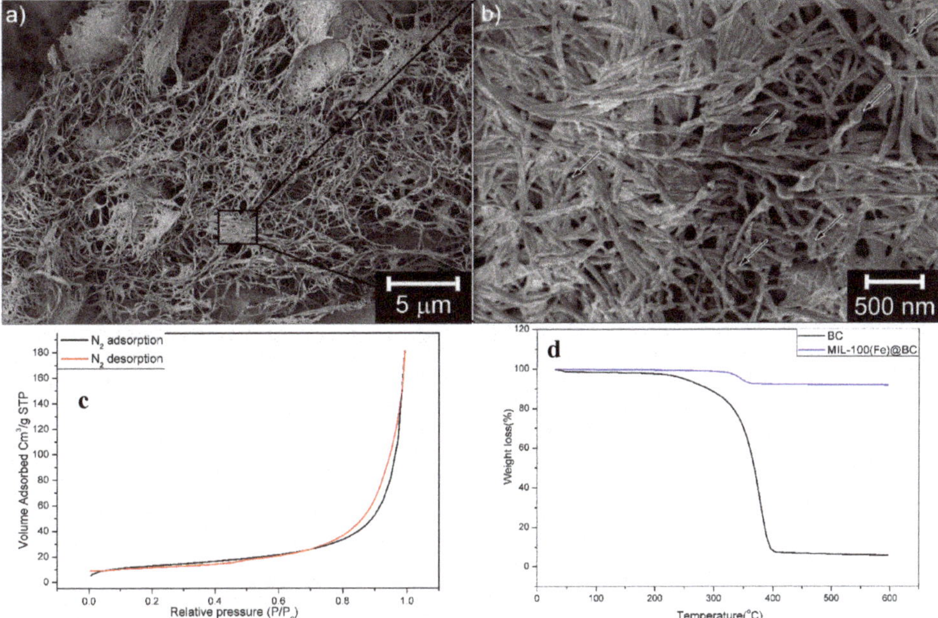

Figure 1. (**a,b**) SEM image of hybrid MIL-100(Fe)@BC nanocomposite (BTC/Fe(III) = 120), (**c**) Brunauer-Emmett-Teller (BET) analysis of MIL-100(Fe)@BC (BTC/Fe(III) = 120), and (**d**) thermogravimetric analysis (TGA) of bacterial cellulose (BC) and MIL-100(Fe)@BC.

Figure 1d shows the thermogravimetric analysis (TGA) of the MOF-cellulose hybrid material and the BC. A gradual decrease in weight of ca. 0.3 wt% occurred up to ca. 180 °C, which was synonymous with structurally bonded water entrapped in the porous material. Further condensation and densification (ca. 1 wt%) of the material occurred from ca. 180 °C until the degradation temperature of the cellulose was reached at ca. 350 °C. The sharp weight loss to 91.8 wt% (of the materials original weight) at 350 °C could thus be correlated to a total loading of bacterial cellulose equivalent with ca. 8 wt%. At temperatures above 400 °C, degradation and evaporation of carbon residuals occurred. The density of the entire cylindrical sample (Scheme 1) could further be derived from the volume of the sample and its weight. The low density of MIL-100(Fe)@BC was 47.8 mg cm^{-3}, implying that the nanocomposite was very porous with a large accessible surface area.

Owing to their high specific surface area, MOFs are intended for various applications [34]. Herein, we evaluate the synthesized MIL-100(Fe)@BC nanocomposite in water purification to demonstrate the adsorption ability of MOF-cellulose materials as interpenetrating networks with an integrated cellulose phase. Arsenic is recognized as one of the most hazardous metal ions in drinking water and is listed by the World Health Organization among the top 10 major public health concerns. In this context, the adsorption capacity of the MIL-100(Fe)@BC nanocomposite was tested for the removal of As(III) from aqueous solutions. In the tests, a small MIL-100(Fe)@BC nanocomposite (0.18 g) was soaked into 50 mL of an aqueous solution containing As(III) (20 ppm) (See Experimental Section for more details). The adsorption capacities (q_t in mg g^{-1}) and removal (%) of As(III) at different times (t) are shown in Figure 2a. High adsorption capacity for As(III) adsorption onto MIL-100(Fe)@BC nanocomposite was observed (removal efficiency $\approx$ 85% within 72 h).

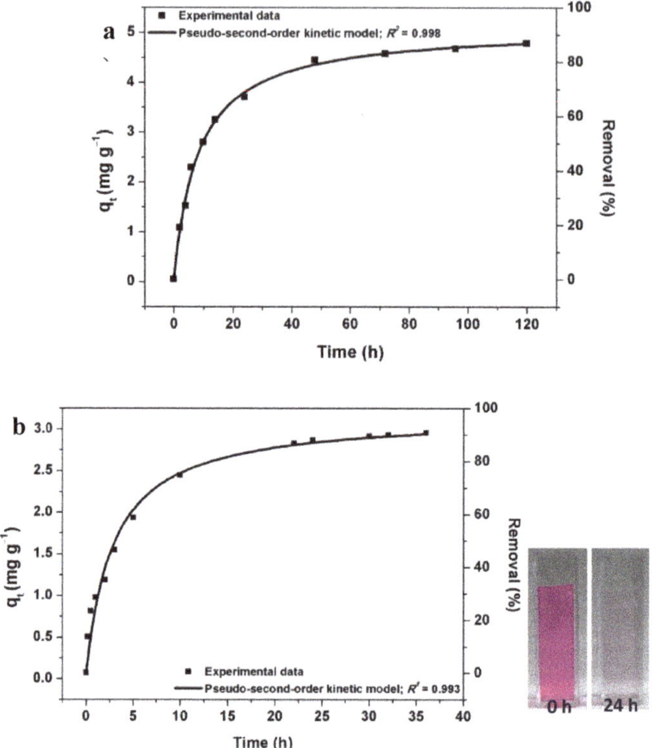

Figure 2. Adsorption capacity versus time, and the nonlinear pseudo-second-order fitting for the adsorption of As(III) (**a**); and Rhodamine B (**b**) on MIL-100(Fe)@BC nanocomposite, respectively. The inset shows a photograph of the Rhodamine B aqueous solutions before and after the adsorption process.

Two kinetic models, the pseudo-first-order and pseudo-second-order models, were used to investigate the mechanism of As(III) adsorption on the MIL-100(Fe)@BC nanocomposite (Figure S3a and Figure 2a, respectively). The nonlinear pseudo-second-order kinetic model fitted well to the experimental data (Figure 2a), and the experimental $q_{t,exp}$ values and the calculated $q_{t,cal}$ values obtained from the pseudo-second-order kinetic model are in good agreement (Table S1). The calculated values of the pseudo-second-order parameters are listed in Table S1. The fitting results indicate that the BC network is not hindering the accessibility of the MIL-100(Fe) pores and most of the MIL-100(Fe) particles are functional in the adsorption of As(III) onto the MIL-100(Fe)@BC nanocomposite as a chemical process via surface complexation [35]. The total concentration of Fe in the solution was measured by ICP-OES after the adsorption experiments in order to investigate the stability of the material. No Fe was detected, indicating a high stability of the synthesized MIL-100(Fe)@BC nanocomposite under the employed conditions.

Removal of organic dyes from aqueous solution is another significant challenge in the field of water purification. Among several organic dyes, Rhodamine B was chosen for the present study due to its extensive use as a colorant in the textile and food industries [36]. Rhodamine B was used as a model tracer to monitor the adsorption capacity and kinetics for adsorption onto the MIL-100(Fe)@BC nanocomposite. Herein, a small amount of the nanocomposite MIL-100(Fe)@BC (0.18 g) was added to 40 mL of an aqueous solution containing Rhodamine B (10 ppm), and the dye concentration in the solution at certain time was determined by UV-Vis (see Experimental Section for more details). The adsorption experiments were carried out at room temperature. The adsorption capacity (q_t) and removal percentage of Rhodamine B as a function of time (*t*) are shown in Figure 2b. The adsorption capacities for Rhodamine B increased with increasing time, and reached equilibrium within 24 h, achieving a removal percentage of 85%. The high adsorption capacity of the MIL-100(Fe)@BC nanocomposite results from the hierarchical porosity and small MIL-100(Fe) crystals integrated within the BC network, which offer a high number of external surface sites for the adsorption.

The kinetics for Rhodamine B adsorption onto the MIL-100(Fe)@BC nanocomposite was studied using pseudo-first-order and pseudo-second-order models (Figure S3b and Figure 2b, respectively). The obtained experimental data for the adsorption process, determined by UV-Vis at 554, fitted very well with the pseudo-second-order kinetic model (Figure 2b), which indicate that the adsorption was controlled by intraparticle diffusion [37]. The calculated values of the pseudo-second-order parameters are listed in Table S1. After soaking the MIL-100(Fe)@BC nanocomposite with the Rhodamine B aqueous solution, the pink dye solution gradually faded into colorless over time, and its UV-Vis maximum absorption beak at 554 nm reduced significantly (Figure S4). The inset in Figure 2b visually confirms the color change of Rhodamine B aqueous solution before and after the adsorption process by MIL-100(Fe)@BC nanocomposite. The adsorption mechanism can be attributed to the electrostatic interactions between MIL-100(Fe)@BC nanocomposite and Rhodamine B [38].

A comparison of the obtained results for As(III) and Rhodamine B removal using MIL-100(Fe)@BC nanocomposite with other reported systems is summarized in Table 1.

Table 1. Comparison of As(III) and Rhodamine B removal with other reported adsorbents.

Adsorbents	q_e (mg g^{-1})	Ref.
As(III) Removal		
MIL-53(Al)-graphene oxide	65.0	[39]
Surfactant-modified montmorillonite	1.48	[40]
Magnetic pinecone biomass	18.02	[41]
Zn-MOF	49.50	[42]
MIL-100(Fe)	120	[35]
MIL-100(Fe)@BC	4.81	This work
Rhodamine B Removal		
Hypercross-linked polymeric adsorbent	2.1	[43]
Mango leaf power	3.31	[44]
Zn-MOF	3.750	[45]
Fe$_3$O$_4$/MIL-100(Fe)	28.36	[38]
MIL-100(Fe)@BC	2.77	This work

4. Conclusions

A MIL-100(Fe)@BC nanocomposite was synthesized by an environmentally friendly method using water as solvent. The bacterial cellulose (BC) acted as structural support during the lyophilisation of the MIL-100(Fe), which resulted in a flexible and light weight material suitable for membrane adsorption processes. It was demonstrated that the size of the loaded MIL-100(Fe) particles on the BC can be tailored by changing the initial ratio of MIL-100(Fe) precursors. The synthesized nanocomposite is efficient in removal of As(III) and Rhodamine B from aqueous solutions. The kinetic studies revealed that the adsorption of As(III) and Rhodamine B was best fitted to the pseudo-second order model. The synthetic approach provides a simple strategy to prepare the MOFs@BC nanocomposites for various applications, in particular for water purification. Further studies on how to immobilize other types of MOFs using BC could be explored.

Polymers **2020**, *12*, 1104

Supplementary Materials: The following are available online at http://www.mdpi.com/2073-4360/12/5/1104/s1, Figure S1: Photograph (inset) and TEM image of MIL-100(Fe)@BC nanocomposite (BTC/Fe(III) = 25), Figure S2: XRD patterns of (a) BC and (b) MIL-100(Fe)@BC nanocomposite (BTC/Fe(III) = 120), Figure S3: Plots of pseudo-first-order kinetics models for the adsorption of As(III) (a) and Rhodamine B (b) using MIL-100(Fe)@BC nanocomposite (BTC/Fe(III) = 120), Figure S4: UV-Vis spectra for the adsorption of Rhodamine B using MIL-100(Fe)@BC nanocomposite (BTC/Fe(III) = 120) at different time, and Table S1: Kinetic parameters for the adsorption of As(III) and Rhodamine B using MIL-100(Fe)@BC nanocomposite.

Author Contributions: R.M.A., A.F.A.-M. and Q.W.: Investigation; R.T.O. and K.F.: Validation; R.M.A. and A.F.A.-M.: Writing—original draft; R.T.O. and K.F.: Writing—Review & Editing; R.M.A., A.F.A.-M., R.T.O. and K.F.: Methodology. All authors have read and agreed to the published version of the manuscript.

Funding: This work was funded by a start-up grant at KTH Royal Institute of Technology.

Conflicts of Interest: The authors declare no conflict of interest.

References

1. Yin, F.; Deng, X.; Jin, Q.; Yuan, Y.; Zhao, C. The impacts of climate change and human activities on grassland productivity in Qinghai Province, China. *Front. Earth Sci.* **2014**, *8*, 93–103. [CrossRef]
2. Geisse, R.A.; Ngule, M.C.; Genna, T.D. Removal of lead ions from water using thiophene-functionalized metal–organic frameworks. *Chem. Commun.* **2020**, *56*, 237–240. [CrossRef] [PubMed]
3. Li, D.; Tian, X.; Wang, Z.; Guan, Z.; Li, X.; Qiao, H.; Ke, H.; Luo, L.; Wei, Q. Multifunctional adsorbent based on metal-organic framework modified bacterial cellulose/chitosan composite aerogel for high efficient removal of heavy metal ion and organic pollutant. *Chem. Eng. J.* **2020**, *383*, 123127. [CrossRef]
4. Huang, L.; He, M.; Chen, B.; Hu, B. Magnetic Zr-MOFs nanocomposites for rapid removal of heavy metal ions and dyes from water. *Chemosphere* **2018**, *199*, 435–444. [CrossRef]
5. Chen, Q.; Luo, Z.; Hills, C.; Xue, G.; Tyrer, M. Precipitation of heavy metals from wastewater using simulated flue gas: Sequent additions of fly ash, lime and carbon dioxide. *Water Res.* **2009**, *43*, 2605–2614. [CrossRef] [PubMed]
6. Wang, Z.; Feng, Y.; Hao, X.; Huang, W.; Feng, X. A novel potential-responsive ion exchange film system for heavy metal removal. *J. Mater. Chem. A* **2014**, *2*, 10263–10272. [CrossRef]
7. Tofighy, A.M.; Mohammadi, T. Divalent heavy metal ions removal from contaminated water using positively charged membrane prepared from a new carbon nanomaterial and HPEI. *Chem. Eng. J.* **2020**, *388*, 124192. [CrossRef]
8. Li, H.; Eddaoudi, M.; O'Keeffe, M.; Yaghi, O.M. Design and synthesis of an exceptionally stable and highly porous metal-organic framework. *Nature* **1999**, *402*, 276. [CrossRef]
9. David, F.; Sonia, A.; Catherine, P. Metal-Organic frameworks: Opportunities for catalysis. *Angew. Chem. Int. Ed.* **2009**, *48*, 7502–7513.
10. Lee, J.; Farha, O.K.; Roberts, J.; Scheidt, K.A.; Nguyen, S.T.; Hupp, J.T. Metal-Organic framework materials as catalysts. *Chem. Soc. Rev.* **2009**, *38*, 1450–1459. [CrossRef]
11. Zhu, L.; Liu, X.-Q.; Jiang, H.-L.; Sun, L.-B. Metal-Organic frameworks for heterogeneous basic catalysis. *Chem. Rev.* **2017**, *117*, 8129–8176. [CrossRef] [PubMed]
12. Wu, M.-X.; Yang, Y.-W. Metal-Organic framework (MOF)-based drug/Cargo delivery and cancer therapy. *Adv. Mater.* **2017**, *29*, 1606134. [CrossRef] [PubMed]
13. Alkordi, M.H.; Belmabkhout, Y.; Cairns, A.; Eddaoudi, M. Metal–organic frameworks for H2 and CH4 storage: Insights on the pore geometry–sorption energetics relationship. *IUCrJ* **2017**, *4*, 131–135. [CrossRef] [PubMed]
14. Tian, T.; Zeng, Z.; Vulpe, D.; Casco, M.E.; Divitini, G.; Midgley, P.A.; Silvestre-Albero, J.; Tan, J.-C.; Moghadam, P.Z.; Fairen-Jimenez, D. A sol-gel monolithic metal-organic framework with enhanced methane uptake. *Nat. Mater.* **2017**, *17*, 174. [CrossRef]
15. Zheng, H.; Zhang, Y.; Liu, L.; Wan, W.; Guo, P.; Nyström, A.M.; Zou, X. One-pot synthesis of metal–organic frameworks with encapsulated target molecules and their applications for controlled drug delivery. *J. Am. Chem. Soc.* **2016**, *138*, 962–968. [CrossRef]
16. Chen, B.; Yang, Z.; Zhu, Y.; Xia, Y. Zeolitic imidazolate framework materials: Recent progress in synthesis and applications. *J. Mater. Chem. A* **2014**, *2*, 16811–16831. [CrossRef]

17. Li, J.-R.; Sculley, J.; Zhou, H.-C. Metal-organic frameworks for separations. *Chem. Rev.* **2012**, *112*, 869–932. [CrossRef]

18. Qiu, S.; Xue, M.; Zhu, G. Metal–organic framework membranes: From synthesis to separation application. *Chem. Soc. Rev.* **2014**, *43*, 6116–6140. [CrossRef]

19. Abdel-Magied, A.F.; Abdelhamid, H.N.; Ashour, R.M.; Zou, X.; Forsberg, K. Hierarchical porous zeolitic imidazolate frameworks nanoparticles for efficient adsorption of rare-earth elements. *Microporous Mesoporous Mater.* **2019**, *278*, 175–184. [CrossRef]

20. Almeida Paz, F.A.; Klinowski, J.; Vilela SM, F.; Tomé JP, C.; Cavaleiro JA, S.; Rocha, J. Ligand design for functional metal–organic frameworks. *Chem. Soc. Rev.* **2012**, *41*, 1088–1110. [CrossRef]

21. Chen, Y.; Huang, X.; Zhang, S.; Li, S.; Cao, S.; Pei, X.; Zhou, J.; Feng, X.; Wang, B. Shaping of metal–organic frameworks: From fluid to shaped bodies and robust foams. *J. Am. Chem. Soc.* **2016**, *138*, 10810–10813. [CrossRef] [PubMed]

22. Liu, C.; Zhang, J.; Zheng, L.; Zhang, J.; Sang, X.; Kang, X.; Zhang, B.; Luo, T.; Tan, X.; Han, B. Metal-organic framework for emulsifying carbon dioxide and water. *Angew. Chem. Int. Ed.* **2016**, *55*, 11372–11376. [CrossRef] [PubMed]

23. Bétard, A.; Fischer, R.A. Metal—organic framework thin films: From fundamentals to applications. *Chem. Rev.* **2012**, *112*, 1055–1083. [CrossRef] [PubMed]

24. Denny, M.S.; Cohen, S.M. In situ modification of metal—organic frameworks in mixed-matrix membranes. *Angew. Chem. Int. Ed.* **2015**, *54*, 9029–9032. [CrossRef]

25. Moon, R.J.; Martini, A.; Nairn, J.; Simonsen, J.; Youngblood, J. Cellulose nanomaterials review: Structure, properties and nanocomposites. *Chem. Soc. Rev.* **2011**, *40*, 3941–3994. [CrossRef]

26. Olsson, R.T.; Azizi Samir MA, S.; Salazar-Alvarez, G.; Belova, L.; Ström, V.; Berglund, L.A.; Ikkala, O.; Nogués, J.; Gedde, U.W. Making flexible magnetic aerogels and stiff magnetic nanopaper using cellulose nanofibrils as templates. *Nat. Nanotechnol.* **2010**, *5*, 584. [CrossRef]

27. Lee, K.Y. *Nanocellulose and Sustainability: Production, Properties, Applications, and Case Studies*; CRC Press: Boca Raton, FL, USA, 2018.

28. Matsumoto, M.; Kitaoka, T. Ultraselective gas separation by nanoporous metal-organic frameworks embedded in gas-barrier nanocellulose films. *Adv. Mater.* **2016**, *28*, 1765–1769. [CrossRef]

29. Zhu, H.; Yang, X.; Cranston, E.D.; Zhu, S. Flexible and porous nanocellulose aerogels with high loadings of metal–organic-framework particles for separations applications. *Adv. Mater.* **2016**, *28*, 7652–7657. [CrossRef]

30. Au-Duong, A.-N.; Lee, C.-K. Flexible metal–organic framework-bacterial cellulose nanocomposite for iodine capture. *Cryst. Growth Des.* **2018**, *18*, 356–363. [CrossRef]

31. Férey, G.; Serre, C.; Mellot-Draznieks, C.; Millange, F.; Surblé, S.; Dutour, J.; Margiolaki, I. A hybrid solid with giant pores prepared by a combination of targeted chemistry, simulation, and powder diffraction. *Angew. Chem. Int. Ed.* **2004**, *43*, 6296–6301. [CrossRef]

32. Roman, M.; Winter, W.T. Effect of sulfate groups from sulfuric acid hydrolysis on the thermal degradation behavior of bacterial cellulose. *Biomacromolecules* **2004**, *5*, 1671–1677. [CrossRef] [PubMed]

33. Cravillon, J.; Nayuk, R.; Springer, S.; Feldhoff, A.; Huber, K.; Wiebcke, M. Controlling Zeolitic imidazolate framework nano- and microcrystal formation: Insight into crystal growth by time-resolved in situ static light scattering. *Chem. Mater.* **2011**, *23*, 2130–2141. [CrossRef]

34. Pettinari, C.; Marchetti, F.; Mosca, N.; Tosi, G.; Drozdov, A. Applications of metal-organic frameworks. *Polym. Int.* **2017**, *66*, 731–744. [CrossRef]

35. Georgiou, Y.; Perman, J.A.; Bourlinos, A.B.; Deligiannakis, Y. Highly efficient arsenite [As(III)] adsorption by an [MIL-100(Fe)] metal–organic framework: Structural and mechanistic insights. *J. Phys. Chem. C* **2018**, *122*, 4859–4869. [CrossRef]

36. Mohammadi, M.; Hassani, A.J.; Mohamed, A.R.; Najafpour, G.D. Removal of Rhodamine B from aqueous solution using Palm shell-based activated carbon: Adsorption and kinetic studies. *J. Chem. Eng. Data* **2010**, *55*, 5777–5785. [CrossRef]

37. Plazinski, W.; Dziuba, J.; Rudzinski, W. Modeling of sorption kinetics: The pseudo-second order equation and the sorbate intraparticle diffusivity. *Adsorption* **2013**, *19*, 1055–1064. [CrossRef]

38. Haochi, L.; Xiaohui, R.; Ligang, C.J. Synthesis and characterization of magnetic metal–organic framework for the adsorptive removal of Rhodamine B from aqueous solution. *Ind. Eng. Chem.* **2016**, *34*, 278–285.

39. Tonoy, C.; Lei, Z.; Junqing, Z.; Srijan, A. Removal of Arsenic(III) from aqueous solution using metal organic framework-graphene oxide nanocomposite. *Nanomaterials* **2018**, *8*, 1062.

40. Samira, S.; Ghasem, A.; Faramarz, M.; Abdolreza, K.; Kazem, G.; Ehsan, N. Application of surfactant-modified montmorillonitefor As (III) removal from aqueous solutions: Kinetics and isotherm study. *Desalin. Water Treat.* **2018**, *115*, 236–248.

41. Agnes, P.; Eliazer, B.N.; Augustine, E.O. Enhanced Arsenic (III) adsorption from aqueous solution by magnetic pine cone biomass. *Mater. Chem. Phys.* **2019**, *222*, 20–30.

42. Jian, M.; Liu, B.; Zhang, G.; Liu, R.; Zhang, X. Adsorptive removal of arsenic from aqueous solution by zeolitic imidazolate framework-8 (ZIF-8) nanoparticles. *Colloids Surf. Physicochem. Eng. Asp.* **2015**, *465*, 67–76. [CrossRef]

43. Huang, J.-H.; Huang, K.-L.; Liu, S.-Q.; Wang, A.-T.; Yan, C. Adsorption of Rhodamine B and methyl orange on a hypercrosslinked polymeric adsorbent in aqueous solution. *Colloid Surf. A Physicochem. Eng. Asp.* **2008**, *330*, 55. [CrossRef]

44. Khan, T.A.; Sharma, S.; Ali, I. Adsorption of Rhodamine B dye from aqueous solution onto acid activated mango (*Magnifera indica*) leaf powder: Equilibrium, kinetic and thermodynamic studies. *Toxicol. Environ. Health Sci.* **2011**, *3*, 286–297.

45. Zhang, J.; Li, F.; Sun, Q. Rapid and selective adsorption of cationic dyes by a unique metal-organic framework with decorated pore surface. *Appl. Surf. Sci.* **2018**, *440*, 1219–1226. [CrossRef]

Article

Development of a Highly Proliferated Bilayer Coating on 316L Stainless Steel Implants

Fatemeh Khosravi [1], Saied Nouri Khorasani [1,*], Shahla Khalili [1], Rasoul Esmaeely Neisiany [2,*], Erfan Rezvani Ghomi [3], Fatemeh Ejeian [4], Oisik Das [5,*] and Mohammad Hossein Nasr-Esfahani [4,*]

[1] Department of Chemical Engineering, Isfahan University of Technology, Isfahan 8415683111, Iran; f.khosravi@ce.iut.ac.ir (F.K.); shahla.khalili65@gmail.com (S.K.)

[2] Department of Materials and Polymer Engineering, Faculty of Engineering, Hakim Sabzevari University, Sabzevar 9617976487, Iran

[3] Department of Mechanical Engineering, Center for Nanofibers and Nanotechnology, National University of Singapore, Singapore 119260, Singapore; erfanrezvani@u.nus.edu

[4] Department of Cellular Biotechnology, Cell Science Research Center, Royan Institute for Biotechnology, ACECR, Isfahan 8159358686, Iran; fatemeh.eje@gmail.com

[5] Material Science Division, Department of Engineering Sciences and Mathematics, Luleå University of Technology, 97187 Luleå, Sweden

* Correspondence: saied@cc.iut.ac.ir (S.N.K.); r.esmaeely@hsu.ac.ir (R.E.N.); oisik.das@ltu.se (O.D.); mh.nasr-esfahani@royaninstitute.org (M.H.N.-E.)

Received: 1 April 2020; Accepted: 28 April 2020; Published: 1 May 2020

Abstract: In this research, a bilayer coating has been applied on the surface of 316 L stainless steel (316LSS) to provide highly proliferated metallic implants for bone regeneration. The first layer was prepared using electrophoretic deposition of graphene oxide (GO), while the top layer was coated utilizing electrospinning of poly (ε-caprolactone) (PCL)/gelatin (Ge)/forsterite solutions. The morphology, porosity, wettability, biodegradability, bioactivity, cell attachment and cell viability of the prepared coatings were evaluated. The Field Emission Scanning Electron Microscopy (FESEM) results revealed the formation of uniform, continuous, and bead-free nanofibers. The Energy Dispersive X-ray (EDS) results confirmed well-distributed forsterite nanoparticles in the structure of the top coating. The porosity of the electrospun nanofibers was found to be above 70%. The water contact angle measurements indicated an improvement in the wettability of the coating by increasing the amount of nanoparticles. Furthermore, the electrospun nanofibers containing 1 and 3 wt.% of forsterite nanoparticles showed significant bioactivity after soaking in the simulated body fluid (SBF) solution for 21 days. In addition, to investigate the in vitro analysis, the MG-63 cells were cultured on the PCL/Ge/forsterite and GO-PCL/Ge/forsterite coatings. The results confirmed an excellent cell adhesion along with considerable cell growth and proliferation. It should be also noted that the existence of the forsterite nanoparticles and the GO layer substantially enhanced the cell proliferation of the coatings.

Keywords: biocomposites; nanofibers; electrospinning; cell culture; graphene oxide

---

1. Introduction

At present, numerous types of bone diseases, e.g., bone fractures, bone infections, bone cancers, and genetic diseases are rising due to increasing prevalence of physical inactivity, obesity and lack of safe exercising [1]. It is reported that over 20 million people suffer from bone disorders and clinical troubles annually, making this an global issue [2]. Traditional bone regeneration methods were based on utilizing autograft and allograft [3]. There are serious drawbacks for using bone substituents from the patient's iliac crest, including limited donor tissue, donor site illness and increased risk for infections or disease transmission, which highlights the importance of engineered implants [1,4]. Emerging tissue engineering strategies provide a remarkable opportunity for the regeneration of injured tissues through the fabrication of the artificial constructs [5,6]. Such structures must afford a suitable microenvironment for cell attachment and proliferation to stimulate the damaged tissue formation [7]. Furthermore, biocompatibility, biodegradability, and porosity of the structures directly affect their treatment performance [8]. Currently, different types of materials such as metals, polymers, and ceramics are used as biomedical implants [9]. The metallic implants such as stainless steel, cobalt, and titanium alloys are mainly exploited due to the excellent mechanical properties and superior corrosion resistance in orthopedic targets, while polymer and ceramic-based implants exhibit weak and brittle properties [10]. Among the different types of metallic implants, the surgical grade 316LSS is the most common bone-implant offering high mechanical properties, low cost, and availability [11]. Regarding the 316LSS properties, the biggest drawback is the release of the metal ions, e.g., iron, nickel, and chromium in the biological environment, making it pernicious in nature [12,13].

In order to overcome the aforementioned issue, several surface modification procedures have been applied. Based on the literature, the composite coatings method using polymers and ceramic components is considered as the most popular strategy for this purpose [14,15]. Poly (ε-caprolactone) (PCL) is a well-known synthetic polymer composed of semi-crystalline linear polyester, which is approved by the U.S. Food and Drug Administration (FDA) as a biomedical material [16]. Although PCL exhibits significant mechanical strength and biocompatibility, it is inherently hydrophobic which negatively affects its biological properties such as cell adhesion and proliferation [17,18]. Therefore, a combination of PCL with a natural hydrophilic polymer such as Ge was utilized as an ideal coating for bone regeneration [19–21]. Ge has been utilized widely in medical applications as a natural biopolymer derived from partial hydrolysis of collagen. In addition to its biocompatibility, low cost, availability the suitable hydrophilicity of Ge-based materials promote cell attachment and proliferation of the blends comprised of Ge [22].

Since rapid biodegradability and weak mechanical properties are considered as the key drawbacks of Ge, it is normally used for tissue engineering combined with artificial polymers such as PCL to fulfill the mechanical properties requirement [23,24]. Yao et al. [25] fabricated PCL/Ge nanofibrous scaffolds containing various polymer ratios for tissue engineering application. The essays of cellular behaviors indicated that the blend of PCL/Ge had higher adhesion and proliferation in comparison with pure PCL and Ge. Additionally, the PCL/Ge having the ratio of 2:1 showed the best cell spreading, viability and cytoskeleton organization. Fanaee et al. [26] prepared PCL/Ge nanofiber mats with a 70/30 weight ratio containing bioactive glass particles via electrospinning for bone tissue engineering application. The results of in vitro tests confirmed no considerable cytotoxicity as well as good cell adhesion for the prepared nanofibers comprised of PCL and Ge. Moreover, bioceramics are exploited to generate osteocunductive feature for these artificial constructs. Various types of bioceramics and bioglasses such as HA, alumina, zirconia, phosphates, and forsterite have been used to stimulate cell growth and/or bone cell formations by releasing active ions in cell microenvironment [27–31]. Recently, forsterite (Mg_2SiO_4) has been highly recommended as an osteocunductive biomaterial for use in bone regeneration applications, based on its remarkable mechanical properties and biocompatibility. It is worthwhile to note that forsterite enhances cell proliferation and bone regeneration by releasing Mg ion after implantation. Moreover, a higher degradation rate of forsterite composite scaffolds is reported because of its low degree of crystallinity [32,33].

GO is one of the most efficient derivatives of graphene, which has abundant hydroxyls, epoxides and carboxyl functional groups on its surface [34,35]. GO possesses many benefits such as solubility in water and some polar solvents, excellent biocompatibility, good mechanical properties, and high flexibility. It is also a potential biomaterial for cell proliferation enhancement because of its superior biocompatibility. GO nanosheets were incorporated into PCL nanofibers in order to investigate cell behavior of two types of cells such as mMSCs and PC12-L on the PCL/GO [36]. The results showed that GO incorporation substantially improved the cell attachment, spreading and proliferation of the prepared scaffolds. Therefore, the shortcomings of 316LSS—which include releasing of ions such as nickel and chromium—can be addressed using these two materials to improve the biocompatibility and corrosion resistance of 316LSS.

In our previous research, a bilayer coating of GO and polymeric nanofibrous composite was prepared via electrophoretic deposition (EPD) and electrospinning, whereby the corrosion resistance of 316LSS significantly improved [37]. The central aim of this research is evaluating the cellular behavior of that nanocomposite and the bilayer coating. In other words, the effects of GO layer and forsterite concentration on the bioactivity of the nanofibers were assessed.

2. Materials and Methods

2.1. Surface and Coatings Preparation

The 316LSS substrate was cut into rectangular samples with a dimension of $2 \times 1 \times 0.4$ cm^3. Before the EPD process, the samples were mechanically polished using SiC papers with 80, 120, 240 and 320 grit-size. Then, the samples were rinsed with deionized (DI) water and were sonicated in acetone to remove any remained grease on the surface of the samples followed by drying at room temperature.

To apply GO coating on the surface of the samples, firstly different amounts of GO nanopowder (Nanosany Corporation, Mashhad, Iran) were dispersed in DI water by ultra-sonication to obtain a homogenous suspension. To reach a uniform coating, different EPD variables such as voltage and deposition time were optimized as discussed in our previous study [37]

To perform the electrospinning process, solutions containing PCL (average Mw = 80,000, Sigma, St. Louis, MO, USA), gelatin (type B bovine skin, Mw = 50,000–100,000) with 1 and 3 weight percent of forsterite nanoparticles were prepared using formic acid and acetic acid (1:3 *v/v*) as solvents. In order to prepare the solvents, forsterite nanoparticles were first ultrasonically dispersed in adequate solvents. Afterward, the solutions were prepared by dissolving the PCL and gelatin in the solvents and magnetically stirred at room temperature for more than six hours. The solutions were then electrospun with a constant gap distance of 15 cm, applied voltage range of 12–26 kV, and feed rates of 0.1–0.5 mL/h.

2.2. Characterization of the Nanofibrous Layer

The morphology of the electrospun samples and distribution of the nanoparticles in the PCL/gelatin nanofibers were evaluated by FESEM (Quanta 450 FEG, Graz, Austria) and Energy Dispersive X-ray (EDS, Octane Elite EDS, Graz, Austria), respectively.

The porosity of the electrospun layer was determined based on the analysis of the nanofiber FESEM micrographs, utilizing image J software (Image J, National Institutes of Health, Bethesda, MD, USA). The surface area of pores (S_p) and the total surface area of the samples (S_t) were measured. Moreover, the porosity percent was calculated according to Equation (1) [38]:

$$\%P = \frac{S_p}{S_t} \times 100 \tag{1}$$

Brookfield DV-II viscometer (Middleboro, MA, USA) and JENWA 3540 conductivity meter (Burlington, NJ, USA) were used to measure the conductivity and viscosity of the electrospinning solutions, respectively. The viscosity was measured at 25 °C and the rotational speed of 6 rpm. Water contact angle measurements were carried out with a drop shape analyzer (Sessile Drop-G10, Tehran, Iran) to investigate the surface wettability and hydrophilicity of the GO layer and electrospun nanofibers.

The degradation rate of the samples was determined by measuring the weight loss of the samples based on ASTM-F1635 after 21 days of immersion in PBS at 37 °C and pH = 7.4. The weight loss percentage was calculated according to Equation (2). In the equation, the W_0, W_t refer to the weights of the coated samples before and after immersion, respectively. In addition, W_s is the weight of the 316LSS substrate.

$$\%\text{Weight loss} = \frac{w_0 - w_t}{w_0 - w_s} \times 100 \tag{2}$$

Since the pH changes indicate the release of the alkaline ions and HA formation [39], the pH value of the solutions was measured during the soaking time using an electrolyte-type pH meter.

2.3. Bioactivity Investigations of the Coatings

The bioactivity of the coatings was investigated according to the amount of HA formed on the substrates after soaking in SBF. The SBF solution was prepared according to the Kokubo et al. method [40]. The substrates were immersed in SBF at 37 °C in a stable water bath for 21 days. X-ray diffraction (ASENWARE, AW-XDM 300, Shenzhen, China), using monochromatized CuKα radiation generated from 40 kV and 30 mA and ranging from 10° to 80°, was employed to confirm the crystalline phase of the formed HA on the coated substrates. The morphology of the HA was evaluated by FESEM images after 3, 7, 14 and 21 days soaking in the SBF.

2.4. In-vitro Cell Behavior of the Coatings

The MG-63 cells were cultured in Dulbecco's modified Eagle's medium (DMEM) complemented with 10% FBS (Gibco, Biosciences, Dublin, Ireland), 1% Glutamax, 1% penicillin/streptomycin and 1% Non-essential Amino Acid (NEAA). The seeded cells were incubated at 37 °C and carbon dioxide amount of 5%. The nanofibrous coatings were electrospun on circular disks based on previous work [41]. All the coatings were sterilized under UV over 15 min on each side, immersed in 70% ethanol for 12 h and then washed with amphotericin/gentamicin/penicillin and PBS for 15 min. After that, the electrospun coated substrates were placed in 24-well plates and MG-63 cells, at a density of 30,000 cells, were seeded on the surfaces of each sample. The cell morphology and adhesion on the seeded nanofibers were evaluated by FESEM images after 1 and 7 days of cell seeding. The cultured cells on the coatings were fixed by 2.5 v% glutaraldehyde solution (Sigma, St. Louis, MO, USA) in PBS and dehydrated through various concentrations of ethanol (0, 25, 50, 75 and 100 v%).

The MTS tests were also performed on the coatings after 1, 3, 5 and 7 days' cell seeding to evaluate the viability of the samples. The seeded samples were washed and placed in an incubator with 10% of MTS reagent under 37 °C and 5% carbon dioxide. After 3.5 h of incubation, the aliquots were transferred into a 96-well plate. Then, the absorbance of the samples at 429 nm was quantified using a spectrophotometric plate reader (Awareness Technology Inc., Palm City, FL, USA).

3. Results and Discussion

3.1. Characterization of the Electrospun Nanofibers

Figure 1 indicates the FESEM micrographs of the electrospun nanofibers. In addition, the viscosity and conductivity of the solutions along with the average fiber diameter of the electrospun nanofibers were measured and summarized in Tables 1 and 2, respectively. Table 2 reveals that the incorporation of 1 wt.% forsterite to the PCL/Ge composition decreased the fiber diameter from 167 to 148 nm. Increasing the forsterite content to 3 wt.% increased the average fiber diameter to 171 nm. The addition of the forsterite nanopowder increased the conductivity and the surface charge density of the solution, which caused the diameter reduction. On the other hand, the higher amount of forsterite had a dominant effect on the solution viscosity, leading to an increase in the nanofiber diameter. These results are in agreement with previous researches [42,43].

The morphology and the corresponding EDS analysis of the nanoparticles distribution in the PCL/Ge nanofibers are presented in Figure 2. It can be observed that the nanoparticles are uniformly distributed on the coatings. It can be also discerned that the dispersion of the nanoparticles in the nanofibers with 1 wt.% forsterite is better than the 3 wt.% loaded sample. The low amount of agglomeration in the PCL/Ge/forsterite-3 sample can be attributed to the strong surface energy among the nanoparticles [44].

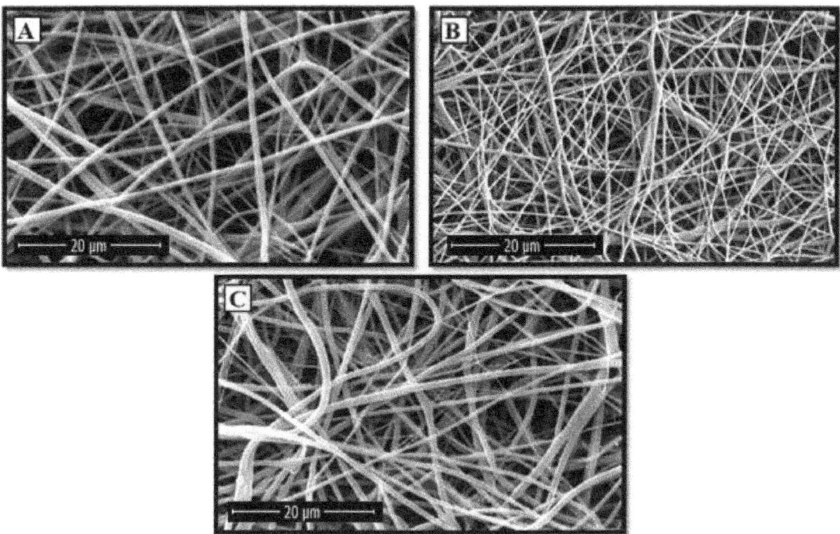

Figure 1. FESEM micrographs of the prepared PCL/Ge nanofibers containing (**A**) 0%, (**B**) 1%, and (**C**) 3 wt.% forsterite nanoparticles.

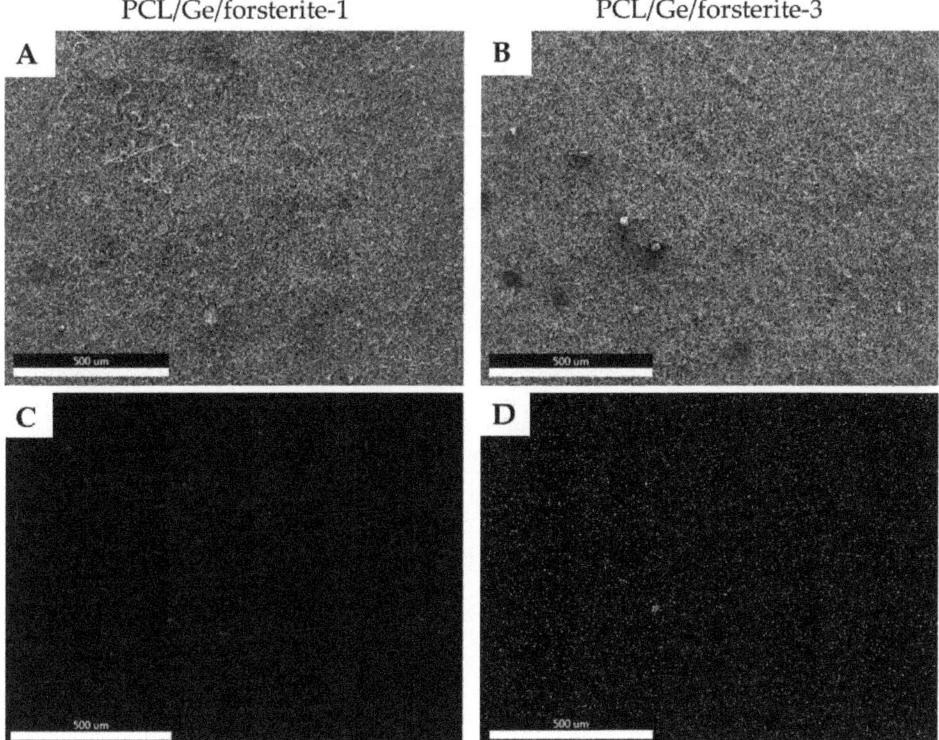

Figure 2. FESEM micrographs (**A**) and (**B**), and the distribution map of Mg element (**C**) and (**D**) of the electrospun PCL/Ge nanofibers with 1% and 3% forsterite nanoparticles.

Table 1. Physical properties of the solutions.

Nanofiber Composition	Viscosity (cP)	Conductivity (µS/cm)
PCL/Ge	910 ± 32	271 ± 13
PCL/Ge/forsterite-1	980 ± 24	288 ± 10
PCL/Ge/forsterite-3	1400 ± 100	290 ± 20

Table 2. Morphology characteristic of the electrospun scaffolds.

Nanofiber Composition	Fiber Diameter (nm)	Porosity (%)	Weight Loss (%)
PCL/Ge	167 ± 29	77.4 ± 0.2	12.0 ± 0.2
PCL/Ge/forsterite-1	148 ± 36	71.1 ± 0.1	15.0 ± 0.2
PCL/Ge/forsterite-3	171 ± 43	82.6 ± 0.2	17.9 ± 0.1

Since the porosity influences the scaffold's cell adhesion and proliferation, it is essential to consider this scaffold characteristic during the tissue engineering [45]. It was reported that the porosity of the electrospun nanofibers are mostly controlled by the diameter of the nanofibers [46]. The porosity of the samples was measured based on the FESEM micrographs (Figure 1) and reported in Table 2. The porosity of the electrospun scaffolds was reduced by introducing 1 wt.% of nanoparticles and then increased at the nanoparticles content of 3 wt.%. Therefore, the effect of the amount of nanoparticles on the porosity was similar to the fiber diameter. The lowest porosity content was present in the coatings having thinner nanofibers. On the other hand, the highest porosity was assigned to the PCL/Ge/forsterite-3 nanofibers at 82.6% ± 0.2%, which had a thicker fiber diameter. Generally, all of the samples illustrated porosity above 70%, which is apt for medical applications [47]. Therefore, it is anticipated that all of the electrospun mats would have a high potential for cell attachment and proliferation.

Figure 3 shows the obtained results of the wettability analysis by measuring the water contact angle for the PCL/Ge, PCL/Ge/forsterite-1 and PCL/Ge/forsterite-3 coatings. The relaxation time of the water droplet was 10 s. According to Sup Kim et al. [48], the contact angle of PCL was reported to be 120°, hence it is clear that the incorporation of gelatin increases the hydrophilicity of PCL nanofibers, which is due to the existence of amine and carboxylic groups in gelatin [49]. As can be seen in Figure 3, when the forsterite nanoparticles content increased from 1 to 3 wt.%, the water contact angle of the nanofibers decreased from 53.59° to 35.55°. Therefore, the hydrophilicity of the nanofibers is affected by the concentration of the nanoparticles. As a result, it is expected that cells would show higher extent adhesion on PCL/Ge/forsterite-3 nanofibers due to increased hydrophilicity.

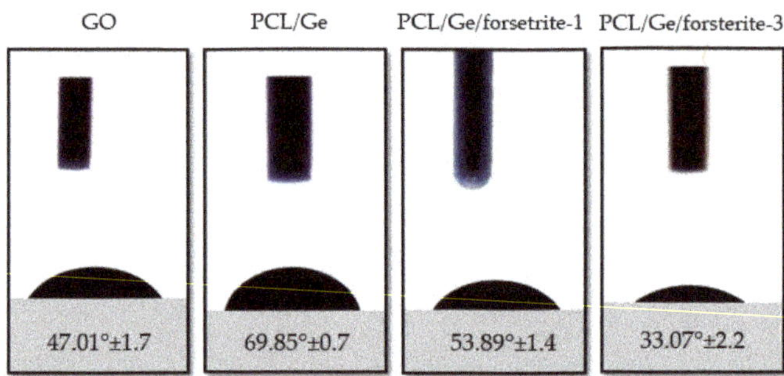

Figure 3. The water contact angles of GO layer and PCL/Ge nanofibers containing 0, 1 and 3 wt.% forsterite.

The weight loss percentage of the samples was measured after 21 days' soaking in PBS at 37 °C and pH = 7.4. The results are summarized in Table 2. The weight loss of the PCL/Ge nanofibers increased from ca. 12% to ca. 18% by increasing the forsterite content from 0 to 3 wt.%. Therefore, increasing the forsterite content increased the hydrophilicity, porosity and the weight loss of the scaffolds. The degradation of the coating can be associated not only with the hydrolysis of gelatin but also the diffusion of the nanoparticles from the surface of the nanofibrous coating to the solution [50]. Moreover, due to the highly crystalline phase of PCL, its weight loss is considered to be negligible [51].

The pH changes of the PBS solutions containing the scaffolds were assessed and depicted in Figure 4. It is clear that the pH values of the solution reduced from 7.4 to 6.9 in the PBS solution of the PCL/Ge nanofibers within 21 days. Releasing of acidic products from the degradation of the PCL and gelatin is responsible for the decrease in pH values [52]. In contrast to PCL/Ge, the amount of pH in the solution containing PCL/Ge/forsterite with 1 and 3 wt.% nanoparticles increased to 8.1 and 8.2, respectively, during the first week. The release of the Mg ions from forsterite incorporated into the scaffolds increased the alkalinity of the solution and consequently the pH value [52]. Moreover, the pH slightly decreased in the next week due to polymer degradation and remained constant.

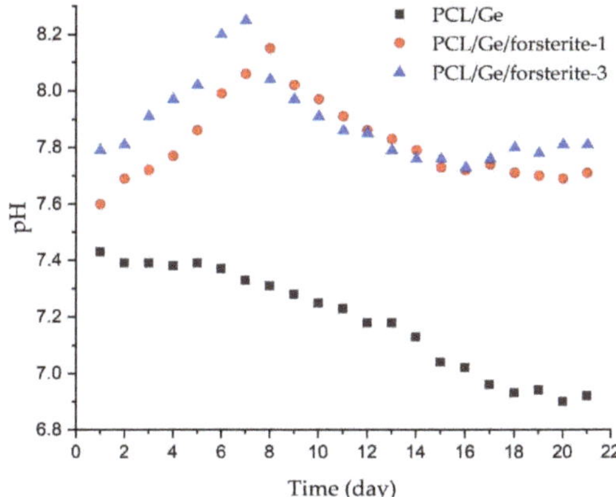

Figure 4. The pH values of the PBS solutions containing PCL/Ge with 0, 1 and 3 wt.% forsterite during 21 days immersion.

3.2. Bioactivity of the Electrospun Scaffolds

Osteoconductivity is a crucial characteristic of the implants utilized in bone tissue engineering applications to predict bone regeneration during implant employment [41]. The osteoconductivity of the electrospun nanofibers was evaluated by studying the capability of the coated samples in the creation of bone-like apatite. The PCL/Ge nanofibers containing various concentrations of the nanoparticle (1 and 3 wt.%) were soaked in SBF for 3, 7, 14 and 21 days. Figure 5 demonstrates the FESEM images of the PCL/Ge/forsterite with 1 and 3 wt.% after immersion in SBF. The bone-like apatite deposition was found to form on the surface of the resulting electrospun structure after three days in SBF and noticeably increased within 21 days. The presence of the nanoparticle in the scaffold structure led to the formation of the silanol (–Si–OH) groups, which contributed to Ca-P nucleation. As a result, the interaction of phosphate and carbonate groups in SBF and positively charged positions of the Ca-P nucleation caused phosphate layer formation during the immersion time [53]. From Figure 5, it can be seen that the HA formation on the electrospun structures having 3 wt.% of forsterite is significantly higher than the PCL/Ge/forsterite-1 structure, especially in three and seven days' immersion. Figure 6 presents the WAXS profiles of the samples PCL/Ge/forsterite-1 and PCL/Ge/forsterite-3 after 21 days soaking in SBF. From the diffractograms, the peaks at 2Θof 21.4° and 23.8° can be ascribed to the existence of PCL in the structure [26]. In addition, the peaks at 26.7°, 31.7°, 43.6°, 45.5°, 50.7°, and 74.6° can be related to the created HA on the electrospun substrates containing nanoparticles. The observed peaks for HA were also reported for HA in previous works [14,54]. The XRD patterns also confirm the formation of HA after 21 days' incubation of samples in SBF.

PCL/Ge/forsterite-1   PCL/Ge/forsterite3

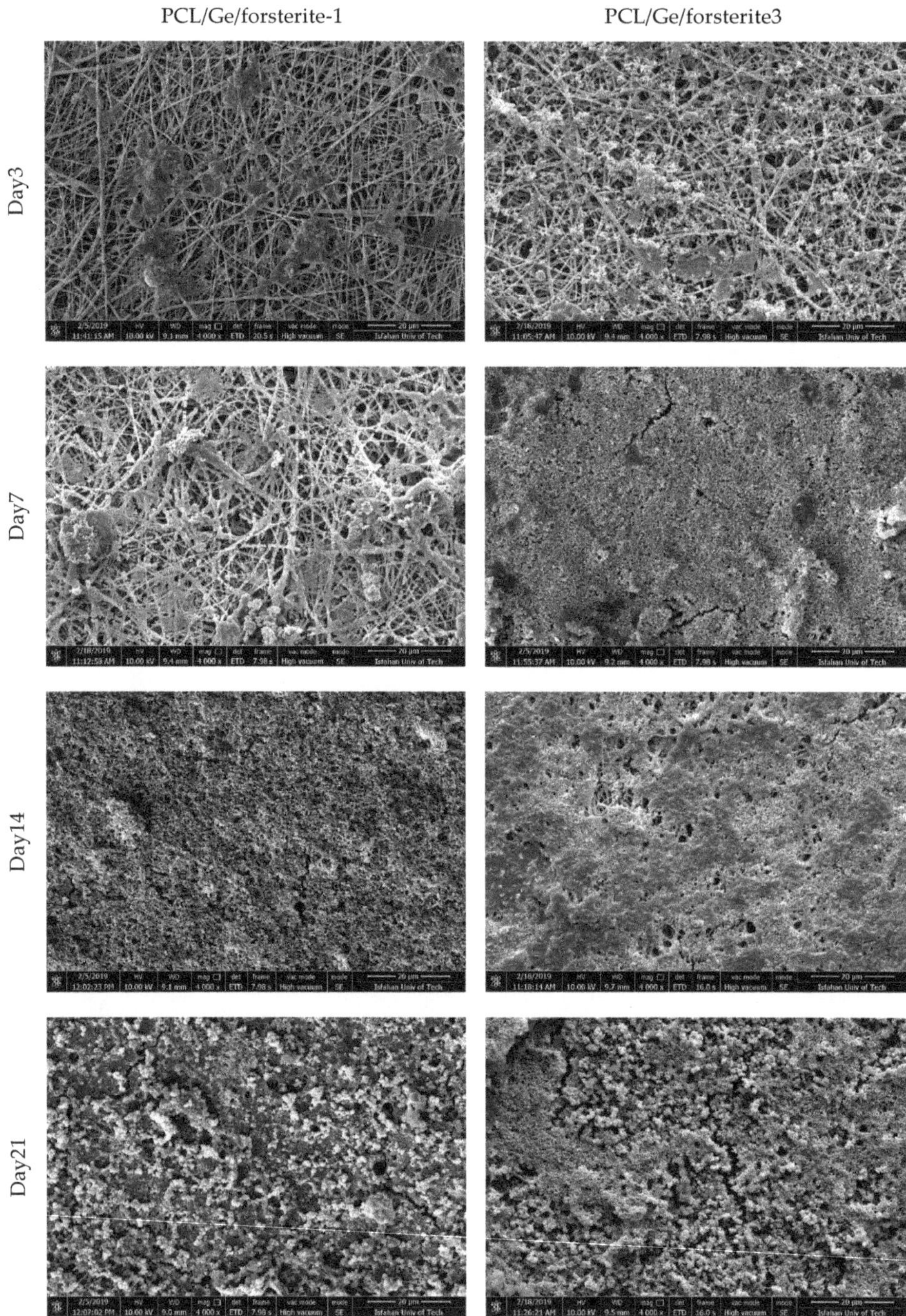

Figure 5. FESEM micrographs of the PCL/Ge/forsterite with 1 and 3 wt.% after 3, 7, 14, and 21 days immersion in the SBF solution.

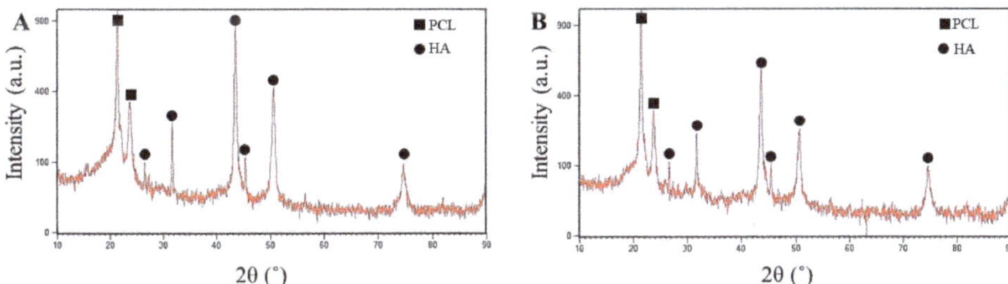

Figure 6. XRD patterns of PCL/Ge nanofibers containing (**A**) 1 and (**B**) 3% of forsterite after 21 days of immersion in SBF.

3.3. Cell Culture Studies

Cell attachment and proliferation are the results of efficient cell-material interactions [55]. The cell morphology on the PCL/Ge and GO-PCL/Ge electrospun structures containing 1 and 3 wt.% forsterite nanoparticles after one and seven days of MG-63 cells seeding is shown in Figure 7. It can be observed that the cells were well attached and spread on all the samples due to the proper interactions between the cells and the coatings. Specific cellular adhesion and well-spread morphology were higher for the GO-PCL/Ge structure rather than PCL/Ge after seven days. The better performance of the bilayer structure can be attributed to the presence of GO layer (with OH and COOH groups on the surface) and high hydrophilicity of the coatings. The existence of the surface roughness, as the intrinsic property of GO [36], and GO functional groups assisted the serum protein adsorption as well as cell attachment [56]. Additionally, the cells showed better growth on the structures containing a higher amount of the nanoparticles. The MTS results are illustrated in Figure 8 where the progressive growth of the cells confirms the non-cytotoxicity of the coatings [57]. All the bilayer structures with various amounts of the nanoparticles showed effective compatibility and interactions with the cells. The GO-PCL/Ge with 3 wt.% nanoparticles showed a noticeably higher cell viability in comparison with other nanostructures. The higher hydrophilicity of the coatings in the presence of the GO layer and the larger amount of the nanoparticles increased the cell viability. Besides, adsorption of the serum protein is affected by surface oxygen-containing groups of GO as its intrinsic feature [58]. Another reasonable explanation for the good cell growth is the conductivity of GO having oxygenated groups on its structure. Although GO is a poor conductor compared to graphene, it has higher conductivity than PCL/Ge nanofibrous layer [59].

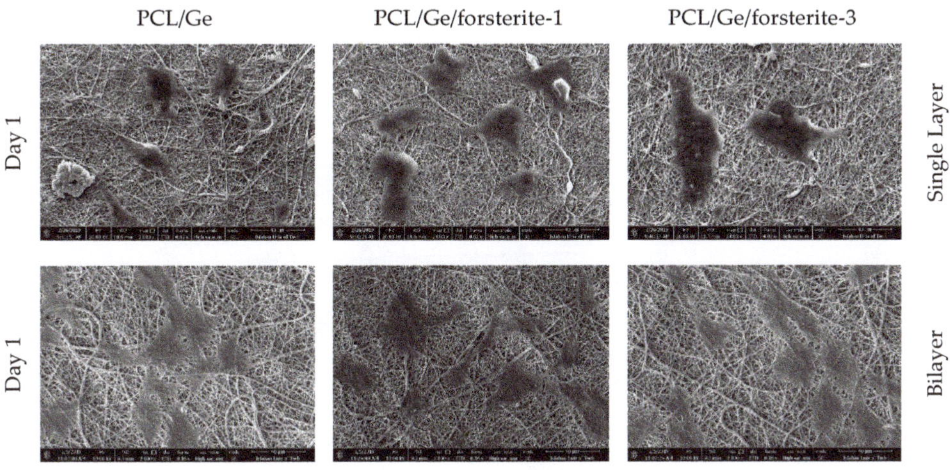

Figure 7. *Cont.*

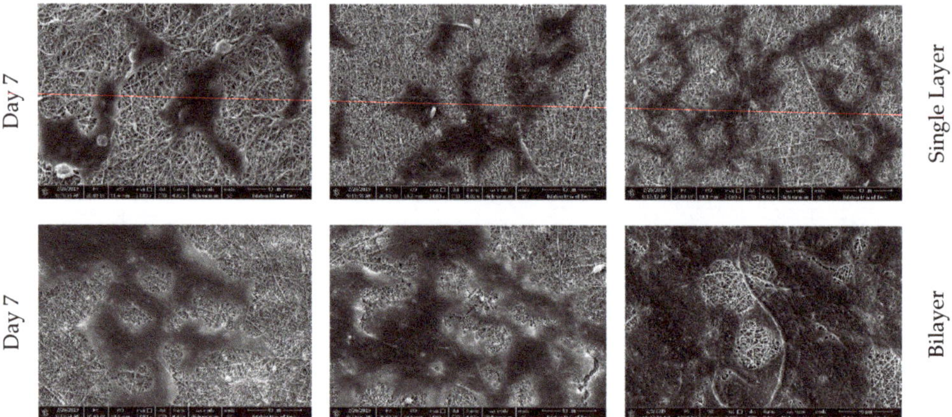

Figure 7. Morphology of the MG-63 cells on PCL/Ge/forsterite nanofibers with 1 and 3 wt.% and GO-PCL/Ge/forsterite with 1 and 3 wt.% after one and seven days of cell culture.

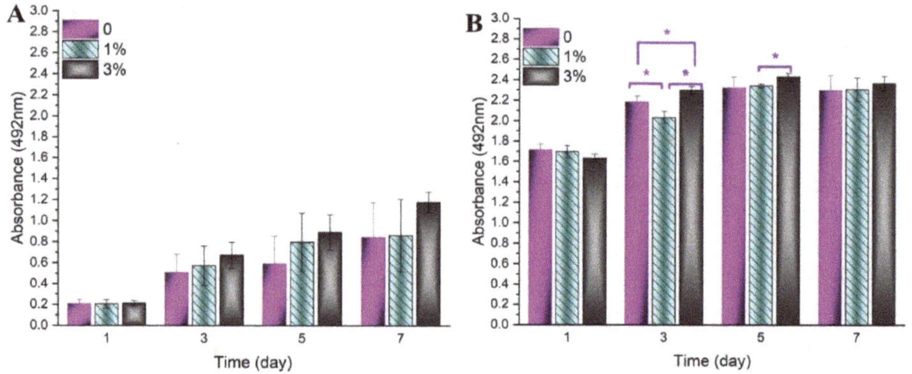

Figure 8. The MTS results of PCL/Ge structures containing 0, 1 and 3 wt.% forsterite nanoparticles, (**A**) without the GO layer and (**B**) with the GO layer (*significant difference at $p_{value} < 0.05$).

4. Conclusions

To recapitulate, a bilayer bioactive coating containing GO layer and nanofibrous PCL/Ge/forsterite was applied on 316LSS to develop a potential platform for bone implant application.

- Characterization of the nanofiber layer revealed the formation of a uniform beadless nanofibrous layer on the surface of the GO layer. It was also indicated that the forsterite nanoparticles were well-distributed on the top layer. The presence of gelatin and forsterite nanoparticles increased the wettability and biodegradation rate of the top layer (electrospun nanofibrous layer) which marks a development in bilayer coating in bone implant applications.
- The bioactivity results indicated the formation of HA on the surface of the nanofiber structures which was subsequently confirmed by XRD. the incorporation of the forsterite nanoparticles increased the bioactivity of the samples, especially after 14 and 21 days of soaking in the SBF solution.
- The PCL/Ge/forsterite and GO-PCL/Ge/forsterite coatings were found to be non-cytotoxic structures with an ability to enhance cell attachment and proliferation. Furthermore, the enhanced adhesion and growth of MG63 cells on bilayer coatings in comparison with nanocomposite coatings revealed the beneficial biocompatibility and hydrophilicity of GO due to functional groups on its surface as well as high surface roughness.

Author Contributions: Methodology, F.K., S.K., R.E.N., E.R.G., and F.E.; validation, F.K., S.K., R.E.N., E.R.G., F.E., and O.D.; formal analysis, F.K., S.N.K., S.K., R.E.N., E.R.G., F.E., and M.H.N.-E.; investigation, F.K., S.K., R.E.N., E.R.G., and F.E.; data curation, F.K., S.K., R.E.N., E.R.G., O.D., writing—original draft preparation, F.K., S.K., R.E.N., E.R.G., and F.E; writing—review and editing, F.K., S.N.K., S.K., R.E.N., E.R.G., F.E., O.D. and M.H.N.-E.; supervision, S.N.K., and M.H.N.-E. All authors have read and agreed to the published version of the manuscript.

Funding: This research received no external funding.

Conflicts of Interest: The authors declare no conflict of interest. The funders had no role in the design of the study; in the collection, analyses, or interpretation of data; in the writing of the manuscript, or in the decision to publish the results.

References

1. De Witte, T.-M.; Fratila-Apachitei, L.E.; Zadpoor, A.A.; Peppas, N.A. Bone tissue engineering via growth factor delivery: From scaffolds to complex matrices. *Regen. Biomater.* **2018**, *5*, 197–211. [CrossRef] [PubMed]
2. Habibovic, P. Strategic Directions in Osteoinduction and Biomimetics. *Tissue Eng. Part A* **2017**, *23*, 1295–1296. [CrossRef] [PubMed]
3. Ricciardi, B.F.; Bostrom, M.P. Bone graft substitutes: Claims and credibility. *Semin. Arthroplast.* **2013**, *24*, 119–123. [CrossRef]
4. Burg, K.J.L.; Porter, S.; Kellam, J.F. Biomaterial developments for bone tissue engineering. *Biomaterials* **2000**, *21*, 2347–2359. [CrossRef]
5. Rezwan, K.; Chen, Q.Z.; Blaker, J.J.; Boccaccini, A.R. Biodegradable and bioactive porous polymer/inorganic composite scaffolds for bone tissue engineering. *Biomaterials* **2006**, *27*, 3413–3431. [CrossRef]
6. Esmaeely Neisiany, R.; Enayati, M.S.; Sajkiewicz, P.; Pahlevanneshan, Z.; Ramakrishna, S. Insight into the Current Directions in Functionalized Nanocomposite Hydrogels. *Front. Mater.* **2020**, *7*, 25. [CrossRef]
7. Bose, S.; Vahabzadeh, S.; Bandyopadhyay, A. Bone tissue engineering using 3D printing. *Mater. Today* **2013**, *16*, 496–504. [CrossRef]
8. Loh, Q.L.; Choong, C. Three-Dimensional Scaffolds for Tissue Engineering Applications: Role of Porosity and Pore Size. *Tissue Eng. Part B Rev.* **2013**, *19*, 485–502. [CrossRef]
9. Meinert, K.; Uerpmann, C.; Matschullat, J.; Wolf, G.K. Corrosion and leaching of silver doped ceramic IBAD coatings on SS 316L under simulated physiological conditions. *Surf. Coat. Technol.* **1998**, *103–104*, 58–65. [CrossRef]
10. Sivaraj, D.; Vijayalakshmi, K. Novel synthesis of bioactive hydroxyapatite/f-multiwalled carbon nanotube composite coating on 316L SS implant for substantial corrosion resistance and antibacterial activity. *J. Alloys Compd.* **2019**, *777*, 1340–1346. [CrossRef]
11. Yao, H.; Li, J.; Li, N.; Wang, K.; Li, X.; Wang, J. Surface Modification of Cardiovascular Stent Material 316L SS with Estradiol-Loaded Poly (trimethylene carbonate) Film for Better Biocompatibility. *Polymers* **2017**, *9*, 598. [CrossRef] [PubMed]
12. Gopi, D.; Prakash, V.C.A.; Kavitha, L. Evaluation of hydroxyapatite coatings on borate passivated 316L SS in Ringer's solution. *Mater. Sci. Eng. C* **2009**, *29*, 955–958. [CrossRef]
13. González, M.B.; Saidman, S.B. Electrodeposition of polypyrrole on 316L stainless steel for corrosion prevention. *Corros. Sci.* **2011**, *53*, 276–282. [CrossRef]
14. Madhan Kumar, A.; Rajendran, N. Electrochemical aspects and in vitro biocompatibility of polypyrrole/TiO2 ceramic nanocomposite coatings on 316L SS for orthopedic implants. *Ceram. Int.* **2013**, *39*, 5639–5650. [CrossRef]
15. Chang, S.-H.; Hsiao, Y.-C. Surface and Protein Adsorption Properties of 316L Stainless Steel Modified with Polycaprolactone Film. *Polymers* **2017**, *9*, 545. [CrossRef]
16. Ghasemi-Mobarakeh, L.; Prabhakaran, M.P.; Morshed, M.; Nasr-Esfahani, M.-H.; Ramakrishna, S. Electrospun poly(ε-caprolactone)/gelatin nanofibrous scaffolds for nerve tissue engineering. *Biomaterials* **2008**, *29*, 4532–4539. [CrossRef]
17. Gautam, S.; Dinda, A.K.; Mishra, N.C. Fabrication and characterization of PCL/gelatin composite nanofibrous scaffold for tissue engineering applications by electrospinning method. *Mater. Sci. Eng. C* **2013**, *33*, 1228–1235. [CrossRef]
18. Ren, K.; Wang, Y.; Sun, T.; Yue, W.; Zhang, H. Electrospun PCL/gelatin composite nanofiber structures for effective guided bone regeneration membranes. *Mater. Sci. Eng. C* **2017**, *78*, 324–332. [CrossRef]
19. Rajzer, I.; Menaszek, E.; Kwiatkowski, R.; Planell, J.A.; Castano, O. Electrospun gelatin/poly(ε-caprolactone) fibrous scaffold modified with calcium phosphate for bone tissue engineering. *Mater. Sci. Eng. C* **2014**, *44*, 183–190. [CrossRef]

20. Lee, J.-h.; Lee, Y.J.; Cho, H.-j.; Kim, D.W.; Shin, H. The incorporation of bFGF mediated by heparin into PCL/gelatin composite fiber meshes for guided bone regeneration. *Drug Deliv. Transl. Res.* **2015**, *5*, 146–159. [CrossRef]

21. Cirillo, V.; Guarino, V.; Ambrosio, L. Design of Bioactive Electrospun Scaffolds for Bone Tissue Engineering. *J. Appl. Biomater. Funct. Mater.* **2012**, *10*, 223–228. [CrossRef] [PubMed]

22. Rezvani Ghomi, E.; Khalili, S.; Nouri Khorasani, S.; Esmaeely Neisiany, R.; Ramakrishna, S. Wound dressings: Current advances and future directions. *J. Appl. Polym. Sci.* **2019**, *136*, 47738. [CrossRef]

23. Chong, E.J.; Phan, T.T.; Lim, I.J.; Zhang, Y.Z.; Bay, B.H.; Ramakrishna, S.; Lim, C.T. Evaluation of electrospun PCL/gelatin nanofibrous scaffold for wound healing and layered dermal reconstitution. *Acta Biomater.* **2007**, *3*, 321–330. [CrossRef] [PubMed]

24. Khalili, S.; Khorasani, S.N.; Neisiany, R.E.; Ramakrishna, S. Theoretical cross-link density of the nanofibrous scaffolds. *Mater. Des. Process. Commun.* **2019**, *1*, e22. [CrossRef]

25. Yao, R.; He, J.; Meng, G.; Jiang, B.; Wu, F. Electrospun PCL/Gelatin composite fibrous scaffolds: Mechanical properties and cellular responses. *J. Biomater. Sci. Polym. Ed.* **2016**, *27*, 824–838. [CrossRef]

26. Fanaee, S.; Labbaf, S.; Enayati, M.H.; Baharlou Houreh, A.; Esfahani, M.-H.N. Creation of a unique architectural structure of bioactive glass sub-micron particles incorporated in a polycaprolactone/gelatin fibrous mat; characterization, bioactivity, and cellular evaluations. *J. Biomed. Mater. Res. Part A* **2019**, *107*, 1358–1365. [CrossRef] [PubMed]

27. Pang, X.; Zhitomirsky, I. Electrophoretic deposition of composite hydroxyapatite-chitosan coatings. *Mater.Charact.* **2007**, *58*, 339–348. [CrossRef]

28. Kheirkhah, M.; Fathi, M.; Salimijazi, H.R.; Razavi, M. Surface modification of stainless steel implants using nanostructured forsterite (Mg2SiO4) coating for biomaterial applications. *Surf. Coat. Technol.* **2015**, *276*, 580–586. [CrossRef]

29. Vallet-Regí, M.; Salinas, A.J. 6—Ceramics as bone repair materials. In *Bone Repair Biomaterials*, 2nd ed.; Pawelec, K.M., Planell, J.A., Eds.; Woodhead Publishing: Sawston, UK; Cambridge, UK, 2019; pp. 141–178. [CrossRef]

30. Enayati, M.S.; Neisiany, R.E.; Sajkiewicz, P.; Behzad, T.; Denis, P.; Pierini, F. Effect of nanofiller incorporation on thermomechanical and toughness of poly (vinyl alcohol)-based electrospun nanofibrous bionanocomposites. *Theor. Appl. Fract. Mech.* **2019**, *99*, 44–50. [CrossRef]

31. Kouhi, M.; Jayarama Reddy, V.; Fathi, M.; Shamanian, M.; Valipouri, A.; Ramakrishna, S. Poly (3-hydroxybutyrate-co-3-hydroxyvalerate)/fibrinogen/bredigite nanofibrous membranes and their integration with osteoblasts for guided bone regeneration. *J. Biomed. Mater. Res. Part A* **2019**, *107*, 1154–1165. [CrossRef]

32. Devi, K.B.; Tripathy, B.; Roy, A.; Lee, B.; Kumta, P.N.; Nandi, S.K.; Roy, M. In Vitro Biodegradation and In Vivo Biocompatibility of Forsterite Bio-Ceramics: Effects of Strontium Substitution. *ACS Biomater.Sci. Eng.* **2019**, *5*, 530–543. [CrossRef]

33. Tavangarian, F.; Emadi, R. Improving degradation rate and apatite formation ability of nanostructure forsterite. *Ceram. Int.* **2011**, *37*, 2275–2280. [CrossRef]

34. Dreyer, D.R.; Park, S.; Bielawski, C.W.; Ruoff, R.S. The chemistry of graphene oxide. *Chem. Soc. Rev.* **2010**, *39*, 228–240. [CrossRef] [PubMed]

35. Cobos, M.; De-La-Pinta, I.; Quindós, G.; Fernández, J.M.; Fernández, D.M. Synthesis, Physical, Mechanical and Antibacterial Properties of Nanocomposites Based on Poly(vinyl alcohol)/Graphene Oxide–Silver Nanoparticles. *Polymers* **2020**, *12*, 723. [CrossRef] [PubMed]

36. Song, J.; Gao, H.; Zhu, G.; Cao, X.; Shi, X.; Wang, Y. The preparation and characterization of polycaprolactone/graphene oxide biocomposite nanofiber scaffolds and their application for directing cell behaviors. *Carbon* **2015**, *95*, 1039–1050. [CrossRef]

37. Khosravi, F.; Nouri Khorasani, S.; Rezvani Ghomi, E.; Kichi, M.K.; Zilouei, H.; Farhadian, M.; Esmaeely Neisiany, R. A bilayer GO/nanofibrous biocomposite coating to enhance 316L stainless steel corrosion performance. *Mater. Res. Exp.* **2019**, *6*, 086470. [CrossRef]

38. Grove, C.; Jerram, D.A. jPOR: An ImageJ macro to quantify total optical porosity from blue-stained thin sections. *Comput. Geosci.* **2011**, *37*, 1850–1859. [CrossRef]

39. Diba, M.; Kharaziha, M.; Fathi, M.H.; Gholipourmalekabadi, M.; Samadikuchaksaraei, A. Preparation and characterization of polycaprolactone/forsterite nanocomposite porous scaffolds designed for bone tissue regeneration. *Compos. Sci. Technol.* **2012**, *72*, 716–723. [CrossRef]

40. Kokubo, T.; Takadama, H. How useful is SBF in predicting in vivo bone bioactivity? *Biomaterials* **2006**, *27*, 2907–2915. [CrossRef]

41. Masoudi Rad, M.; Nouri Khorasani, S.; Ghasemi-Mobarakeh, L.; Prabhakaran, M.P.; Foroughi, M.R.; Kharaziha, M.; Saadatkish, N.; Ramakrishna, S. Fabrication and characterization of two-layered nanofibrous membrane for guided bone and tissue regeneration application. *Mater. Sci. Eng. C* **2017**, *80*, 75–87. [CrossRef]

42. Haider, A.; Haider, S.; Kang, I.-K. A comprehensive review summarizing the effect of electrospinning parameters and potential applications of nanofibers in biomedical and biotechnology. *Arab. J. Chem.* **2018**, *11*, 1165–1188. [CrossRef]

43. Sharifi, A.; Khorasani, S.N.; Borhani, S.; Neisiany, R.E. Alumina reinforced nanofibers used for exceeding improvement in mechanical properties of the laminated carbon/epoxy composite. *Theor. Appl. Fract. Mech.* **2018**, *96*, 193–201. [CrossRef]

44. Shi, H.; Liu, F.; Yang, L.; Han, E. Characterization of protective performance of epoxy reinforced with nanometer-sized TiO_2 and SiO_2. *Prog. Org. Coat.* **2008**, *62*, 359–368. [CrossRef]

45. Madhan Kumar, A.; Nagarajan, S.; Ramakrishna, S.; Sudhagar, P.; Kang, Y.S.; Kim, H.; Gasem, Z.M.; Rajendran, N. Electrochemical and in vitro bioactivity of polypyrrole/ceramic nanocomposite coatings on 316L SS bio-implants. *Mater. Sci. Eng. C* **2014**, *43*, 76–85. [CrossRef] [PubMed]

46. Neisiany, R.E.; Enayati, M.S.; Kazemi-Beydokhti, A.; Das, O.; Ramakrishna, S. Multilayered Bio-Based Electrospun Membranes: A Potential Porous Media for Filtration Applications. *Front. Mater.* **2020**, *7*, 67. [CrossRef]

47. Meng, Z.X.; Wang, Y.S.; Ma, C.; Zheng, W.; Li, L.; Zheng, Y.F. Electrospinning of PLGA/gelatin randomly-oriented and aligned nanofibers as potential scaffold in tissue engineering. *Mater. Sci. Eng. C* **2010**, *30*, 1204–1210. [CrossRef]

48. Kim, M.S.; Jun, I.; Shin, Y.M.; Jang, W.; Kim, S.I.; Shin, H. The Development of Genipin-Crosslinked Poly(caprolactone) (PCL)/Gelatin Nanofibers for Tissue Engineering Applications. *Macromol. Biosci.* **2010**, *10*, 91–100. [CrossRef]

49. Xue, J.; He, M.; Liang, Y.; Crawford, A.; Coates, P.; Chen, D.; Shi, R.; Zhang, L. Fabrication and evaluation of electrospun PCL–gelatin micro-/nanofiber membranes for anti-infective GTR implants. *J. Mater. Chem. B* **2014**, *2*, 6867–6877. [CrossRef]

50. Nie, L.; Wu, Q.; Long, H.; Hu, K.; Li, P.; Wang, C.; Sun, M.; Dong, J.; Wei, X.; Suo, J.; et al. Development of chitosan/gelatin hydrogels incorporation of biphasic calcium phosphate nanoparticles for bone tissue engineering. *J. Biomater. Sci. Polym. Ed.* **2019**, *30*, 1636–1657. [CrossRef]

51. Bartnikowski, M.; Dargaville, T.R.; Ivanovski, S.; Hutmacher, D.W. Degradation mechanisms of polycaprolactone in the context of chemistry, geometry and environment. *Prog. Polym. Sci.* **2019**, *96*, 1–20. [CrossRef]

52. Kharaziha, M.; Fathi, M.H.; Edris, H. Development of novel aligned nanofibrous composite membranes for guided bone regeneration. *J. Mech. Behav. Biomed. Mater.* **2013**, *24*, 9–20. [CrossRef] [PubMed]

53. Jokar, M.; Darvishi, S.; Torkaman, R.; Kharaziha, M.; Karbasi, M. Corrosion and bioactivity evaluation of nanocomposite PCL-forsterite coating applied on 316L stainless steel. *Surf. Coat. Technol.* **2016**, *307*, 324–331. [CrossRef]

54. Bavya Devi, K.; Singh, K.; Rajendran, N. Sol–gel synthesis and characterisation of nanoporous zirconium titanate coated on 316L SS for biomedical applications. *J. Sol-Gel Sci. Technol.* **2011**, *59*, 513. [CrossRef]

55. Kumbar, S.G.; Nukavarapu, S.P.; James, R.; Nair, L.S.; Laurencin, C.T. Electrospun poly(lactic acid-co-glycolic acid) scaffolds for skin tissue engineering. *Biomaterials* **2008**, *29*, 4100–4107. [CrossRef]

56. Yang, J.; Wan, Y.; Tu, C.; Cai, Q.; Bei, J.; Wang, S. Enhancing the cell affinity of macroporous poly(L-lactide) cell scaffold by a convenient surface modification method. *Polym. Int.* **2003**, *52*, 1892–1899. [CrossRef]

57. Mahlooji, E.; Atapour, M.; Labbaf, S. Electrophoretic deposition of Bioactive glass—Chitosan nanocomposite coatings on Ti-6Al-4V for orthopedic applications. *Carbohydr. Polym.* **2019**, *226*, 115299. [CrossRef]

58. Ku, S.H.; Park, C.B. Myoblast differentiation on graphene oxide. *Biomaterials* **2013**, *34*, 2017–2023. [CrossRef]

59. Ryu, S.; Kim, B.-S. Culture of neural cells and stem cells on graphene. *Tissue Eng. Regen. Med.* **2013**, *10*, 39–46. [CrossRef]

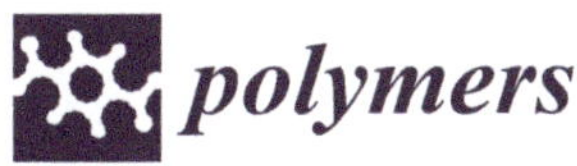 

Article

Performance of Straw/Linear Low Density Polyethylene Composite Prepared with Film-Roll Hot Pressing

Lei Zhang, Huicheng Xu and Weihong Wang *

Key Lab of Bio-based Material Science and Technology (Ministry of Education), College of Material Science and Engineering, Northeast Forestry University, Harbin 150040, China; zl857910712@163.com (L.Z.); Hynner@163.com (H.X.)
* Correspondence: weihongwang2001@nefu.edu.cn; Tel.: +86-451-8219-0932

Received: 18 December 2019; Accepted: 17 March 2020; Published: 9 April 2020

Abstract: Thermoplastic composites are usually prepared with the extrusion method, and straw reinforcement material must be processed to fiber or powder. In this study, film-roll hot pressing was developed to reinforce linear low density polyethylene (LLDPE) with long continuous straw stems. The long straw stems were wrapped with LLDPE film and then hot pressed and cooled to prepare straw/LLDPE composite. Extruded straw fiber/LLDPE composite was prepared as a control. The mechanical properties of these LLDPE-based composites were evaluated. The hot pressed straw/LLDPE composite provided higher tensile strength, tensile modulus, flexural strength, flexural modulus, and impact strength than the traditional extruded straw/LLDPE composite, by 335%, 107%, 68%, 57%, and 181%, respectively, reaching 35.1 MPa, 2.65 GPa, 3.8 MPa, 2.15 GPa, and 25.1 KJ/m^2. The density of the hot pressed straw/LLDPE composite (0.83 g/cm^3) was lower than that of the extruded straw/LLDPE composite (1.31 g/cm^3), and the former had a higher ratio of strength-to-weight. Scanning electron microscopy indicated that the orientation of the straws in the composite was better with the new method. Differential scanning calorimetry tests revealed that in hot pressed straw/LLDPE composite, straw fibers have a greater resistance to the melting of LLDPE than extruded composite. Rotary rheometer tests showed that the storage modulus of the hot pressed straw/LLDPE was less affected by frequency than that of the extruded composite, and the better elastic characteristics were pronounced at 150 °C. The hot pressed straw/LLDPE composite absorbed more water than the extruded composite and showed a potential ability to regulate the surrounding relative humidity. Our results showed that straw from renewable sources can be used to produce composites with good performance.

Keywords: bio-composite; linear low density polyethylene; performance; straws

---

1. Introduction

Natural fiber reinforced composite is a kind of biomass composite. It uses plant-based natural fibers (such as straw, bamboo, jute, and sisal, etc.) as reinforcement, polymers such as polyethylene (PE) and polypropylene (PP) as the matrix, prepared by blending extrusion or molding [1,2]. Natural fiber reinforced composites are widely used in automotive interiors, panel or wall panels, and sports equipment because of their low cost, safety, non-toxicity, renewability, wide source and excellent performance [3,4].

Straw is a type of renewable, abundant natural fiber. The world's crops can provide approximately two billion tons of straw per year [5]. In the past, the main deposal method of straw was incineration, but given the serious associated environmental problems and increasing environmental awareness, straw recycling has become an international priority [6].

Researchers have made some achievements in straw fiber reinforced composites. Nyambo et al. [7] used maleic acid-grafted polyurethane (PU-g-MA) to improve the interfacial adhesion between wheat straw and polyurethane (PU). They found that the addition of 3 phr and 5 phr PU-g-MA significantly increased the tensile strength (20%) and flexural strength (14%) of straw/PU composites, and proved that the increase in strength was due to the well combination of fibers and matrix. Xiao et al. [8] treated the straw with NaOH solution, blended the straw, polyethylene, stearic acid and maleic anhydride, then hot pressed to manufacture the straw/PP composite. The composite has low water absorption and good acid and alkali resistance; Zhang et al. [9] investigated the effects of different straw treatment methods, the particle size of straw powder, and the mass fraction of straw on the mechanical properties of straw/PP composites. The results show that when the straw is treated with the silane coupling agent KH570, the mechanical properties of the straw/PP composite are the best when the particle size of the straw powder is 60 mesh and the mass fraction is 50%; Zabihzadeh et al. [10] investigated the effect of maleic acid grafted polyethylene on the mechanical properties of straw/high density polyethylene (HDPE) composites. It was found that compared with no addition, adding 2% MAPE can increase the tensile strength of the composite by 43%, increase the tensile modulus by 116%, and increase the impact strength by 12%. Even with the addition of 1%, there is a clear improvement.

Straw fiber reinforced composites are a branch of Wood Plastic Composites (WPC). The extrusion molding process, one of the main molding processes of WPC, refers to a processing method in which natural fiber powder, thermoplastic, and various additives are melted in a high temperature, high pressure extruder to be fully mixed, plasticized, finally passed through the top mold of the machine continuously. Due to its advantages of continuous production and high production efficiency, extrusion molding is widely used in industrial production. As of 2017, China's WPC output was close to three million tons, accounting for two-thirds of the world's total output, and the China's production, consumption and exports ranked first in the world, which is precisely due to the development and improvement of extrusion molding processes over the years. The extrusion molding process is mainly divided into one-step extrusion and two-step extrusion. In two-step molding process, the raw materials are first pelletized in a twin-screw extruder, then extruded in single-screw extruder to prepare the composite, and the one-step process skips the pelletizing stage. The two-step process is simpler, flexible, and easy to adjust, which is the most commonly used molding process for enterprises and research units. However, the extrusion molding is easily affected by many factors due to the complicated process. The combined effect of many factors caused many uncertainties among the variables [11].

In extrusion molding process, straw is mostly used in form of short fiber or powder in composite structures. The interface bonding conditions of long fiber reinforced composites are very different from that of short fiber reinforced composites [12]. For example, the friction at the fiber/plastic interface of the former is much larger than that of the latter [13]. The stem of the straw itself has good tensile strength. Nevertheless, this strength often decreases when the straw is used in powder form. We know that orienting short fibers and powders is difficult. Therefore, owing to long straw integrity, long straw reinforced composites should theoretically have high strength. In addition, polypropylene (PP) and high density polyethylene are commonly used as matrixes because of their high strength and suitable processing temperatures [14,15], but the toughness of HDPE and PP-based composites is poor [16]. Therefore, other matrix options must be developed. Linear low density polyethylene (LLDPE) has excellent properties, such as tear strength and environmental stress crack resistance, in addition to the properties of general polyolefins [17]. LLDPE has a disadvantage of poor stiffness besides, it potentially could be remedied by straws.

The main aim of this study is to develop a novel method to prepare long straw stem reinforced LLDPE composites, whose properties are compared with those of short fiber reinforced LLDPE composites.

2. Materials and Methods

2.1. Materials

Straws were obtained from the suburb of Harbin, China. LLDPE film (Film, 10H01) and LLDPE particles (Hytrel, 22402) were purchased from Runwen Packaging Materials Co., Ltd., Shanghai, China. Talcum powder (2000 mesh) was produced by Liangjiang Titanium Chemical Products Co., LTD, Shanghai, China, and was used to enhance the stiffness of straw-plastic composite. Maleic anhydride grafted polyethylene (MAPE) with a grafting rate of 0.9% was purchased from Rizhisheng Fine Chemical Co., Ltd., Nantong, China, and was used as the coupling agent. PE wax, also from Shanghai Liangjiang Titanium Chemical Co., Ltd. and zinc stearate, purchased from Natural Oil Chemical Co., Ltd., Pasir Gudang Town, Johor Bahru, Malaysia, were used as lubricants.

2.2. Preparation of Straw/LLDPE Composite

The straw was first oven dried at 103 °C in a DHG-9140A drier (Yiheng Scientific Instrument Co., Ltd., Shanghai, China) to decrease its moisture content to below 3%. Then the straw was used to prepare composite through two molding processes.

2.2.1. Extrusion Molding

Dry straws were cut to 1 cm lengths. Then the short straws, LLDPE particles, talcum, MAPE, PE wax and zinc stearate were weighed in a ratio of 60:25:10:3:1:1. These materials were mixed in a SHR-10A high-speed mixer (Tonghe Rubber & Plastic Machinery Co., Ltd., Zhangjiagang, China) for 5 min. The mixture was pelleted in a JSH30 twin-screw extruder (Nanjing Rubber & Plastic Machinery Factory in Nanjing, China) at 140 °C and then shattered in a GL-01 pulverizer (Evian Machinery Co., Ltd., Shanghai, China). The pulverized raw materials were fed into a BHMS single-screw extruder (Nanjing Saiwang Technology Development Co., Ltd., Nanjing, China). The extruded lumber had a rectangular cross section of 40 mm in width and 4 mm in thickness.

2.2.2. Film-Roll Hot Press Molding

MAPE and talcum powder were evenly spread on the LLDPE film. Long dry straws were spread in parallel on the film. The proportions of straw, LLDPE, talcum and MAPE were 60%, 27%, 10% and 3%, respectively. The film was rolled up, and the straw stems were enveloped. The rolls were first pre-heated without pressure for 4 min and then hot pressed for 5 min under 10 MPa pressure in a SY01 hot press (Shanghai Board Equipment Technology Co., Ltd., Shanghai, China). In this process, the temperature was set to 140 °C. After hot pressing, the panel was cooled. A gauge was used to control the thickness of the straw/LLDPE composite. The size of the result panel was 165mm × 165mm × 4 mm. The detailed steps are shown in Figure 1. Straw particles and straw stems are shown in Figure 2.

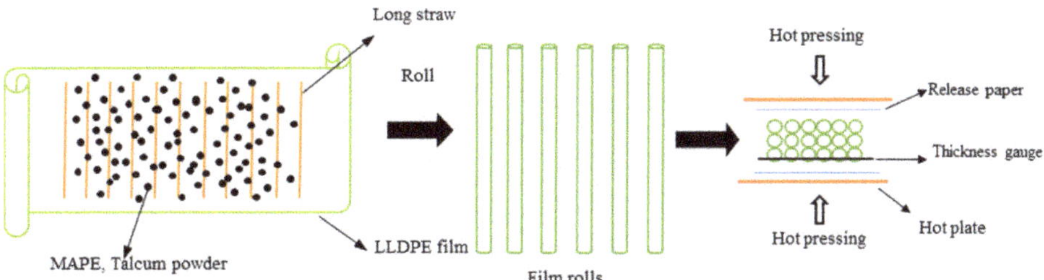

Figure 1. Preparation of straw/LLDPE composite through the film-roll hot pressing process.

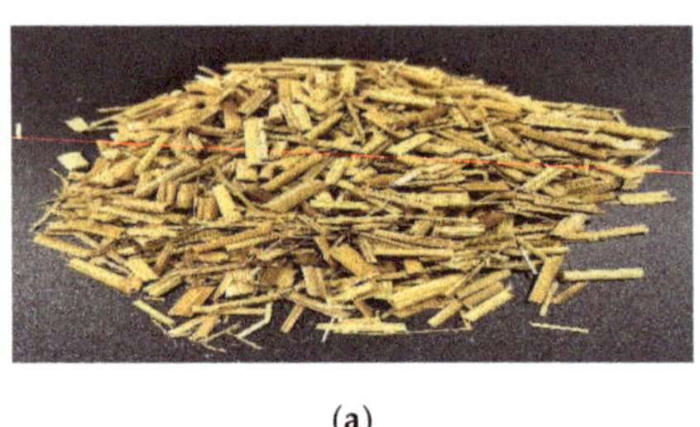 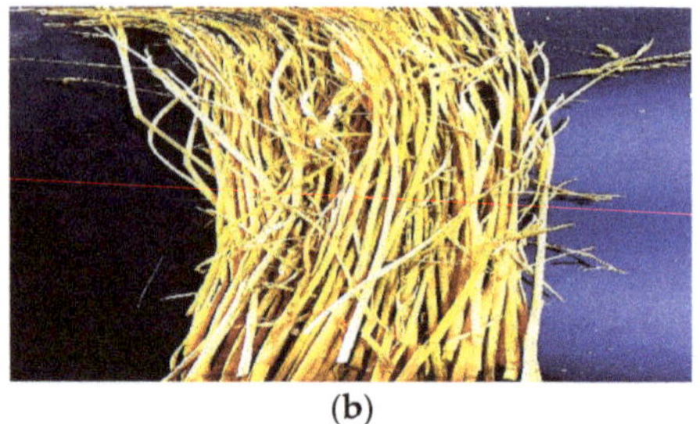

(a) (b)

Figure 2. Straw particle (**a**), straw stem (**b**).

2.3. Characterization of Straw/LLDPE Composite

2.3.1. Mechanical Property Tests

The unnotched impact strength was examined on the basis of GB/T 1043.1–2008 ("Plastics, Determination of Charpy Impact Properties, Part 1: Noninstrumented Impact Test") with a JC-5 Charpy Impact Tester (Chengde Precision Testing Machine Co., Ltd., Chengde, Hebei, China). Specimens of 80mm × 10mm × 4 mm with a span length of 60 mm were analyzed. The striking velocity of the tests was 2.9 m/s, and the pendulum energy was 2 J. Six replicates of each preparation were tested to determine the impact strength.

The tensile tests were carried out in accordance with the method of GB/T 1040.2–2006 ("Plastics–Determination of tensile properties–Part 2: Test conditions for molding and extrusion plastics"). The test piece had a dumbbell shape, and its gauge length was 50 mm. The length of the specimen is parallel to the fiber orientation. The length of the test piece was 165 mm, the width of the narrow portion was 13 mm, and the thickness was 4 mm. The tests were carried out with a loading speed of 5 mm/min. The clamp stretches the specimen along the fiber orientation. Six replicates of each preparation were tested to obtain values for the tensile modulus and strength.

Flexural tests were carried out in accordance with the procedure of GB/T 1449–2005 ("Fiber-Reinforced Plastic composite- Determination of Flexural Properties"). The specimens of 80mm × 13mm × 4 mm had a span length of 64 mm. A loading speed of 2 mm/min was used for testing. The extension direction of the probes is perpendicular to the fiber orientation. Six replicates of each preparation were tested to obtain values for the flexural modulus and flexural strength. The tensile and flexural tests are completed by a CMT5504 mechanical testing machine (MTS industrial Systems (China) Co., LTD., Shanghai, China).

2.3.2. Density Tests

The composites were cut into 50 mm × 35 mm test pieces, and the length, width and thickness of the sample and the sample quality were measured according to GB/T 17657–2013 "Physical Testing Methods for Artificial Board and Finished Panels." The composite density was characterized according to the ratio of mass to volume. Three replicates of each preparation were tested. The density of each replicate was measured once, then averaged the measurement of each replicate.

2.3.3. Water Absorption Performance

The material was formed into a test piece of 76.2mm × 25.4mm × 4 mm, and water absorption performance tests were carried out in accordance with the standard ASTM D570 "Standard Test Method for Water Absorption of Plastics." The test pieces were dried in an oven for 24 h at 50 °C and then completely immersed in water at 24 °C. After 24 h, the test pieces were removed and weighed immediately. Vernier calipers were used to measure the thickness of test pieces before and after soaking. Three replicates of each preparation were tested. The mass and thickness of each replicate were measured three times and then averaged.

The water absorption and thickness expansion were calculated according to Equations (1) and (2), respectively:

$$c = \frac{m_1 - m_0}{m_0} \times 100\% \tag{1}$$

$$T = \frac{t_1 - t_0}{t_0} \times 100\% \tag{2}$$

where c denotes the water absorption mass fraction, T denotes the thickness expansion ratio, m_0 and t_0 denote the mass and thickness of the test piece after drying, respectively, and m_1 and t_1 denote the mass and thickness of the test piece after immersion, respectively.

2.3.4. Differential Scanning Calorimetry (DSC) Analysis

A differential scanning calorimeter (DSC Q100, TA Instruments, New Castle, PA, USA) was used to detect the melting behavior of the straw/LLDPE composite. The temperature of the straw/LLDPE composite sample was reduced to −70 °C, then heated to 180 °C at a rate of 10 °C min^{-1}. The entire testing process was performed under a nitrogen atmosphere.

2.3.5. Interfacial Morphological Observations by Scanning Electric Microscopy (SEM)

SEM was used to characterize the internal structural changes. The straw/LLDPE composite samples were frozen in liquid nitrogen for 10 min and then broken. The broken surfaces were sputter-coated with gold and then observed under a scanning electron microscope (FEI Quanta 200, FEI Co, Hillsboro, TX, USA) operated at an acceleration voltage of 12.5 kV. Three samples of each preparation were observed.

2.3.6. Rotating Rheological Tests

The dynamic rheological properties of straw/LLDPE composite were tested with a rotary rheometer (AR2000ex, TA Instruments, New Castle, PA, USA). Dynamic frequency scanning tests were conducted. The frequency range was 628.3 rad/s to 0.01 rad/s, and the strain was fixed at 0.05%. The above operations were carried out at 150 °C.

3. Results

3.1. Mechanical Property Analysis

The mechanical properties of extruded and hot pressed straw/LLDPE composite are shown in Figure 3.

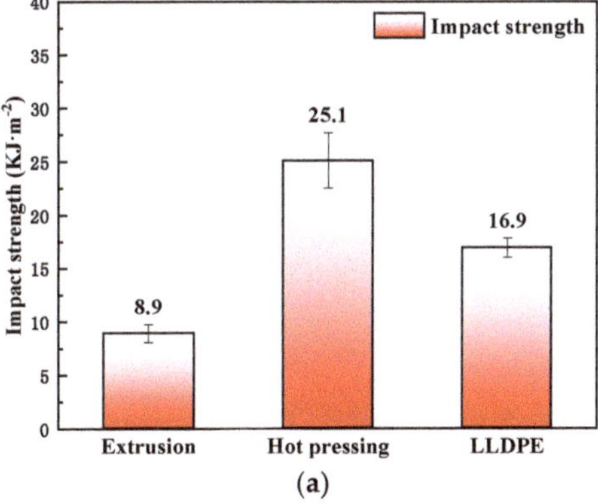

Figure 3. *Cont.*

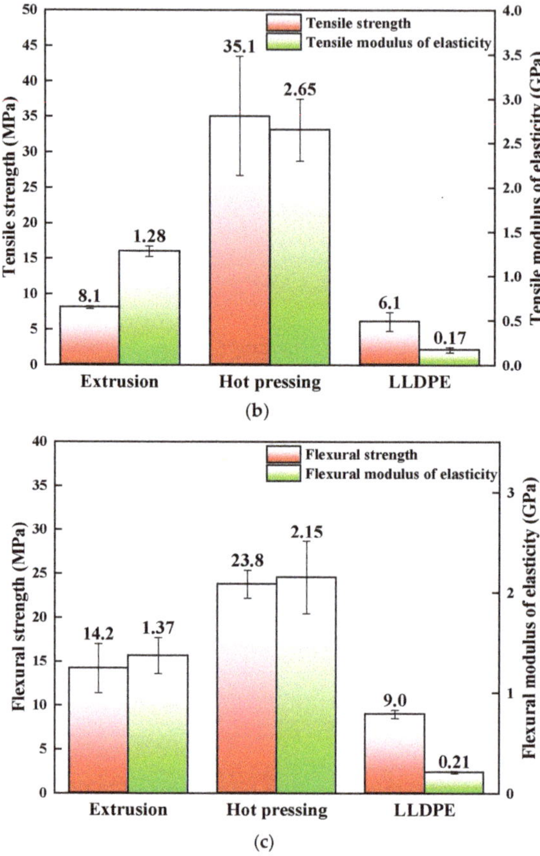

Figure 3. Impact strength (**a**), tensile properties (**b**) and bending properties (**c**) of hot pressed and extruded straw/LLDPE composite.

As shown in Figure 3, the mechanical properties of hot pressed straw/LLDPE composite were significantly higher than those of the extruded composite. Compared with the extruded panel, the hot pressed panel showed 181% greater impact strength (Figure 3a), 335% greater tensile strength, 107% greater tensile modulus (Figure 3b), 68% greater bending strength and 57% greater flexural modulus (Figure 3c). These results might be due to several factors.

Fiber orientation is known to positively influence the strength of reinforced composites. In the extruded straw/LLDPE composite, the straw fibers were randomly distributed, whereas in the hot pressed straw/LLDPE composite, the straw stems were highly oriented and remained at full length. Under loading, more energy was needed to overcome the interface bonding between the straws and the LLDPE as well as to break the straw stems themselves [18]. However, in extrusion preparation, the straw was in the forms of short fibers or powder, whose specific ratios were small and whose lengths were shorter than the critical length. If the length is less than the critical length, straw fibers are not be snapped but may be pulled out (the fibers slip from the matrix), i.e., straw fibers fail to fully exert the fiber strength within critical length and cannot play a reinforcing role (only as a filling material).

Because the length of straw stems is longer than straw fibers, the energy consumed for fiber extraction in the composite is higher, so the impact strength of hot pressed composite is higher. In addition, the fiber ends where stress is concentrated is the crack initiation point. The longer the fiber, the fewer the fiber ends in the composite. This is also the reason for its high impact strength.

3.2. Melting Performance Analysis

The melting performance of extruded and hot pressed straw/LLDPE composites is shown in Figure 4. The crystallinity of LLDPE was calculated according to Equation (3):

$$X_C = \frac{\Delta H_f (m_c / m_{LLDPE})}{\Delta H_f^0} \tag{3}$$

where ΔH_f denotes the melting enthalpy, m_c denotes the mass of the sample, m_{LLDPE} denotes the mass of LLDPE in the sample and ΔH_f^0 denotes the melting enthalpy of LLDPE with 100% crystallization, 293 J/g [19].

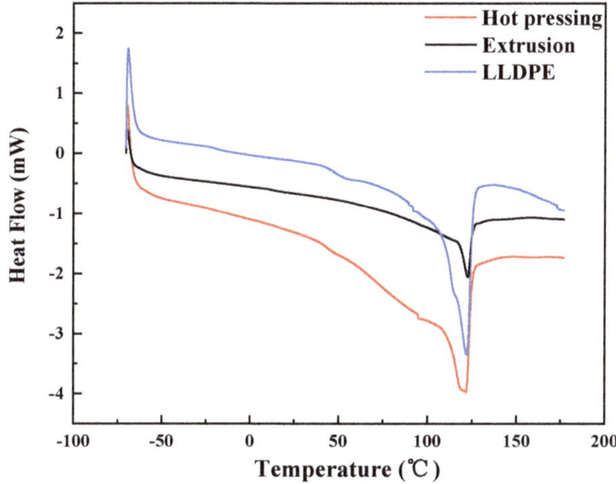

Figure 4. DSC image of hot pressed and extruded straw/LLDPE composite.

The data in Table 1 show the enthalpy and melting temperatures of samples. The composite manufactured by hot pressing and extrusion showed clear melting peaks at 121.97 °C and 122.61 °C, respectively, values similar to that of LLDPE, at 122.28 °C. The endothermic peaks might possibly be caused by the melting of LLDPE [20]. The peak of the hot pressed composite was wider and higher than that of the extruded composite. As shown in Table 1, the melting enthalpy of the hot pressed straw/LLDPE composite was much higher than that of the extruded composite. This shows that in hot pressed straw/LLDPE composite, straw fibers have a greater resistance to the melting of LLDPE, so that the melting process requires more heat than extruded composite [21]. This is because the melting enthalpy is related to the crystallinity. Compared with long straw stems, short straw fibers are more easily dispersed uniformly in the composite, and the contact area with LLDPE molecules increases, which plays a role in diluting LLDPE, so it can reduce the interaction between LLDPE molecules to a greater extent. The lower crystallinity, the fewer the heat required for heating and melting, i.e., the lower melting enthalpy. Pore structure of long straw in hot pressed composite would be expected to store more thermal energy, thereby resulting in slow progress of melting.

Table 1. Melting enthalpy and temperature of straw/LLDPE composite.

Sample	X_c (%)	T_m (°C)	ΔH_f (J/g)
LLDPE	17.41	122.28	51.01
hot pressing	35.76	121.97	62.87
extrusion	6.94	122.61	12.20

X_c—crystallinity; T_m—melting temperature; ΔH_f—melting enthalpy.

3.3. Water Absorption Performance Analysis

The water resistance test results of hot pressed and extruded straw/LLDPE composites are shown in Figure 5.

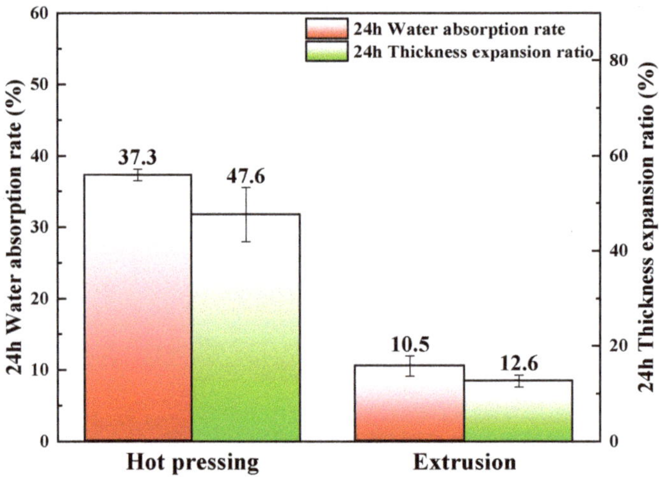

Figure 5. Water absorption mass fractions and thickness expansion ratios of hot pressed and extruded straw/LLDPE composite (immersed for 24 h).

Both the water absorption rate and the thickness expansion ratio of hot pressed straw/LLDPE composite were higher than those of the extruded composite, because in the hot pressing process, LLDPE matrix cannot completely fill in the cavities of the straw stem, and consequently the area of the straw contacting water is larger. In the extrusion process, almost all surfaces of the tiny straw particles were covered with LLDPE matrix [22]. Evenly distributing the raw material can effectively improve water resistance [23]. Possible effective measures include breaking the stem along the length of straw, which can take advantage of long fibers and also facilitate LLDPE matrix to enter the straw cavity during hot pressing, reducing the contact area between straw and water. In addition, a layer of LLDPE can be coated on the outside before hot pressing to seal the material.

3.4. Morphology of the Fracture Surface

3.4.1. Interface Bonding

The interfacial bonding and fiber orientation of the straw/LLDPE composite manufactured through the two methods were characterized by SEM (Figure 6).

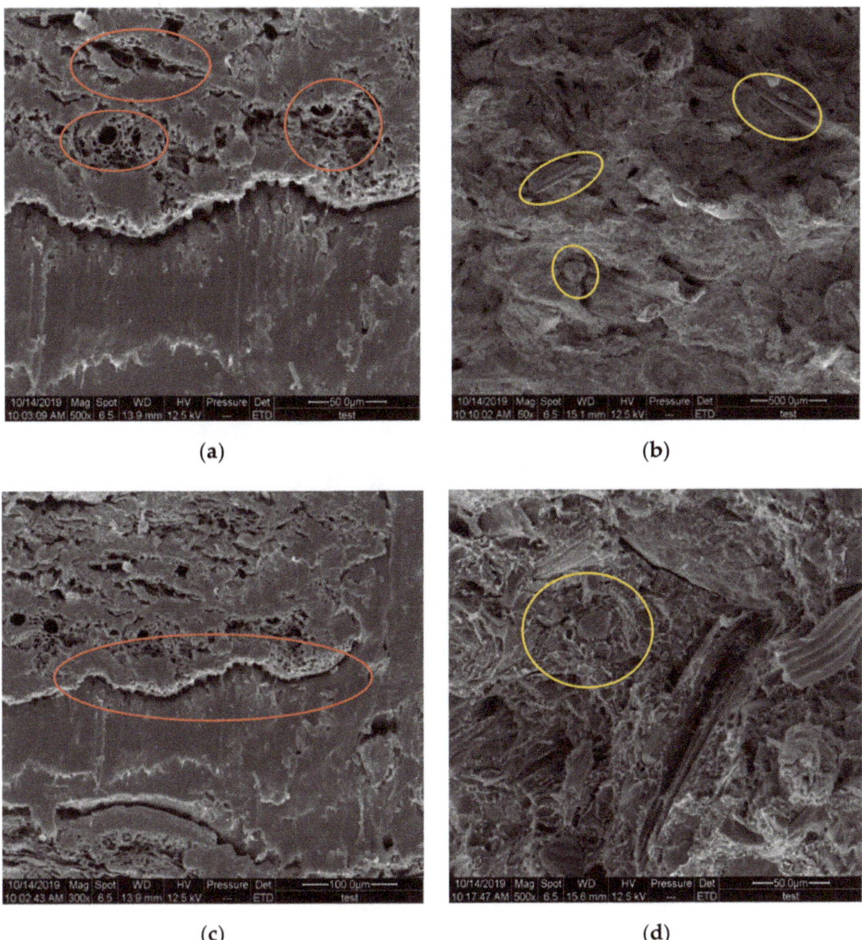

Figure 6. Microscopic morphology of the fracture surface of hot pressed (**a,c**) and extruded (**b,d**) straw/LLDPE composite.

In the hot pressed composite, only the cross section of the straw could be seen (circles in Figure 6a), thus indicating that the direction of the straw was well fixed. In Figure 6b, the circled straw fibers in the extruded composite are randomly distributed. As further shown in circles of Figure 6c, a crack is present at the interface location between the straw and the LLDPE matrix in the hot pressed composite. In contrast, the straw and LLDPE matrix are relatively tightly combined in the extruded composite (circles in Figure 6d). In addition, the hollow structure of the straw in the hot pressed composite is not filled, thereby contributing to the higher impact strength. The holes left by straws that had been pulled out indicated that the straw fibers did not break in tensile loading (Figure 6b,d). However, the straw stems fully broke, and no holes were left in the matrix (Figure 6a,c), thus indicating that the long straw stems contributed to the strength. These structural results explained the differences in both the strength and water absorption of the composite. The hollow structures of the long straws that remained also explained the low density of the hot pressed straw/LLDPE composite (Table 2) [24]. In addition, the extrusion process can uniformly mix the raw materials and tightly combine them, thereby increasing the density of the composite.

Table 2. Density of extruded and hot pressed straw/LLDPE composite.

Manufactured Process	Extrusion	Hot Pressing
Density/(g·cm^{-3})	1.31	0.83

3.4.2. Analysis of Tensile Fracture Mechanism

In the straw/LLDPE composite, the main fracture forms are fiber fracture, interface detachment and matrix fracture. In Figure 7, the broken straw fiber (Circle 1) and the tip-shaped LLDPE (Circle 2) can be seen. In Figure 8, we can see the holes (The circled part) left by short fibers pull-out, indicating that the strength of the straw itself is not fully exerted in the extruded composite.

Figure 7. Optical micrograph of fracture surface of the hot pressed straw/LLDPE composite.

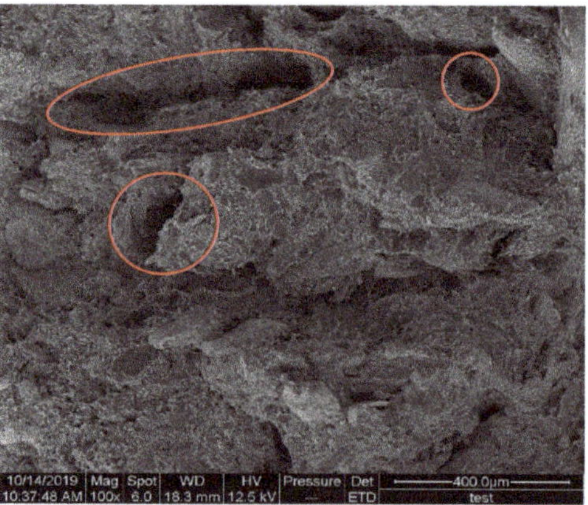

Figure 8. SEM of fracture surface of the extruded straw/LLDPE composite.

Tensile stress-strain curve of hot pressed and extruded straw/LLDPE composites is shown in Figure 9. Comparing Figure 9a,b, it can be seen that the stress-strain curve of the extruded composite is relatively smooth, thus the fracture process is relatively gentle, which indicates that the tougher LLDPE plays a major bearing role. Therefore, combining Figures 8 and 9b shows that the mechanical strength of the extruded composite mainly comes from the LLDPE matrix. In the curve image of the hot pressed composite, the stress drops suddenly. This is due to the sudden separation of the straw fiber and LLDPE matrix interface and the sudden breaking of the straw. Combining Figures 7 and 9a, it can be known that the mechanical strength of the hot pressed composite mainly comes from the combination of the fiber matrix and the strength of the straw, and the matrix itself has less effect.

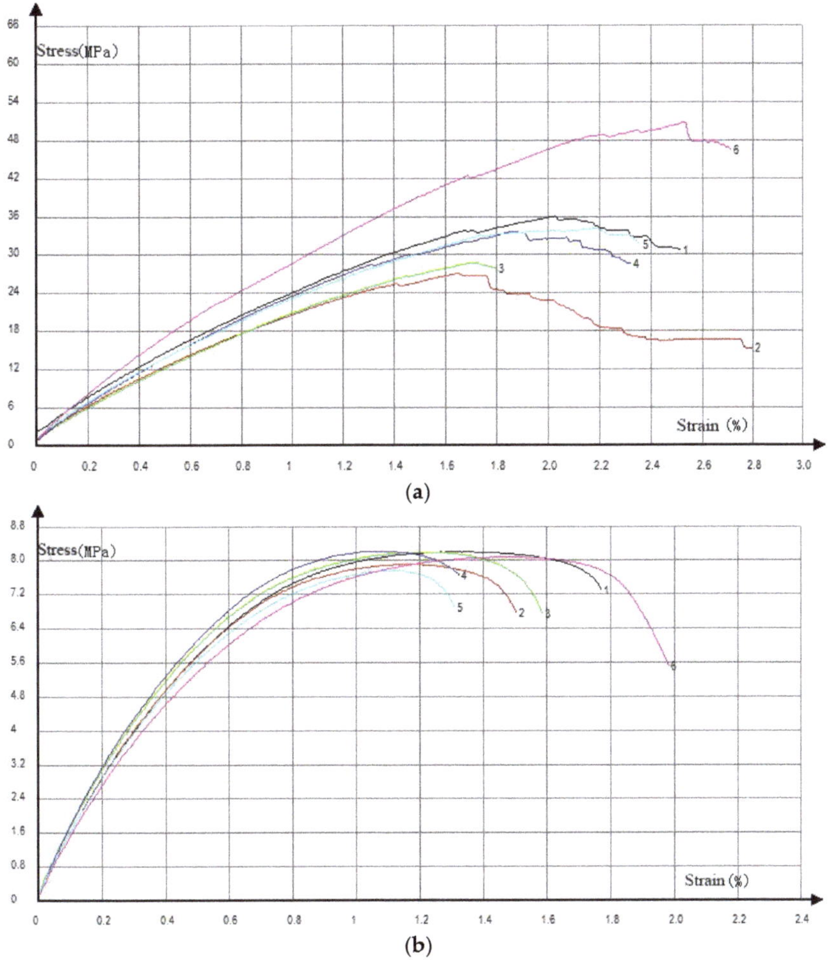

Figure 9. Tensile stress-strain curves of specimen 1 to 6 of hot pressed (**a**) and extruded (**b**) straw/LLDPE composite.

3.5. Dynamic Rheological Performance Analysis

Plots of the storage modulus, loss modulus, loss tangent value and complex viscosity vs ω of straw/LLDPE composite is shown in Figure 10. Figure 10a shows the relationship between the storage modulus (G′) and the angular frequency. By increasing the test temperature (from 130 °C to 150 °C), a large increase in storage modulus can be seen for the hot pressed composite, which is not the case for extruded composite. This may be because at a temperature closer to the T_m of LLDPE (T_m of LLDPE measured by DSC is 122 °C), the LLDPE melt is still relatively hard and the molecular chain flexibility is poor. In this case, the uneven fiber distribution has a negative effect on the deformation of the matrix. In contrast to the changes in G′ of hot pressed straw/LLDPE composite with frequency, a modulus platform in the low frequency region was observed for the extruded straw/LLDPE composite. This phenomenon, so-called solid-like behavior [25–27], occurs because of the formation of three-dimensional ordered structures such as agglomerates, skeletons and networks inside the system [28]. The viscoelastic behavior of the low ω region is the motion response of the long-chain segment of the polymer or even the entire macromolecular chain, and the three-dimensional ordered structure limits the long-term movement of macromolecular motion units [25,26,29]. The hot pressed straw/LLDPE composite showed very little change as the frequency increased. The slight modulus increase occurred because the parallel straws acted as a skeleton and prevented the LLDPE from sliding. Similarly, as shown in Figure 10b, both the hot pressed and extruded straw/LLDPE composites showed stable loss modulus with frequency variation.

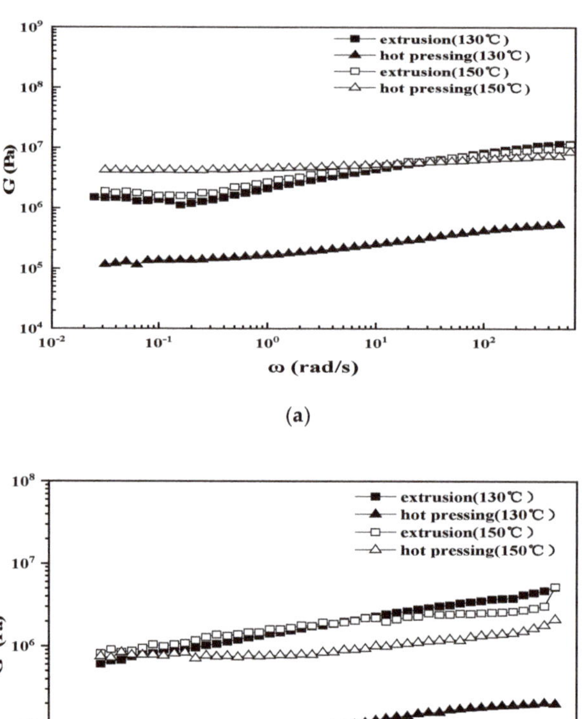

Figure 10. *Cont.*

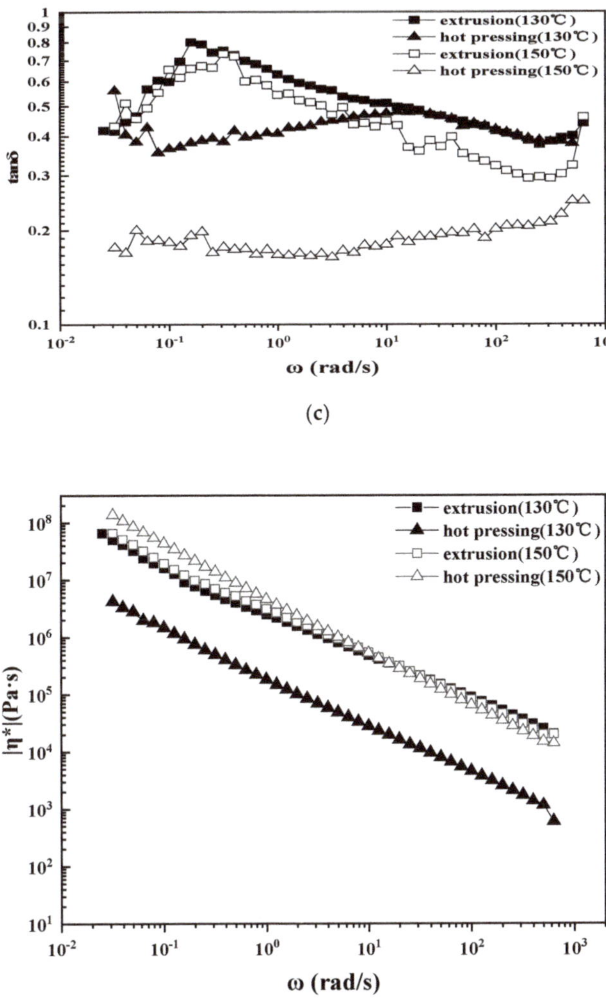

(c)

(d)

Figure 10. Plots of the storage modulus (**a**), loss modulus (**b**), loss tangent value (**c**) and complex viscosity (**d**) vs ω of straw/LLDPE composite.

Figure 10c shows the relationship between the loss tangent value of the straw/LLDPE melting and the angular frequency. The loss tangent value is the ratio of the loss modulus to the storage modulus. As shown in Figure 10c, the tan δ of the extruded straw/LLDPE melt showed a sharp peak at around 0.2 rad/s, whereas the hot pressed straw/LLDPE melt did not show a clear peak. In the G′—ω curve, the tanδ peak appears along with the modulus platform; that is, the tanδ peak is also a characteristic of solid-like behavior [30]. The change in the value of tanδ indicates a change in viscoelasticity, thus demonstrating that the viscoelasticity of the hot pressed composite is fairly stable [31].As shown in Figure 10c, tanδ was less than 1 over the entire scanning frequency range, thus indicating that the straw/LLDPE melting exhibited elasticity. We conclude that hot pressed straw/LLDPE composite had the clearest elastic characteristics, according to its low tanδ value and high G′ value [32].

Figure 10d shows the relationship between the complex viscosity ($|\eta^*|$) and the angular frequency.

As shown in the figure, as the frequency increases, the viscosity shows a downward trend, i.e., the phenomenon of shear thinning occurs. This is because the viscosity is the ratio of stress to strain rate, and according to Power-Law Equation

$$\sigma = K \cdot \gamma^{n} \tag{4}$$

We can get Equation

$$\eta = \frac{\sigma}{\gamma} = K \cdot \gamma^{n-1} \tag{5}$$

where η denotes the viscosity, σ denotes the stress, γ denotes the strain rate, K and n are both constants. High-molecular polymers such as PE are pseudoplastic liquids, and the n value of the pseudoplastic fluid is less than 1, so when the strain rate increases, the viscosity of the system decreases. The opposite of pseudoplastic fluid is dilatant liquid, such as corn paste, whose n value is more than 1, and the viscosity will increase as the strain rate increases [33].

There is always a certain speed gradient between the various liquid layers when the polymer flows. If a large molecule passes through several liquid layers with different flow rates at the same time, each part of the same macromolecule must advance at different speeds. This situation obviously cannot last. Therefore, during the flow, each long-chain molecule always tries to make itself all enter the liquid layer with the same flow rate. The parallel distribution of liquid layers with different flow rates results in the orientation of the macromolecules in the flow direction, which causes the viscosity to decrease with increasing frequency during the flow. With the increase of experimental temperature, the kinetic energy of the molecule of hot pressed composite is increased, but also increases the degree of intermolecular collision, which makes the viscosity increase.

In dynamic tests, the relationship between dynamic viscosity and loss viscosity, the Cole-Cole curve, can give information about the various relaxation processes in heterogeneous polymer systems. As shown in Figure 11, the right end of the curve of the extruded composite is slightly upturned, so-called "tailing" phenomenon occurs. This shows that there are two relaxation mechanisms in the system of the extruded composite. This may be because LLDPE is more prone to entanglement in the extruded composite system, and this structure relaxes very slowly.

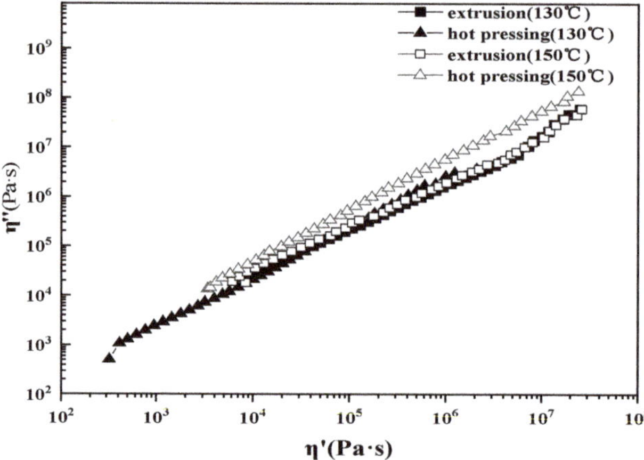

Figure 11. Loss viscosity of straw/LLDPE composite as a function of storage viscosity.

3.6. Physical Drawing of Extruded and Hot Pressed Straw/LLDPE Composites

The physical picture of the extruded and hot pressed composites is shown in Figure 12.

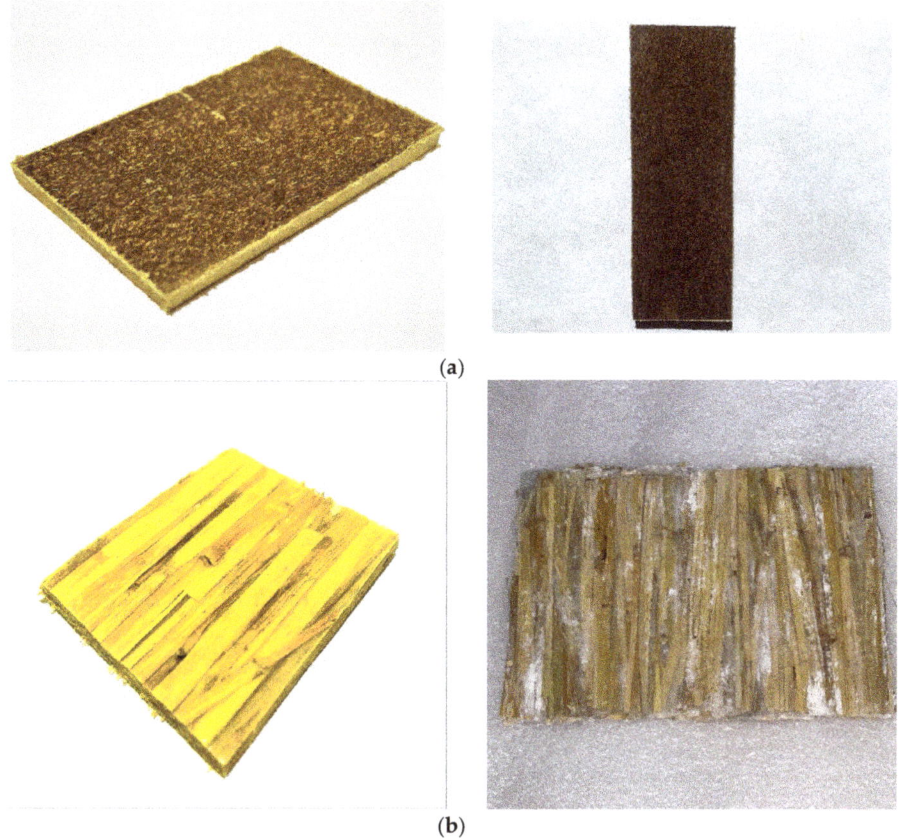

Figure 12. Extruded (**a**) and hot pressed (**b**) straw/LLDPE composites.

4. Conclusions

This study developed a new method for constructing straw-plastic composite and compared it with conventional extrusion methods. Tests of the mechanical properties verified that the hot pressed long straw stem reinforced LLDPE composite had relatively higher strength and modulus. Microstructural observations showed better fiber orientation of the hot pressed straw/LLDPE composite, and this factor had the greatest influence on the mechanical properties of the straw/LLDPE composite. According to the results of DSC, straw fibers have a greater resistance to the melting of LLDPE in hot pressed straw/LLDPE composite, so that the melting process requires more heat than extruded composite. The results from dynamic rheological analysis indicated that the storage modulus of the straw/LLDPE melt manufactured by hot pressing was more stable than that of the extruded composite. The elastic characteristics of the hot pressed straw/LLDPE melt were more pronounced than those of the extruded composite. The hot pressed straw/LLDPE composite had higher water absorption, thus indicating its ability to regulate the surrounding relative humidity. Straw/LLDPE composites are expected to be applicable in interior decoration materials, owing to their high strength-to-weight ratio and the absence of chemical emission such as those from adhesives.

Author Contributions: W.W. and L.Z. conceived and designed the experiments; L.Z. and H.X. performed the experiments; W.W. and L.Z. analyzed the data; L.Z. wrote the paper; W.W. revised the paper. All authors have read and agreed to the published version of the manuscript.

Funding: This research received no external funding.

Acknowledgments: The authors gratefully acknowledge the support of the National Natural Science Foundation of China (No. 31670573), and the Innovation Team Sustainable Development Special Project of the Central University Fundamental Research Business Expenses Special Fund Project (No. 2572017ET05).

Conflicts of Interest: The authors declare no conflict of interest.

References

1. Militky, J.Í.; Jabbar, A. Comparative evaluation of fiber treatments on the creep behavior of jute/green epoxy composites. *Compos. Part B Eng.* **2015**, *80*, 361–368. [CrossRef]
2. Bengtsson, M.; Baillif, M.L.; Oksman, K. Extrusion and mechanical properties of highly filled cellulose fibre–polypropylene composites. *Compos. Part A* **2007**, *38*, 1922–1931. [CrossRef]
3. Graupner, N.; Herrmann, A.S.; Müssig, J. Natural and man-made cellulose fibre-reinforced poly(lactic acid) (PLA) composites: An overview about mechanical characteristics and application areas. *Compos. Part A Appl. Sci. Manuf.* **2009**, *40*, 810–821. [CrossRef]
4. Dicker, M.P.M.; Duckworth, P.F.; Baker, A.B.; Francois, G.; Weaver, P.M. Green composites: A review of material attributes and complementary applications. *Compos. Part A Appl. Sci. Manuf.* **2014**, *56*, 280–289. [CrossRef]
5. Liu, Z.; Huang, F.; Li, B. Investigating contribution factors to China's grain output increase in period of 2003 to 2011. *Trans. Chin. Soc. Agric. Eng.* **2013**, *29*, 1–8.
6. Li, Y.; Cheng, C.; Xu, L. Design and experiment of baler for 4L-4.0 combine harvester of rice and wheat. *Trans. Chin. Soc. Agric. Eng.* **2016**, *32*, 29–35.
7. Nyambo, C.; Mohanty, A.K.; Misra, M. Effect of Maleated Compatibilizer on Performance of PLA/Wheat Straw-Based Green Composites. *Macromol. Mater. Eng.* **2011**, *296*, 710–718. [CrossRef]
8. Xiao, Y.; Fu, M. Forming Technology of PP/Straw composites by hot pressing. *Eng. Plast. Appl.* **2005**, *1*, 34–37.
9. Zhang, D.; He, C.; Liu, J. Mechanical properties of straw-powder/PP composites. *Trans. Chin. Soc. Agric. Eng.* **2010**, *26*, 380–384.
10. Zabihzadeh, M.; Dastoorian, F.; Ebrahimi, G. Effect of MAPE on Mechanical and Morphological Properties of Wheat Straw/HDPE Injection Molded Composites. *J. Reinf. Plast. Compos.* **2010**, *29*, 123–131. [CrossRef]
11. Koster, L. Influencing factors and parameters in the extrusion process. *KGK Rubberpoint.* **2005**, *58*, 362–365.
12. Mandell, J.F.; Mcgarry, F.J.; Huang, D.D.; Li, C.G. Some effects of matrix and interface properties on the fatigue of short fiber-reinforced thermoplastics. *Polym. Compos.* **2010**, *4*, 32–39. [CrossRef]
13. Kakisawa, H.; Honda, K.; Kagawa, Y. Effect of wear on interface frictional resistance in fiber-reinforced composite: Model experimental. *Mater. Sci. Eng. A* **2000**, *284*, 226–234. [CrossRef]
14. Park, C.B. *Polymeric Foams: Science and Technology*; CRC Press: Boca Raton, FL, USA, 2006.
15. Machado, J.S.; Santos, S.; Pinho, F.F.S.; Luis, F.; Alves, A.; Simoes, R.; Rodrigues, J.C. Impact of high moisture conditions on the serviceability performance of wood plastic composite decks. *Mater. Design.* **2016**, *103*, 122–131. [CrossRef]
16. Sudar, A.; Renner, K.; Moczo, J.; Lummerstorfer, T.; Burgstaller, C.; Jerabek, M.; Gahleitner, M.; Doshev, P.; Pukanszky, B. Fracture resistance of hybrid PP/elastomer/wood composites. *Compos. Struct.* **2016**, *141*, 146–154. [CrossRef]
17. Yilmazer, U. Effects of the processing conditions and blending with linear low-density polyethylene on the properties of low-density polyethylene films. *J. Appl. Polym. Sci.* **2010**, *42*, 2379–2384. [CrossRef]
18. Phelps, J.H.; Iii, C.L.T. An anisotropic rotary diffusion model for fiber orientation in short- and long-fiber thermoplastics. *J. Non-Newton. Fluid Mech.* **2009**, *156*, 165–176. [CrossRef]
19. Wilfong, D.L.; Knight, G.W. Crystallization mechanisms for LLDPE and its fractions. *J. Polym. Sci. Part A Polym. Chem.* **1990**, *28*, 861–870. [CrossRef]
20. Tubbs, F.R. Physiological Studies in Plant Nutrition: II. The Effect of Manurial Deficiency upon the Mechanical Strength of Barley Straw. *Ann. Botany.* **1930**, *44*, 147–160. [CrossRef]
21. Prasad, A. A quantitative analysis of low density polyethylene and linear low density polyethylene blends by differential scanning calorimetery and fourier transform infrared spectroscopy methods. *Polym. Eng. Sci.* **2010**, *38*, 1716–1728. [CrossRef]

Polymers **2020**, *12*, 860

22. Huang, X.; Li, Q.; Yang, M.; Luo, Y.; Ye, X.; Yang, Y. Thermal Decomposition and Kinetics Analysis of Larger Particle Neodymium Oxalate. *J. Chin. Soc. Rare Earths* **2016**, *34*, 83–92.
23. Haque, A.; Jeelani, S. Environmental Effects on the Compressive Properties: Thermosetting vs. Thermoplastic Composites. *J. Reinf. Plast. Compos.* **1992**, *11*, 146–157. [CrossRef]
24. Czel, G.; Czigany, T. A Study of Water Absorption and Mechanical Properties of Glass Fiber/Polyester Composite Pipes – Effects of Specimen Geometry and Preparation. *J. Compos. Mater.* **2008**, *42*, 2815–2827. [CrossRef]
25. Aranguren, M.I. Effect of reinforcing fillers on the rheology of polymer melts. *J. Rheol.* **1992**, *36*, 1165–1182. [CrossRef]
26. Wu, G.; Song, Y.; Zheng, Q.; Du, M.; Zhang, P. Dynamic rheological properties for HDPE/CB composite melts. *J. Appl. Polym. Sci.* **2003**, *88*, 2160–2167. [CrossRef]
27. Romani, F.; Corrieri, R.; Braga, V.; Ciardelli, F. Monitoring the chemical crosslinking of propylene polymers through rheology. *Polymer* **2002**, *43*, 1115–1131. [CrossRef]
28. Prashantha, K.; Soulestin, J.; Lacrampe, M.F.; Krawczak, P.; Dupin, G.; Claes, M. Masterbatch-based multi-walled carbon nanotube filled polypropylene nanocomposites: Assessment of rheological and mechanical properties. *Compos. Sci. Technol.* **2009**, *69*, 1756–1763. [CrossRef]
29. Wang, M.; Wang, W.; Liu, T.; Zhang, W.D. Melt rheological properties of nylon 6/multi-walled carbon nanotube composites. *Compos. Sci. Technol.* **2008**, *68*, 2498–2502. [CrossRef]
30. Dong, Q.; Zheng, Q.; Du, M.; Zhang, M.Q. Temperature-dependence of dynamic rheological properties for high-density polyethylene filled with graphite. *J. Mater. Sci.* **2005**, *40*, 3539–3541. [CrossRef]
31. Liang, J.Z.; Li, R.K.Y.; Tjong, S.C. Effects of glass bead size and content on the viscoelasticity of filled polypropylene composites. *Polym. Test.* **2000**, *19*, 213–220. [CrossRef]
32. Unwin, A.P.; Hine, P.J.; Ward, I.M.; Guseva, O.A.; Schweizer, T.; Fujita, M.; Tanaka, E.; Gusev, A.A. Predicting the visco-elastic properties of polystyrene/SIS composite blends using simple analytical micromechanics models. *Compos. Sci. Technol.* **2017**, *142*, 302–310. [CrossRef]
33. Masao, D.; See, H. *Introduction to Polymer Physics*; South China University of Technology Press: Guangzhou, China, 2011.

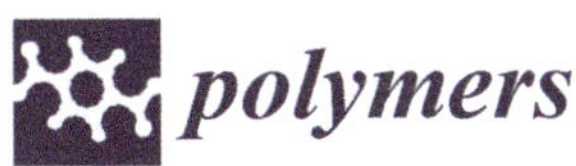

Article

Facile Construction of Superhydrophobic Surfaces by Coating Fluoroalkylsilane/Silica Composite on a Modified Hierarchical Structure of Wood

Jiajie Wang, Yingzhuo Lu, Qindan Chu, Chaoliang Ma, Lianrun Cai, Zhehong Shen * and Hao Chen *

School of Engineering, Zhejiang A&F University, Hangzhou 311300, China; laowangwjj0126@163.com (J.W.); tqy033201@163.com (Y.L.); 15306587705@163.com (Q.C.); qq843543030@163.com (C.M.); cailianr21@163.com (L.C.)
* Correspondence: zhehongshen@zafu.edu.cn (Z.S.); haochen@zafu.edu.cn (H.C.)

Received: 15 March 2020; Accepted: 1 April 2020; Published: 4 April 2020

Abstract: Constructing superhydrophobic surfaces by simple and low-cost methods remains a challenge in achieving the large-scale commercial application of superhydrophobic materials. Herein, a facile two-step process is presented to produce a self-healing superhydrophobic surface on wood to improve water and mildew resistance. In this process, the natural hierarchical structure of wood is firstly modified by sanding with sandpaper to obtain an appropriate micro/nano composite structure on the surface, then a fluoroalkylsilane/silica composite suspension is cast and dried on the wood surface to produce the superhydrophobic surface. Due to the full use of the natural hierarchical structure of wood, the whole process does not need complicated equipment or complex procedures to construct the micro/nano composite structure. Moreover, only a very low content of inorganic matter is needed to achieve superhydrophobicity. Encouragingly, the as-obtained superhydrophobic surface exhibits good resistance to abrasion. The superhydrophobicity can still be maintained after 45 abrasion cycles under the pressure of 3.5 KPa and this surface can spontaneously recover its superhydrophobicity at room temperature by self-healing upon damage. Moreover, its self-healing ability can be restored by spraying or casting the fluoroalkylsilane/silica composite suspension onto this surface to replenish the depleted healing agents. When used for wood protection, this superhydrophobic surface greatly improves the water and mildew resistance of wood, thereby prolonging the service life of wood-based materials.

Keywords: superhydrophobic surfaces; self-healing; natural hierarchical microstructures; wood

1. Introduction

The fabrication of superhydrophobic surfaces with water contact angles (CAs) larger than 150° and sliding angles less than 10° has attracted extensive research attention worldwide due to its great potential in both theoretical research and practical applications, such as self-cleaning [1], anti-fouling [2], durable antibacterial uses [3], oil/water separation [4,5], gas sensing and droplet manipulation [6]. The superhydrophobic property of the surface is controlled by its chemical composition and topography. The cooperation of micro/nano scale hierarchical structures with low-surface energy materials has been the main strategy to fabricate superhydrophobic surfaces [7]. Through years of extensive efforts, many chemical and physical methods that generate superhydrophobic surfaces have been developed, such as plasma polymerization/etching [8], chemical vapor deposition [9–12], solvent-mediated phase separation [13], and polymer self-assembly [14–16]. However, some superhydrophobic surfaces are apt to lose their superhydrophobicity during practical applications since the artificial micro/nano hierarchical architectures are susceptible to being damaged after a slight scratch, abrasion, or even brief contact with fingers, which has hampered their widespread application. Recently, although many researchers have reported the successful fabrication of mechanically durable superhydrophobic surfaces [17–23], it is still highly desired to construct superhydrophobic surfaces by simple and low-cost methods to achieve the large-scale commercial application of superhydrophobic materials.

Wood, as a renewable resource, has great potential in decorative fields to substitute steel, stone, glass, minerals, and synthetic resin. Nevertheless, because it contains many hydrophilic groups such as hydroxyl groups, the wood is apt to absorb water, which can result in dimensional instability and attacks by microorganisms including decay fungi and mildew [24–26]. Therefore, it is of great practical significance to construct a high-performance superhydrophobic surface for the wood.

In this work, we have described a facile process to fabricate a robust self-healing superhydrophobic surface on wood with improved water and mildew resistance. In this process, the natural hierarchical structure of wood was firstly modified by sanding with sandpaper to obtain an appropriate micro/nano composite structure on the surface, then a fluoroalkylsilane/silica composite suspension was cast and dried on the wood surface to produce the superhydrophobic surface. Due to the full use of the natural hierarchical structure of wood, the whole process does not need complicated equipment or complex procedures to construct the micro/nano composite structure. Moreover, only a very low content of inorganic matter is needed to achieve superhydrophobicity. Therefore, the as-obtained superhydrophobic surface is transparent and unveils the natural grains and textures of the original surface. More importantly, benefiting from the unique micro/nano composite structure originated from the natural hierarchical structure of wood, the superhydrophobic surface exhibits robust superhydrophobic performance against physical damages. Furthermore, owing to the intrinsic porous structures of wood, the superhydrophobic surface can preserve a large number of fluoroalkylsilane moieties as the healing agent in the wood pores such as cell cavities and grooves. Once the primary top fluoroalkylsilane layer is decomposed or scratched, the internally-preserved healing agent in the wood pores will migrate to the surface to heal the superhydrophobicity at room temperature. Thus, the superhydrophobic surface exhibits good self-healing ability when damaged. And this self-healing ability can be restored by spraying or casting the fluoroalkylsilane/silica composite suspension onto the wood surface to replenish the depleted healing agents. In addition, when used for wood protection, this superhydrophobic surface greatly improves the water and mildew resistance of wood, thereby prolonging the service life of wood-based materials.

2. Materials and Methods

2.1. Materials

Ethanol (AR, ≥99.5%), nano fumed silica (99.8wt%, diameter: 7-40nm) and acetic acid (AR, ≥99.5%) were purchased from Shanghai Chemical Reagent Co. (China). Perfluorooctyltriethoxysilane (KH1322) and Chinese fir wood were kindly donated by Zhejiang Longyou Wood Bond Co. (China). Woodblocks with a size of 40 × 35 × 10 mm (longitudinal × radial × tangential) were obtained from the sapwood of Chinese fir (*Cunninghamia lanceolata*).

2.2. Preparation of Fluoroalkylsilane/Silica Suspension

Typically, 0.2 g of KH1322 (as a fluoroalkylsilane) was dissolved in 10 g of ethanol. The solution was mildly stirred until it became homogeneous and transparent. Then, 4 mg silica nanoparticles (the ratio of silica to KH1322 was 2%) were added into the previous solution and ultrasonically treated for one minute until the resulting fluoroalkylsilane/silica composite suspension was homogeneous.

2.3. Preparation of Superhydrophobic Surfaces

The woodblock was first sanded with 240-grit sandpaper under a pressure of 40 KN· m^{-2} until a visible clean smooth surface was obtained. In this way, the natural hierarchical structure of wood was modified to obtain an appropriate micro/nano composite structure on the surface. Then, the above suspension was cast onto the wood surface (40 mm × 35 mm) and heated in the oven for a certain period of time to form superhydrophobic surfaces. The dosage of KH1322 plus silica on the wood surface was calculated to be about 4 g· m^{-2}. Unless specified, the heating temperature was 120 °C, and the duration of heating was 2 hours.

2.4. Characterization

The surface morphology was observed by scanning electronic microscopy (SEM). The VK-X 3D optical laser microscope system (OPM) was used to characterize the three-dimensional surface morphology. The roughness factor (Ra) was calculated according to JIS B0601:1994. The measurement of the surface wettability was performed using a dynamic contact angle (CA) testing instrument (OCA40, Filderstadt, Germany). CA was recorded at 30 seconds after a droplet of liquids (5 μL) was placed on the surface. The sliding angle (SA) was measured by recording the tilt angle of the sample platform at which a liquid droplet (10 μL) started to roll off the surface. Average water CAs were obtained on radial sections by measuring the same sample at six different positions. Fourier transform infrared spectroscopy (FTIR) measurements were carried out using a Nicolet Nexus 470 spectrometer (Thermo Fisher, Waltham, MA, USA). The chemical composition of the wood surface was measured by X-ray photoelectron spectroscopy (XPS, Perkin-Elmer PHI 5000C ECSA, Waltham, MA, USA). The abrasion resistance of the superhydrophobic surface was evaluated by dragging a piece of 240-grit sandpaper under 500 g of weight in one direction with a speed of 1 cm s^{-1} at a distance of 10 cm per cycle. Digital pictures were captured using a Canon Power Shot A 95 digital camera, and the optical images were obtained with the VK-9710 microscope. Unless specified, the water CA tests and characterizations such as SEM, XPS, FTIR, and OPM were carried on the radial sections of wood.

3. Results and Discussion

3.1. Fabrication of Superhydrophobic Surfaces

The process to fabricate the superhydrophobic surface was shown in Scheme 1. The raw wood was first sanded with 240-grit sandpapers to modify the natural hierarchical structure of the wood to obtain an appropriate micro/nano composite structure on the surface. Then, the homogeneous suspension of KH1322 fluoroalkylsilane and silica was cast onto the surface of the sanded wood and heated in an oven or room-dried for a certain period of time to form the superhydrophobic surface. The resulting material is denoted as coated sanded wood.

Figure 1a displays that the natural hierarchical structure of raw wood is highly uneven with a maximum height of more than 200 μm (Figure 1b) and a roughness factor (Ra) of 38 μm. Moreover, some pores were found to be too large (the inset of Figure 1a) to support the weight of the water droplet that resulted in a homogeneous wetting, namely, Wenzel states [27]. Thus, this natural hierarchical structure is not applicable to directly construct the superhydrophobic surface. As expected, when the fluoroalkylsilane/silica suspension was cast and dried on the surface of this raw wood, the water CA on the resultant surface was only 132°, hence the superhydrophobicity cannot be achieved. Figure 1c demonstrates that a relatively uniform micro/nano hierarchical structure can be obtained by sanding the raw wood with 240-grit sandpapers. Figure 1d manifests that the maximum height and Ra of the modified surface structure of sanded wood decreased to be less than 100 μm and 5.69 μm, respectively. The air in the uniform pores can reduce contact areas between water droplets and surface, resulting in the Cassie–Baxter state [28]. Benefiting from this state, the coated surface obtained by drying the fluoroalkylsilane/silica suspension on the sanded wood exhibited a typical superhydrophobic feature with a water CA of 160° and a sliding angle below 10°.

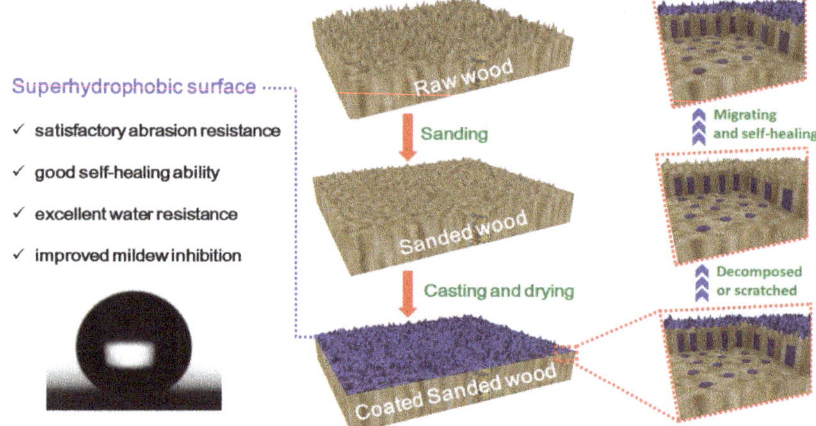

Scheme 1. Schematic procedure for the preparation of the superhydrophobic wood surface (middle) and its self-healing mechanism (right).

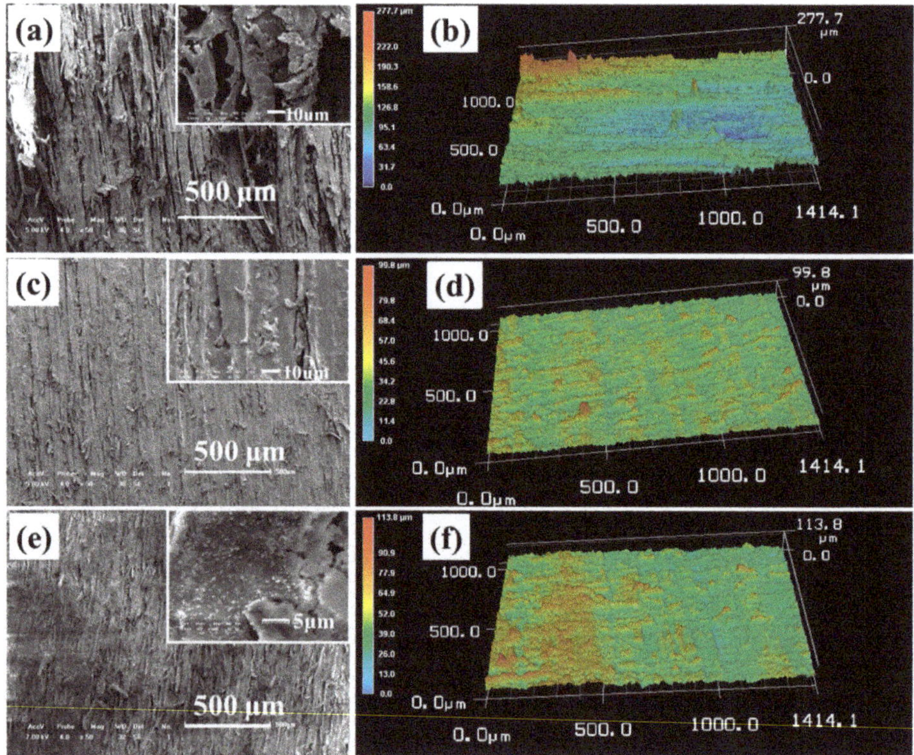

Figure 1. (**a**) Scanning electronic microscopy (SEM) and (**b**) 3D optical laser microscope system (OPM) images of raw wood. (**c**) SEM and (**d**) 3D OPM images of sanded wood. The insets are magnified SEM images with scale bars at 10 μm. (**e**) SEM and (**f**) 3D OPM image of sanded wood coated with KH1322 fluoroalkylsilane and silica.

To understand the detailed formation mechanism of this superhydrophobic surface, its three-dimensional surface morphology was investigated by SEM and OPM. The low magnification SEM image of coated sanded wood in Figure 1e displays a similar micromorphology compared with that of the sanded wood without coating (Figure 1c). However, we can find some new particles on the surface in its high magnification SEM image (the inset of Figure 1e). These particles should be the silane oligomers produced by the KH1322 hydrolysis and silica particles. Their emergence led to an increase in the roughness. As a result, both the maximum height (113.8 μm) and Ra (10.21 μm) of coated sanded wood in the OPM image of Figure 1f have a slight increase compared with the data of the sanded wood without coating (Figure 1d).

Usually, the raw wood surface is hydrophilic due to its naturally porous structure with abundant hydroxyl groups [25]. Thus, the chemical polarity of the surface must be changed to achieve superhydrophobicity. The comparison of FITR spectra in Figure 2 reveals that there are five new absorbance bands for the coated sanded wood compared with the raw wood. Among them, two peaks at 1267 and 617 cm^{-1} can be attributed to the vibrations of CF_3 and CF groups [29]. The other three absorbance peaks at 898, 805 and 467 cm^{-1} are ascribed to the vibrations of Si–O. It is believed that the molecular chains with these hydrophobic groups combine with the uniform micro/nano hierarchical structure of the sanded wood to achieve the construction of the superhydrophobic wood surface.

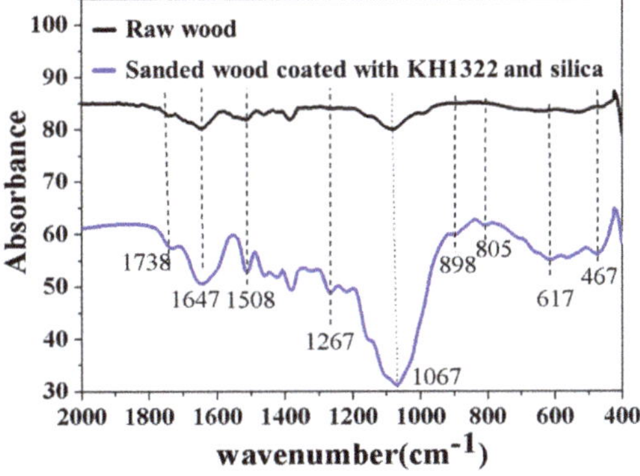

Figure 2. Comparison of Fourier transform infrared spectroscopy (FTIR) spectra between raw wood and sanded wood coated with KH1322 fluoroalkylsilane and silica.

3.2. The Effect of Silica on the Surface Wettability

In this work, the silica was added to prepare the hydrophobic surface. Thus, the effect of silica content on the surface wettability was investigated. It was found that the water CA increased from 152° to 160° when the ratio of silica to KH1322 increased from 0.5% to 2%. However, when the silica ratio further elevated from 2% to 4%, the water CA remained nearly unchanged at 160°. The silica ratio and the total consumption of KH1322 plus silica for superhydrophobic surfaces in the present study were much less than those in previous works [29]. The minimum silica ratio used in this work reached as low as 0.5%, and the total consumption per unit area of KH1322 plus silica was only 4 g·m^{-2}, indicating a low-cost characteristic of the preparation method.

Unlike previously reported method based on silica to prepare the superhydrophobic surface, the silica plays a major role in accelerating the hydrolysis and condensation of KH1322 in this work, rather than building micro-nano morphologies. Therefore, the superhydrophobic surface could be obtained even under a low silica ratio of 0.5%. The detailed reason for the silica ratio affecting the surface wettability can be explained as follows. When the silica particles were added into KH1322 solutions, the hydrolysis of KH1322 was developed more quickly. The hydrophilic molecular chains containing C–O groups reduced and the wood surface was covered by more hydrophobic molecular chains containing Si–O groups, resulting in an increase in the water CAs correspondingly. When the silica feeding ratio further increased to 2%, almost all -OC$_2$H$_5$ groups from KH1322 were hydrolyzed to generate ethanol, which was completely removed from the wood surface during the drying process, thus the C-O groups could hardly be detected. At this point, the number of hydrophobic groups reached the maximum value. In order to confirm the above assumptions, XPS was used to analyze the element change on the wood surfaces. In Figure 3a, the C 1s XPS broad peak of sanded wood coated with KH1322 and 0.5% SiO$_2$ can be fitted to four peaks with binding energies of 283.8, 285.6, 287.3 and 290.9 eV, which can be indexed to the functional groups of Si-C, C-C, C-O, and CF$_2$, respectively [30]. Here, the ratio of hydrophilic carbons (C-O groups) to all carbon atoms was calculated to be 3.46% based on the peak areas in Figure 3a. However, for the sanded wood coated with KH1322 and 2% silica, the hydrophobic C atoms (Si-C, C-C, and CF$_2$ groups) can account for nearly 100% of all C atoms. Hence, the additional increase in the silica ratio to 4% no longer increased the ratio of hydrophobic groups, so the water CAs did not increase further.

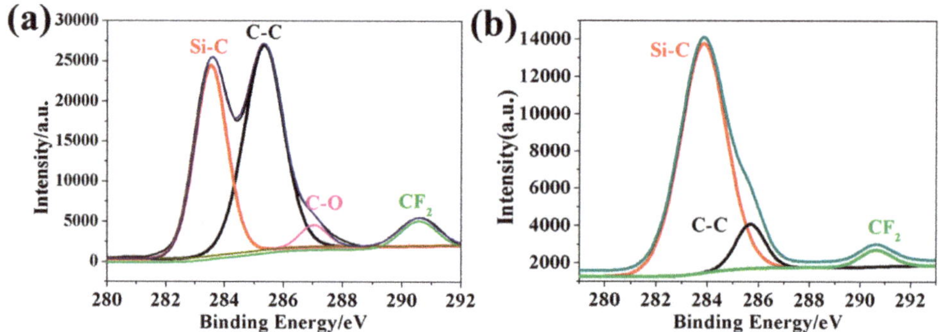

Figure 3. C 1s X-ray photoelectron spectroscopy (XPS) spectra of (**a**) sanded wood coated with KH1322 and 0.5% SiO$_2$ and (**b**) sanded wood coated with KH1322 and 2.0% SiO$_2$.

3.3. The Effect of Drying Temperatures on Surface Wettability

Usually, the drying temperature is a key factor that affects the formation of the coating. Therefore, the influence of drying temperature on the surface wettability was also researched. As shown in Figure 4a, the water CA increased from 157° and 160° to 170° with the elevation of the drying temperature from 100 °C and 120 °C to 130 °C. When the drying temperature was increased again to 140 °C, the water CA decreased slightly from 170° to 167°. The previous CA increase should result from the reduction of hydrophilic C-O groups. It is believed that the increase in the drying temperature from 100 °C to 120 °C accelerated the hydrolysis of KH1322. More -OC$_2$H$_5$ groups of KH1322 reacted with the water to release ethanol vapour. Therefore, the hydrophilic C-O groups reduced correspondingly, as illustrated by the diminished peak intensity of C-O groups at 1271 cm^{-1} (Figure 4b). When the drying temperature was further elevated to 130 °C, almost all -OC$_2$H$_5$ groups of KH1322 were hydrolyzed to generate ethanol vapour. The virtual disappearance of the C-O signal in Figure 4b demonstrates the removal of more hydrophilic groups. Moreover, the adsorption of hydrophobic groups (Si-O at 810 and 470 cm^{-1}) reached the maximum signal. These facilitate the formation of the superhydrophobic surface with a larger water CA. It could be inferred that the drying temperature of 130 °C is high enough to achieve the complete hydrolysis of KH1322. Thus, when the temperature further increased to 140 °C, the ratio of hydrophobic groups on the wood surfaces no longer increased, with the result that the water CA would not increase correspondingly.

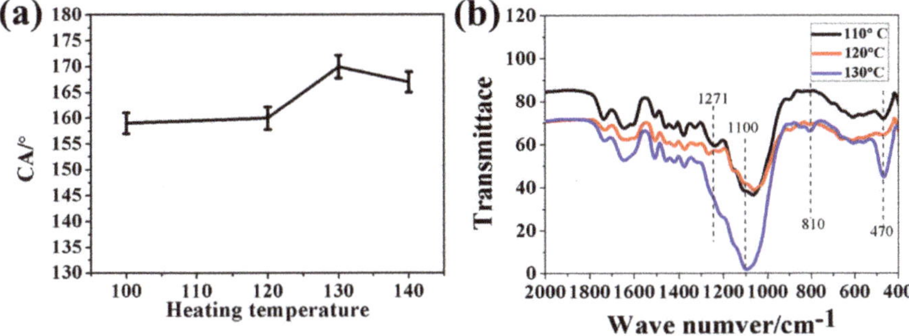

Figure 4. (**a**) The water contact angles (Cas) of the surfaces of coated sanded wood obtained at different drying temperatures, (**b**) Comparison of FTIR spectra of sanded woods coated with KH1322 fluoroalkylsilane and silica obtained at different drying temperatures.

Considering that hours of high-temperature drying may induce damages to some wood, the drying at room temperature (25 °C) was also tried to prepare the hydrophobic surface. It was found that the water CA on the as-obtained surface increased from 75° to 155° when the drying time was extended from 2 hours to 8 hours. This result revealed that the fabrication of a superhydrophobic surface on the wood can also be achieved at room temperature when the drying time is long enough. This suggests that the method can be extended to produce superhydrophobic surfaces for thermally unstable wood products.

3.4. Mechanical Robustness and Self-Healing of Superhydrophobic Surfaces

One obstacle to the widespread practical applications of superhydrophobic surfaces is the lack of enough robustness. To investigate the abrasion resistance of the as-obtained superhydrophobic surface, under the loading of 500 grams (pressure: 3.5 KPa), the hydrophobic surface of the coated sanded wood was oriented toward 240-grit sandpaper and moved a distance of 10 cm (each cycle) in a straight line, as shown in Figure 5a. Then, the water CAs of the resulting wood surface were tested. This process was repeated for 45 cycles, and the test results of water CAs were recorded in Figure 5b. One can see that after 45 cycles, the water CA was able to retain a high value of 153°, and the sliding angle was still less than 10°, indicating that the surface of the coated sanded wood maintained its superhydrophobicity. These results demonstrate the satisfactory abrasion resistance of the as-prepared superhydrophobic surface.

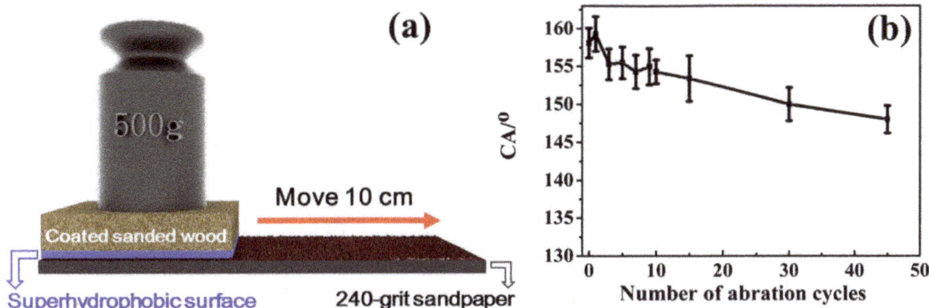

Figure 5. (**a**) Illustration of the abrasion resistance test for the superhydrophobic surface. (**b**) Water CAs as a function of the number of abrasion cycles.

As an important indicator affecting service life, the self-healing ability of the superhydrophobic surface was examined by alkali etching. After it was soaked in alkali solution for two hours, the water CA of the resulting surface decreased from 160° to 0°, indicating that this surface had lost its superhydrophobicity. However, after this damaged surface was exposed in an ambient environment for 8 hours, its water CA and sliding angles returned to 160° and <10°, respectively, demonstrating that the original superhydrophobicity of the damaged surface was restored. The recovery of superhydrophobicity should be attributed to the unique inherent porous structure of wood. This intrinsic porous structure can preserve a large number of fluoroalkylsilane moieties as healing agents in wood pores (such as cell cavities and grooves) beneath the superhydrophobic surface. Once the primary top fluoroalkylsilane layer is decomposed or scratched, the internal preserved healing agents in the cell cavities and grooves will migrate to the wood surface (as illustrated by the right part of Scheme 1) to minimize the surface free energy due to the exposure to the hydrophobic air [31,32]. As a result, the superhydrophobicity is healed at room temperature. In this way, the damaged surface recovered its superhydrophobicity at room temperature when the time duration was long enough for the healing agents in pores to migrate onto the wood surface. As displayed in Figure 6a, the etching–healing process can be repeated for nine cycles without decreasing the superhydrophobicity of the self-healed surface. It is believed that the robust internal porous microstructure of wood is essential for the recovery of superhydrophobicity as shown in Figure 6b,c. In addition, it was found that when the temperature increased to 100 °C and 140 °C, the recovery time of the superhydrophobicity reduced to 2 h and 1 h, respectively. This indicates that this self-healing is temperature-dependent, with a more accelerated self-healing process under higher temperatures and vice versa. At present, the self-healing of the as-constructed superhydrophobic surface can be accomplished at room temperature. Thus, this self-healing can be applied to some wooden decorative materials that are sensitive to heat and ultraviolet radiation. Furthermore, if repeated self-healing runs out of healing agents, the suspension of KH1322 fluoroalkylsilane and silica can be cast and dried onto the surface again to replenish the healing agents.

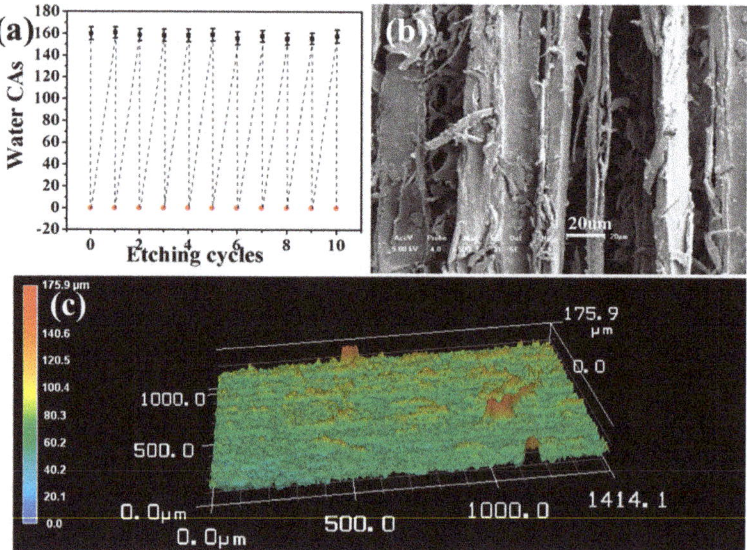

Figure 6. (a) Water CAs of the surface after repeated alkaline etching and self-healing at room temperature. (b) SEM and (c) 3D OPM images of the surface of coated sanded wood after alkaline etching.

3.5. Improvement in Water and Mildew Resistance of Superhydrophobic Surfaces

As a natural biomass material with many hydrophilic groups, raw wood is apt to absorb water and swell in dimension when in contact with water [24]. In order to simulate the rinse effect of rainwater on the surface of outdoor wood products, the as-obtained superhydrophobic surface of coated sanded wood was washed by running water for 20 minutes (Figure 7a). No residual water can be observed on the superhydrophobic surface after this process (Figure 7b). By contrast, excessive water remained on the surface of raw wood after the same treatment (Figure 7c–d). These results indicate the excellent water resistance of the as-constructed superhydrophobic surface.

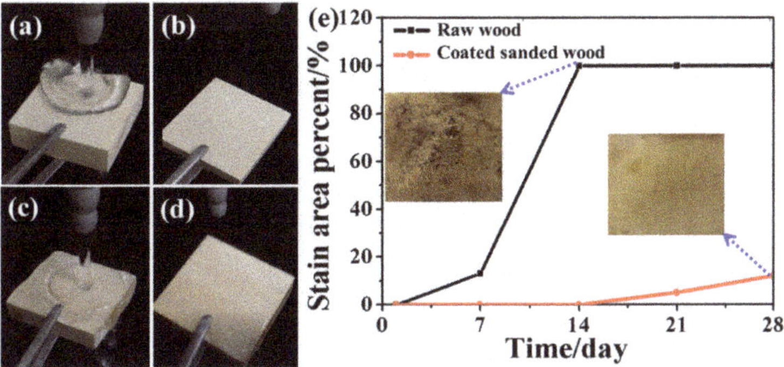

Figure 7. Photographs of the superhydrophobic wood surface (**a**) during water washing and (**b**) after water washing. Photographs of the raw wood surface (**c**) during water washing and (**d**) after water washing. (**e**) Comparison of mildew infection on the surface of raw wood and coated sanded wood in a humid environment.

Wood contains protein, starch, cellulose, etc., which can provide nutrients for microbial growth, so it is easily infected by mildew, especially when the wood surface is wet. Here, the mildew resistance ability of the as-prepared superhydrophobic wood surface was studied and compared with that of the raw wood surface. The results in Figure 7e demonstrate that about 13% of the surface area of the raw wood had been infected by mildew on the 7th day when it was placed in an environment with a humidity of 90% and a temperature of 30 °C, and the percentage of the infected area increased to 100% on the 14th day. By contrast, the coated sanded wood with the superhydrophobic surface was not infected by mildew until the 14th day and only 12% of its surface was infected by mildew on the 28th day under the same environment. The improved mildew resistance could be due to the fact that the superhydrophobic surface repelled water and reduced water adsorption thereby inhibiting the growth of mildew. In addition, this surface can maintain its superhydrophobic ability as shown by only minimal contact angle fluctuations (Figure 8) after a long period of high-intensity UV irradiation which demonstrates the fine UV durability of this superhydrophobic surface. Therefore, it is highly promising to employ this superhydrophobic surface to prolong the service life of wood-based materials.

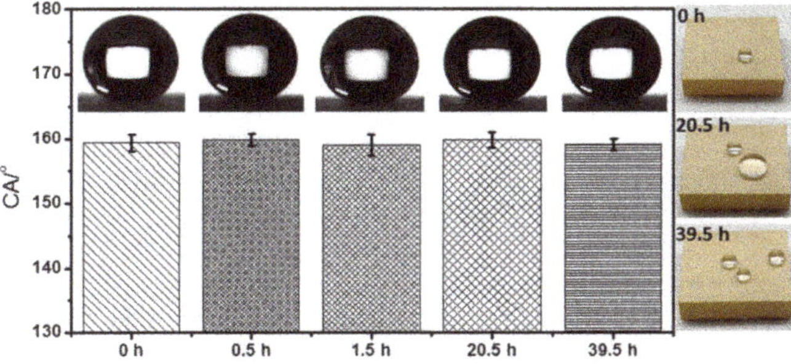

Figure 8. Water CAs and photographs of the coated sanded wood surface after being irradiated with 365 nm UV lamp (11.6 mW·cm^{-2}) for different times.

4. Conclusions

In this work, a novel superhydrophobic surface has been fabricated successfully by casting and drying a composite suspension of KH1322 and silica on the modified hierarchical structure of wood. Due to the full use of the natural hierarchical structure of wood, the whole process needs neither complicated equipment nor complex procedures to construct the micro/nano composite structure. Only a very low content of inorganic matter is needed to achieve superhydrophobicity. Furthermore, the fabrication of a superhydrophobic surface can be achieved at room temperature. More importantly, the as-prepared superhydrophobic surface exhibits a satisfactory resistance to abrasion and is able to self-heal at room temperature upon damage. If the healing agents are depleted, the surface can restore its self-healing ability by casting the fluoroalkylsilane/silica composite suspension to replenish the healing agents. When used for wood protection, this superhydrophobic surface greatly improves both mildew inhibition and water resistance of wood, thereby prolonging the service life of wood-based materials. The excellent performance of the as-constructed superhydrophobic surface, the facile and environmentally friendly fabrication process, and the low cost, make this method highly suitable for the protection of various wood-based materials.

Author Contributions: Conceptualization, Z.S.; Data curation, J.W.; Formal analysis, C.M.; Funding acquisition, Z.S. and H.C.; Investigation, J.W., Y.L., Q.C. and C.M.; Methodology, J.W.; Project administration, Z.S. and H.C.; Resources, Q.C. and L.C.; Supervision, Z.S. and H.C.; Validation, L.C.; Visualization, Y.L.; Writing—original draft, Z.S.; Writing—review and editing, H.C. All authors have read and agreed to the published version of the manuscript.

Funding: This work was supported by the Zhejiang Provincial Key Research and Development Project (2019C02037), National Key Project (2017YFD0601105), Zhejiang Provincial Natural Science Foundation of China (LY20E020004), Zhejiang A&F University Scientific Research Training Program for Undergraduates (KX20180113), Jiangsu North Special Project from Jiangsu Suqian (BN2016176), and 151 Talent Project of Zhejiang Province.

Conflicts of Interest: The authors declare no conflict of interest.

References

1. Lu, Y.; Sathasivam, S.; Song, J.; Crick, C.R.; Carmalt, C.J.; Parkin, I.P. Robust self-cleaning surfaces that function when exposed to either air or oil. *Science* **2015**, *347*, 1132–1135. [CrossRef] [PubMed]
2. Hou, X.; Hu, Y.; Grinthal, A.; Khan, M.; Aizenberg, J. Liquid-based gating mechanism with tunable multiphase selectivity and antifouling behaviour. *Nature* **2015**, *519*, 70–73. [CrossRef] [PubMed]
3. Wu, M.; Ma, B.; Pan, T.; Chen, S.; Sun, J. Silver-nanoparticle-colored cotton fabrics with tunable colors and durable antibacterial and self-healing superhydrophobic properties. *Adv. Funct. Mater.* **2015**, *26*, 569–576. [CrossRef]
4. Zhang, X.; Li, Z.; Liu, K.; Jiang, L. Bioinspired multifunctional foam with self-cleaning and oil/water separation. *Adv. Funct. Mater.* **2013**, *23*, 2881–2886. [CrossRef]
5. Ruan, C.; Ai, K.; Li, X.; Lu, L. A Superhydrophobic sponge with excellent absorbency and flame retardancy. *Angew. Chem. Int. Ed.* **2014**, *53*, 5556–5560. [CrossRef] [PubMed]
6. Liu, Y.; Wang, X.; Fei, B.; Hu, H.; Lai, C.; Xin, J.H. Bioinspired, Stimuli-responsive, multifunctional superhydrophobic surface with directional wetting, adhesion, and transport of water. *Adv. Funct. Mater.* **2015**, *25*, 5047–5056. [CrossRef]
7. Hooda, A.; Goyat, M.S.; Pandey, J.K.; Kumar, A.; Gupta, R. A review on fundamentals, constraints and fabrication techniques of superhydrophobic coatings. *Prog. Org. Coat.* **2020**, *142*, 105557. [CrossRef]
8. Youngblood, J.P.; McCarthy, T.J. Ultrahydrophobic polymer surfaces prepared by simultaneous ablation of polypropylene and sputtering of poly(tetrafluoroethylene) using radio frequency plasma. *Macromolecules* **1999**, *32*, 6800–6806. [CrossRef]
9. Folkers, J.P.; Laibinis, P.E.; Whitesides, G.M. Self-assembled monolayers of alkanethiols on gold: Comparisons of monolayers containing mixtures of short- and long-chain constituents with methyl and hydroxymethyl terminal groups. *Langmuir* **1992**, *8*, 1330–1341. [CrossRef]
10. Sharma, A.; Reiter, G. Instability of thin polymer films on coated substrates: Rupture, dewetting, and drop formation. *J. Colloid Interf. Sci* **1996**, *178*, 383–399. [CrossRef]

11. Shon, Y.-S.; Lee, S.; Colorado, R.; Perry, S.S.; Lee, T.R. Spiroalkanedithiol-Based SAMS reveal unique insight into the wettabilities and frictional properties of organic thin films. *J. Am. Chem. Soc.* **2000**, *122*, 7556–7563. [CrossRef]

12. Miyauchi, M.; Kieda, N.; Hishita, S.; Mitsuhashi, T.; Nakajima, A.; Watanabe, T.; Hashimoto, K. Reversible wettability control of TiO_2 surface by light irradiation. *Surf. Sci.* **2002**, *511*, 401–407. [CrossRef]

13. Mutel, B.; Taleb, A.B.; Dessaux, O.; Goudmand, P.; Gengembre, L.; Grimblot, J. Characterization of mixed zinc-oxidized zinc thin films deposited by a cold remote nitrogen plasma. *Thin Solid Films* **1995**, *266*, 119–128. [CrossRef]

14. Gu, Z.-Z.; Uetsuka, H.; Takahashi, K.; Nakajima, R.; Onishi, H.; Fujishima, A.; Sato, O. Structural color and the lotus effect. *Angew. Chem. Int. Ed.* **2003**, *42*, 894–897. [CrossRef] [PubMed]

15. Wang, J.; Wen, Y.; Feng, X.; Song, Y.; Jiang, L. Control over the wettability of colloidal crystal films by assembly temperature. *Macromol. Rapid Commun.* **2006**, *27*, 188–192. [CrossRef]

16. Wang, J.; Wen, Y.; Hu, J.; Song, Y.; Jiang, L. Fine Control of the wettability transition temperature of colloidal-crystal films: From superhydrophilic to superhydrophobic. *Adv. Funct. Mater.* **2007**, *17*, 219–225. [CrossRef]

17. Deng, X.; Mammen, L.; Butt, H.J.; Vollmer, D. Candle soot as a template for a transparent robust superamphiphobic coating. *Science* **2012**, *335*, 67–70. [CrossRef]

18. Deng, X.; Mammen, L.; Zhao, Y.; Lellig, P.; Müllen, K.; Li, C.; Butt, H.J.; Vollmer, D. Transparent, thermally stable and mechanically robust superhydrophobic surfaces made from porous silica capsules. *Adv. Mater.* **2011**, *23*, 2962–2965. [CrossRef]

19. Latthe, S.; Terashima, C.; Nakata, K.; Sakai, M.; Fujishima, A. Development of sol-gel processed semi-transparent and self-cleaning superhydrophobic coatings. *J. Mater. Chem. A* **2014**, *2*, 5548–5553. [CrossRef]

20. Geng, Z.; He, J.; Xu, L.; Yao, L. Rational design and elaborate construction of surface nano-structures toward highly antireflective superamphiphobic coatings. *J. Mater. Chem. A* **2013**, *1*, 8721–8724. [CrossRef]

21. Chen, L.; Sun, X.; Hang, J.; Jin, L.; Shang, D.; Shi, L. Large-scale fabrication of robust superhydrophobic coatings with high rigidity and good flexibility. *Adv. Mater. Interfaces* **2016**, *3*, 1500718. [CrossRef]

22. Liao, Y.; Zheng, G.; Huang, J.J.; Tian, M.; Wang, R. Development of robust and superhydrophobic membranes to mitigate membrane scaling and fouling in membrane distillation. *J. Membrane Sci.* **2020**, *601*, 117962. [CrossRef]

23. Kobina Sam, E.; Kobina Sam, D.; Lv, X.; Liu, B.; Xiao, X.; Gong, S.; Yu, W.; Chen, J.; Liu, J. Recent development in the fabrication of self-healing superhydrophobic surfaces. *Chem. Eng. J.* **2019**, *373*, 531–546. [CrossRef]

24. Shen, Y.; Yitian, W.; Zhehong, S.; Hao, C. Fabrication of Self-healing superhydrophobic surfaces from water-soluble polymer suspensions free of inorganic particles through polymer thermal reconstruction. *Coatings* **2018**, *8*, 144. [CrossRef]

25. Lewin, M.; Goldstein, I.S. Wood Structure and Composition. *Science* **1991**, *152*, 500–502. [CrossRef]

26. Barbero-López, A.; Chibily, S.; Tomppo, L.; Salami, A.; Ancin-Murguzur, F.J.; Venäläinen, M.; Lappalainen, R.; Haapala, A. Pyrolysis distillates from tree bark and fibre hemp inhibit the growth of wood-decaying fungi. *Ind. Crop. Prod.* **2019**, *129*, 604–610. [CrossRef]

27. Wenzel, R.N. Resistance of solid surfaces to wetting by water. *Ind. Eng. Chem.* **1936**, *28*, 988–994. [CrossRef]

28. Cassie, A.B.D.; Baxter, S. Wettability of porous surfaces. *Trans. Faraday Soc.* **1944**, *40*, 546–551. [CrossRef]

29. Fu, Q.; Wu, X.; Kumar, D.; Ho, J.W.; Kanhere, P.D.; Srikanth, N.; Liu, E.; Wilson, P.; Chen, Z. Development of sol-gel icephobic coatings: Effect of surface roughness and surface energy. *ACS Appl. Mater. Interfaces* **2014**, *6*, 20685–20692. [CrossRef]

30. Esteves, A.C.C.; Luo, Y.; van de Put, M.W.P.; Carcouët, C.C.M.; With, G.D. Self-replenishing dual structured superhydrophobic coatings prepared by drop-casting of an all-in-one dispersion. *Adv. Funct. Mater.* **2014**, *24*, 986–992. [CrossRef]

31. Wang, H.; Xue, Y.; Ding, J.; Feng, L.; Wang, X.; Lin, T. Durable, Self-healing superhydrophobic and superoleophobic surfaces from fluorinated-decyl polyhedral oligomeric silsesquioxane and hydrolyzed fluorinated alkyl silane. *Angew. Chem. Int. Ed.* **2011**, *50*, 11433–11436. [CrossRef] [PubMed]
32. Li, Y.; Chen, S.; Wu, M.; Sun, J. All spraying processes for the fabrication of robust, self-healing, superhydrophobic coatings. *Adv. Mater.* **2014**, *26*, 3344–3348. [CrossRef] [PubMed]

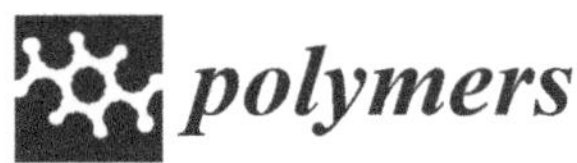 *polymers*

Article

Pyrolytic Kinetics of Polystyrene Particle in Nitrogen Atmosphere: Particle Size Effects and Application of Distributed Activation Energy Method

Lin Jiang [1], Xin-Rui Yang [1], Xu Gao [1], Qiang Xu [1], Oisik Das [2], Jin-Hua Sun [3,*] and Manja Kitek Kuzman [4]

[1] School of Mechanical Engineering, Nanjing University of Science and Technology, Nanjing 210094, China; ljiang@njust.edu.cn (L.J.); yxr1994@njust.edu.cn (X.-R.Y.); gaoxu@njust.edu.cn (X.G.); xuqiang@njust.edu.cn (Q.X.)

[2] The Division of Material Science, Department of Engineering Sciences and Mathematics, Luleå University of Technology, 97187 Luleå, Sweden; oisik.das@ltu.se

[3] State key laboratory of Fire Science, University of Science and Technology of China, Hefei 264000, China

[4] Department of Wood Science and Technology, Biotechnical Faculty, University of Ljubljana, Jamnikarjeva 101, 1000 Ljubljana, Slovenia; manja.kuzman@bf.uni-lj.si

* Correspondence: sunjh@ustc.edu.cn; Tel.: +86-551-63606425

Received: 11 December 2019; Accepted: 8 February 2020; Published: 12 February 2020

Abstract: This work was motivated by a study of particle size effects on pyrolysis kinetics and models of polystyrene particle. Micro-size polystyrene particles with four different diameters, 5, 10, 15, and 50 μm, were selected as experimental materials. Activation energies were obtained by isoconversional methods, and pyrolysis model of each particle size and heating rate was examined through different reaction models by the Coats–Redfern method. To identify the controlling model, the Avrami–Eroféev model was identified as the controlling pyrolysis model for polystyrene pyrolysis. Accommodation function effect was employed to modify the Avrami–Eroféev model. The model was then modified to $f(\alpha) = n\alpha^{0.39n - 1.15}(1 - \alpha)[-\ln(1 - \alpha)]^{1 - 1/n}$, by which the polystyrene pyrolysis with different particle sizes can be well explained. It was found that the reaction model cannot be influenced by particle geometric dimension. The reaction rate can be changed because the specific surface area will decrease with particle diameter. To separate each step reaction and identify their distributions to kinetics, distributed activation energy method was introduced to calculate the weight factor and kinetic triplets. Results showed that particle size has big impacts on both first and second step reactions. Smaller size particle can accelerate the process of pyrolysis reaction. Finally, sensitivity analysis was brought to check the sensitivity and weight of each parameter in the model.

Keywords: particle size; model free; model fitting; avrami–eroféev; DAEM

1. Introduction

To meet the needs of of society, various kinds of advanced materials with different functions have been invented and updated greatly. In the ultrafine materials research area, researchers have tried to generate particles with even smaller diameters. After a normal particle is processed by ultrafine technology, particles will own some unique characteristics, including large specific surface area and high chemical activity. The peculiar physical and chemical characteristics make ultrafine particles the focus of advanced materials nowadays. During the processes of particles' industrial manufacture, storage, and transportation, particles with different sizes behave differently when considering their safety concerns. Therefore, particle size effects are essential influence factors needed to be considered when researchers explore particle thermal safety problems. The chemical kinetics and reaction model can be greatly influenced by particle size [1]. After the block is processed by ultrafine processing technology, particle specific surface area can be greatly increased, which can influence combustible pyrolysis and reaction rates when heating, and even the reaction model and products can be changed [2,3].

So, the work reported here was motivated by a study of particle size effects on pyrolysis behavior, chemical kinetics, and reaction model when surrounded with heating. Micro-size polystyrene particles with four different diameters were selected as typical particle materials. Activation energies were obtained by several different isoconversional methods. The pyrolysis model of each particle size and heating rate was examined by nineteen different reaction model candidates by the Coats–Redfern methods, among which the three best models were then selected, and the reaction model function was then reconstructed by selected models. The particle size effects on kinetics and reaction model could be concluded. To separate step reactions from whole reaction and identify their distributions to kinetics, a distributed activation energy method was then introduced to calculate the weight factor and kinetic triplets.

2. Literature Review

Polystyrene is a commonly-used polymer material in daily life, which is usually employed as thermal insulation materials in extruded or expandable formation, whose kinetics and reaction mechanism have been studied. Jiao et al. studied the kinetics and volatile products of expandable polystyrene and extruded polystyrene with TGA and TGA-MS-FTIR, respectively. They found that the activation energies with conversions of expandable polystyrene are a little higher than extruded polystyrene, which means expandable polystyrene is a little more stable than the extruded one. During the pyrolysis process, small molecules including CO, C_2H_3, C_2H_5, and phenyl were detected [4]. After this, Jiao and Sun explored the reaction mechanism of polystyrene during the pyrolysis process. It was found that two pyrolysis reactions exist during the whole heating process. One is the small pyrolysis of styrene monomers around 275 °C, and the other is breakage of the main chain and large amounts of styrene generation around 430 °C [5]. Cheng et al. compared the thermal degradation behaviors of micron polymethyl methacrylate (PMMA) and polystyrene (PS) by a traditional kinetics method. They found that the particle size diameters can result in the decrease of activation energies, but have no obvious influence on pre-exponential factors [6]. Other researchers have conducted related studies about particle size effects on material pyrolysis behavior. Shen et al. [7] investigated the wood particle size effects on the yield of bio-oil production. Results showed that the yield of bio-oil production can decrease with the particle size increasing, among which the light bio-oil fractions increased and the heavy bio-oil decreased. Marcilla et al. [8] tested different sizes milled powders of almond shells and olive stones. They found that the milling process can provoke the structure damage of both biomasses, and thus cause the difference in thermal behavior. Also, the milling process may cause the increase of mineral substance. Blasi [9] investigated the particle size and heating rate effects on cellulose pyrolysis by means of a computational model. Three main regimes of particle sizes were found to control pyrolysis processes, including thermally thick, thermally thin, and pure kinetic control, which were adjudged by particle size and heating rate conditions. Hanson [10] studied particle size effects on pyrolysis of coal, and found that a smaller particle was more likely to produce char residue larger than itself. For larger particle pyrolysis, it is more likely to produce a fragment. Yu et al. [11] ground the coal sample by a planetary ball mill, and the coal samples were classified into three groups according to different ground particle sizes. They found that particles with different sizes contain different carbon and ash contents, which is resulted by the characteristics of coal's uneven texture and solidity.

Most selected samples of previous pyrolysis studies relating particle size effects were self-ground in the laboratory, among which biomass and coal were mostly employed. During the grinding process, it is hard to form particles with uniform shape and component, as these solids have uneven density and distribution. This can result in that the particles employed in thermal analysis experiments do not have uniform distribution, which can definitely cause thermal analysis profiles fluctuations and bad data repeatability. In this study, the polystyrene sample we used was produced by Suzhou Nanomicro Technology Co., Ltd. The particles were produced with uniform shape and diameter, the diameters of which were 5, 10, 15, and 50 μm. Uniform diameter can guarantee the veracity and reliability of experimental results. More details about experimental sample particle size can be found in Section 4. Most publications including the above reviewed ones preferred to employ a traditional kinetics method when dealing with polymer pyrolysis kinetics problems. However, for the case of polymer pyrolysis, there must be more than one reaction during the pyrolysis procedure. So, in this study, after the traditional kinetics analysis we will introduce distributed activation energy model to explore PS pyrolysis kinetics to distinguish the weight of each sub reaction.

3. Traditional Kinetic Methods

Thermogravimetric analysis (TGA) apparatus can heat the sample with a fixed heating rate and gas flow to blow off the volatiles, and record the instant mass loss. The mass conversion at a certain time can be calculated by instant mass loss divided by total mass loss. The pyrolysis reaction can be expressed by the arithmetic product of two functions, including reaction rate constant and reaction model,

$$d\alpha/dt = A\exp[-E_a/(RT)]f(\alpha) \tag{1}$$

where A, E_a, and R are the pre-exponential factor, the apparent activation energy, and the gas constant, respectively. By TGA testing technique and kinetics calculation methods, the kinetic details can be obtained by measurement and parameterization. After processing natural logarithm to both sides of Equation (1) and then integrating, the reaction rate can yield to

$$g(\alpha) = \frac{A}{\beta} \int_{T_0}^{T} \exp(-\Delta E_a/RT)dT \tag{2}$$

in which the temperature part has no analytical solution. β means heating rate and equals to dT/dt. Many researchers have tried to solve the integration with reasonable approximations, commonly used methods like KAS [12,13], FWO [14,15], and Tang et al. [16,17] methods, among which different approximation solutions were employed to Equation (2) as listed in Table 1.

Table 1. Three commonly used isoconversional methods for activation energy calculation.

Methods	Expression	Description
Flynn–Wall–Ozawa method	$\log\beta = \log(AE_a/Rg(\alpha)) - 2.315 - 0.4567E_a/RT$	Modified general isoconversional equation by Doyle approximation.
Kissinger–Akahira–Sunose	$\ln(\beta/T^2) = \ln(AR/E_ag(a)) - E_a/RT$	Modified general isoconversional equation by Coats-Redfern approximation.
Tang et al.	$\ln(\beta/T^{1.894661}) = \ln[AE_a/Rg(\alpha)] + 3.635041 - 1.894661\ln E_a - 1.001450E_a/RT$	Tang et al. proposed an improved approximation for temperature integral.

Solved by numerical integration, kinetics parameters can be calculated more accurately with appropriate approximations. Vyazovkin et al. [18–20] developed an advanced isoconversional method which contains the temperature integration.

$$I(E_\alpha, T_\alpha) = \int_0^{T_\alpha} \exp\left(\frac{-E_a}{RT}\right)dT \tag{3}$$

$$I = \frac{E_a}{R}p(x) \tag{4}$$

Then the Vyazovkin method equation can be expressed as Equations (3) and (4), where x equals to E_α/RT. At a certain conversional extent, the value of apparent activation can be identified by minimizing the following formula,

$$\Omega(E_a) = \sum_{i=1}^{n} \sum_{\substack{j\neq i}}^{n} \frac{I(E_{a,\alpha}, T_{a,i})\beta_j}{I(E_{a,\alpha}, T_{a,j})\beta_i} \tag{5}$$

The temperature integration can be calculated after a series of transforms. Farjas and Roura [21] derived the six-order Padé approximation, which can give an absolute error less than 10^{-16} for $x > 12$

$$p(x) \approx \frac{\exp(-x)}{x} \times \left(\frac{x^5 + 40x^4 + 552x^3 + 3168x^2 + 7092x + 4320}{x^6 + 42x^5 + 630x^4 + 4200x^3 + 12600x^2 + 15120x + 5040} \right) \tag{6}$$

By Equations (3)–(6), for each conversion the minimization value can be obtained, by this method, a relative dependency between activation energy and conversion range can be obtained.

Model fitting method is a reaction model exploring method using well-known different theoretical reaction models to fit experimental α–T profiles, meanwhile for each model a set of activation energy and pre-exponential factor can be obtained. The Coats–Redfern method is one commonly used model-fitting method, which explores the asymptotic series expansion with the following formula,

$$\ln \frac{g(\alpha)}{T^2} = \ln\left(\frac{AR}{\beta E_a} [1 - (\frac{2RT^*}{E_a})] \right) - \frac{E_a}{RT} \tag{7}$$

where $g(\alpha)$ is the integral form of the reaction model as shown in Table 1, and T^* is the average temperature during all the heating process. For each reaction model as listed in Table 1, plotting $\ln[g(\alpha)/T^2]$ vs. $1/T$ can obtain sets of activation energy and pre-exponential factor. The model which has the best linearity with experimental profile is considered as the real reaction model.

There are nineteen commonly used reaction models in a kinetics area [5,6]. Each model will be used to fit the experimental formation with the obtainment of activation energy and pre-exponential factor. Then according to the fitness of experimental data and theoretical model calculation, one correlation coefficient can be obtained. So, for all nineteen models, there must exist one maximum correlation coefficient. In previous studies, usually the model with the maximum coefficient is identified as the ideal reaction model. However, sometimes the model with the maximum coefficient may be not the real reaction model, which can be checked by model reconstruction with experimental data. So, this model reconstruction [22–26] should be further processed to check if the obtained model can fit experimental profile well, and which procedure is necessary, but is usually ignored in previous related literatures.

The compensation effects means that there must exist one relation between the kinetics parameters that the change of activation energy causing a linear variation of the natural logarithm of the pre-exponential factor. The change of activation energy can be caused by the heating rate or model selection; however, they must be limited to one reaction. When several models are used in the same heating rate, several sets of activation energies and pre-exponential factors can be obtained, then the kinetics compensation effects can be created. The compensation effects between kinetics parameters can be expressed by the following formula,

$$\ln A_i = a + bE_i \tag{8}$$

where i means that the kinetic parameters are obtained from the i-th model, and parameters a and b are kinetics compensation parameters.

All models listed can be examined by Coats–Redfern method, by which nineteen corresponding sets of kinetics parameters can be obtained. Then the calculated activation energy and the pre-exponential factor can be used to evaluate the compensation effect formula parameters a and b. Based on the obtained compensation effects formula, the pre-exponential factor at each conversional extent can be evaluated according to the activation energies obtained by isoconversional methods.

4. Distributed Activation Energy Method

The kinetics methods introduced above belong to traditional kinetics methodology, which usually regards the pyrolysis reaction as one overall reaction, and the activation energy at a certain conversion extent is regarded as global activation energy. However, for polymer pyrolysis reaction, it is unreasonable to take one overall pyrolysis as one step reaction. Considering this reason, distributed activation energy method (DAEM) is adopted to separate the total reaction into several parallel reactions, which was originally adopted to separate the sub-reactions of biomass and coal [27–30]. The idea of distributed activation energy was firstly brought up by Vand [31], and then was developed to solve the pyrolysis problem of coal by Pitt [32].

DAEM assumes that the total reaction can consist of several parallel reaction groups. For each reaction group, it has its own sets of reactions on a molecular level. The decomposition reaction on molecular level can be expressed as,

$$d\left(\frac{m_i(t)}{m_{i*}}\right)/dt = A_i \exp(\frac{-E_i}{RT})\left(\frac{m* - m_i(t)}{m_{i*}}\right) \tag{9}$$

where i means the i_{th} molecular level reaction, $m_i(t)$ means the volatile mass fraction at time t, m_i^* means the total volatile mass fraction, A_i and E_i are the kinetic parameters for this reaction.

Integrating Equation (9) and assuming that the species i is one of the pool reaction group of component j, then we have the following expression of degradation of component j,

$$\alpha_j = 1 - \int_0^\infty \exp[-\int_{T0}^T \frac{A_j}{\beta_j} \exp(-\frac{E_i}{RT})dT]f(E)dE \tag{10}$$

$$\frac{d\alpha_j}{dT} = \int_0^\infty \frac{A_j}{\beta_j} \exp[-\frac{E_i}{RT} - \int_{T0}^T \frac{A_j}{\beta_j} \exp(-\frac{E_i}{RT})dT]f(E)dE \tag{11}$$

a_j means the conversion of component j. $f(E)$ means that the group reaction in component j follows the distribution functions $f(E)$, among which Gaussian distribution function is the earliest and most extensive applied one. The Gaussian distribution can be expressed as

$$f_G(E) = \frac{1}{\sigma\sqrt{2\pi}} \exp(-\frac{(E-E_0)^2}{2\sigma^2}) \tag{12}$$

where the distribution has the center at E_0 and the standard deviation σ. The random distribution is distributed symmetrically at the left and right sides of E_0. For Gaussian distribution, the range between $E_0 - 1.5\sigma$ and $E_0 + 1.5\sigma$ covers 99.7% random distribution. In this study, we consider 60 times standard deviation, which means the integration of Gaussian distribution ranges from $E_0 - 30\sigma$ and $E_0 + 30\sigma$. All equations about DAEM have temperature integration, which cannot be solved accurately in Equations (10)–(12). So, an approximation about temperature integration is also recommended here, here we calculate $p(x)$ the same as Equation (6). By calculating the j_{th} component DAEM mass loss rate, the overall reaction formula can be calculated as a linear reaction combination of all components,

$$\alpha = \sum_{j=1}^M c_j \alpha_j \tag{13}$$

$$d\alpha/dT = \sum_{j=1}^M c_j(d\alpha/dT)_j \tag{14}$$

where c_j means a weight factor equaling to the amount of volatiles formed from the j_{th} pseudo-component decomposition. It should be noted that Gaussian distribution is a symmetric distribution centered at E_0 from the shape of the curve. The distributed activation energy assumed that the total pyrolysis reaction is made of multiple parallel reactions, which is a reasonable assumption for polymer degradation.

5. Experimental

Micro polystyrene particles were provided by Nano-Micro Technology Co., Ltd., Suzhou, China. Four available diameters, 5, 10, 15, and 50 μm, were selected. All particles showed uniform size according to the scanning electron microscopy examining figures (http://en.nanomicrotech.com/). Furthermore, the particle size was double checked by a Laser Diffraction Particle Size Analyzer, SALD-2300, produced by Shimadzu Corporation, Kyoto, Japan. Particle size was identified by the light intensity distribution pattern of scattered light that is irradiated from sample particle surface when laser lights radiate them. The particle size diameters for four particle sizes are shown in Figure 1.

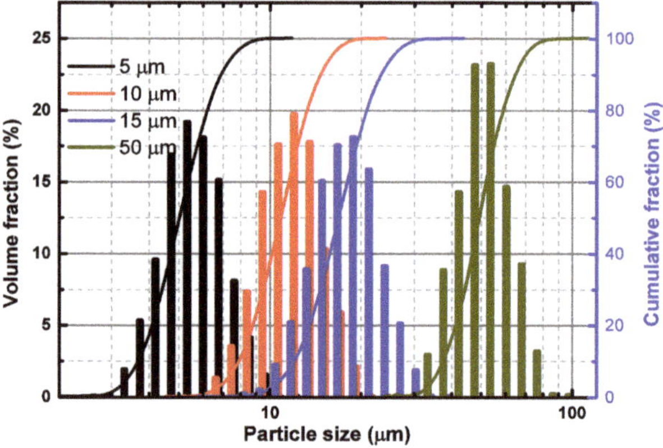

Figure 1. Particle size diameters of polystyrene particles with four different sizes, 5, 10, 15, and 50 μm.

The thermal degradation experiments were conducted on SDT Q600 instrument by TA Instruments (New Castle, USA). Experiments were performed in nitrogen atmosphere with 40 mL min^{-1} flow rate as purge gas and 20 mL min^{-1} as protective gas. Samples were heated in TGA with four heating rates, 3, 5, 7.5, and 10 K·min^{-1} from ambient temperature to 850 °C. An initial sample weight around 3 mg was guaranteed for all testing.

6. Results and Discussion

6.1. Pyrolytic Characteristics Observations

Figure 2 shows the TGA and Differential thermogravimetry (DTG) profiles of polystyrene with four different sizes in nitrogen atmosphere. Detailed thermal pyrolysis temperatures are listed in Table 2. We can find that the pyrolysis profiles of polystyrene with different sizes show similar variations. The DTG curve shows an obvious single peak, which can be identified as a one step reaction. In nitrogen atmosphere, the percentage of heat loss keeps around 90.71 ± 0.80% constantly. From the TGA and DTG curves, we can find that sample sizes cannot cause the change of the reaction process or TGA profiles obviously.

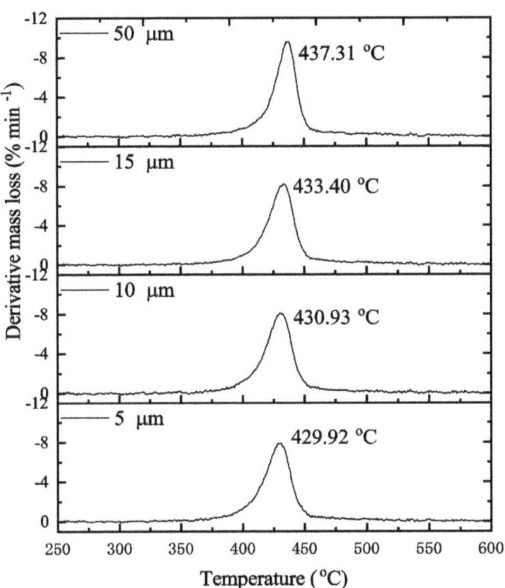

Figure 2. Differential thermogravimetry (DTG) rofiles of polystyrene pyrolysis in nitrogen atmosphere at 3 K·min^{-1} for 5, 10, 15, and 50 μm particle sizes.

Table 2. Characteristic temperature T_0, T_p and T_f for polystyrene pyrolysis determined from thermogravimetric analysis (TGA) profiles at different heating rates.

β (°C·min^{-1})	T_0 (°C)	T_p (°C)	T_f (°C)	α_{max}
5 μm				
3	367	430	535	91
5	351	438	545	92
7.5	369	443	528	91
10	349	447	530	92
10 μm				
3	378	431	572	92
5	378	437	531	91
7.5	381	458	528	90
10	379	460	534	91
15 μm				
3	368	433	524	90
5	369	440	534	91
7.5	386	446	529	90
10	382	449	529	91
50 μm				
3	375	437.31	537.50	90
5	381	444.23	534.04	90
7.5	385	451.87	530.90	89
10	359	455.82	535.81	90

For all the samples with different sizes, DTG curves show similar variations with one single peak, as the particles are produced from the same assignment. With the increase of particle size from 5 to 50 μm, the peak temperature increased monotonically. The 5 μm particle shows the minimum pyrolysis peak temperature and onset temperature, and 50 μm shows the maximum temperatures. For a polystyrene particle with a smaller diameter, it has a larger specific surface area, which means for the same sample masses, a smaller particle has more surface heated than a larger particle. For the TGA experiments in this study, we controlled all testing at the same weight at around 3 mg. Then for the 5 μm particle, its specific surface area is 10 times larger than 50 μm particle. Large specific surface area results in faster heat transfer and shorter time to trigger reaction.

6.2. Kinetics Parameters

The activation energies of polystyrene with four different sample sizes were calculated by five different commonly used isoconversional methods. Then, the dependences of activation energies on conversional extent for different calculated methods can be obtained. Figure 3a shows the activation energy calculation results based on different calculation methods. Five curves show the same variation with increasing conversional extent while Friedman results showed different variation from four other methods. The main reason that caused the deviation by the Friedman method with others is data noise brought during data differential process by total mass to use dα/dt, while the other four methods do not need a derivation step [33–35]. So FWO, KAS, Tang et al., and advanced Vyazovkin methods show almost the same calculation values, which proved the accuracy of the method calculation.

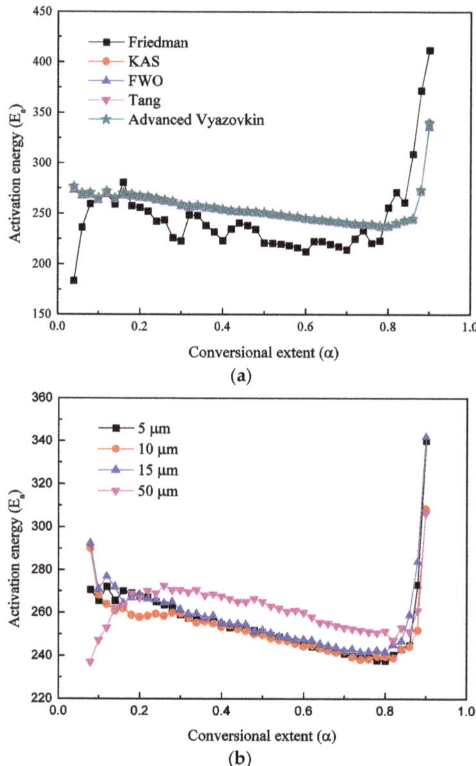

Figure 3. (**a**) Dependencies of the activation energy on extent of polystyrene conversion determined by five iso-conversional methods including KAS, FWO, Tang, Friedman, and advanced Vyazovkin methods. (**b**) Dependencies of the activation energy on conversional extent of four different size polystyrene determined by Vyazovkin methods.

Figure 3b shows the dependencies of activation energies on conversional extent for four different polystyrene particle sizes. The activation energy results were calculated by the advanced Vyazovkin method. The advanced isoconversional method developed by Vyazovkin is a commonly used thermal kinetics method, which excluded the influences of reaction model and needs for differential data to obtain activation energies. From Figure 3b, we can find that the variation tendencies are the same. During the conversional extent 0–0.2, the activation energies fluctuate significantly because a small amount of styrene molecules pyrolyzes and escapes from the main chain. During the conversional extent 0.2–0.85, four size samples show the same variation tendencies. With the increase of conversional extent, the activation energies of all four samples decrease almost linearly, which stage corresponds to the pyrolysis of polystyrene main body. When $\alpha > 0.85$, the activation energies increase rapidly with the increase of conversional extent. During this extent, the mass loss is mainly composed by polystyrene residue, which is hard to pyrolyze continuously and results in a rapid increase of activation energy.

During the main pyrolysis stage, with the increase of conversional extent, activation energies decrease slowly and linearly for all four sizes of samples. The activation energies of 5 and 10 µm are very close to each other for each conversional extent, both of which are smaller than activation energies of 15 µm particle size. The 50 µm size particle shows the maximum activation energies compared with another three sizes, which means that the reaction of 50 µm is the hardest to trigger. This difference on kinetics is mainly caused by their different specific surface area. For all four samples, 50 µm particle sample has the smallest specific surface area, therefore it has the maximum activation energies. The specific surface area of 5 µm particle size is 10 times than 50 µm particle size.

6.3. Model Fitting Method and Compensation Effects

By the isoconversional method calculation, we learned that the main pyrolysis stage ($a = 0.2$–0.85) of four sample sizes has constant activation energies where one existing reaction model may fit well. Isoconversional methods can only calculate the activation energies at a certain conversional extent, but fail to obtain the reaction model. With employment of the Coats–Redfern method, experimental data for four particle sizes can fit with all nineteen models. Then for each tested model, one set of activation energy and pre-exponential factor can be obtained. Three models with best linear coefficients for four sample sizes and heating rates are selected to list in Table 3, considering the linearity coefficient and activation energy appropriateness.

Table 3. Activation energies, pre-exponential, and corresponding linearity coefficient calculated by Coats–Redfern method for the three best models.

		$3\,^\circ$C min^{-1}				$5\,^\circ$C min^{-1}				$7.5\,^\circ$C min^{-1}				$10\,^\circ$C min^{-1}		
	Model	ln$A$	Ea	r^2	Model	ln$A$	Ea	r^2	Model	ln$A$	Ea	r^2	Model	ln$A$	Ea	r^2
	8	73.17	462.33	0.998	8	73.54	466.64	0.999	8	70.47	451.10	0.999	8	66.29	426.76	0.998
5 µm	12	45.99	304.33	0.998	12	46.41	307.16	0.999	12	44.47	296.76	0.998	12	41.76	280.52	0.998
	13	32.32	225.33	0.998	13	32.75	227.42	0.999	13	31.39	219.59	0.998	13	29.41	207.40	0.998
	8	75.06	469.28	0.990	8	71.09	448.69	0.988	8	71.57	453.59	0.987	8	68.97	439.20	0.988
10 µm	12	47.26	308.99	0.990	12	44.76	295.22	0.987	12	45.22	298.44	0.986	12	43.56	288.83	0.988
	13	33.28	228.84	0.989	13	31.52	218.48	0.987	13	31.95	220.87	0.986	13	30.77	213.65	0.987
	8	76.54	478.30	0.993	8	74.02	466.03	0.993	8	72.31	458.14	0.992	8	69.84	444.63	0.992
15 µm	12	48.26	315.00	0.993	12	46.73	306.77	0.992	12	45.71	301.48	0.992	12	44.15	292.45	0.992
	13	34.03	233.35	0.992	13	33.00	227.15	0.992	13	32.32	223.15	0.992	13	31.22	216.36	0.992
	8	80.36	502.22	0.989	8	77.78	489.67	0.984	8	76.64	485.47	0.983	8	75.33	478.41	0.983
50 µm	12	50.81	330.93	0.987	12	49.24	322.52	0.984	12	48.61	319.68	0.982	12	47.82	314.95	0.982
	13	35.96	245.28	0.988	13	34.89	238.94	0.983	13	34.51	236.78	0.982	13	33.99	233.22	0.981

Models 8, 12, 13 means first order model, Avrami–Eroféev model ($n = 1.5$), and Avrami–Eroféev model ($n = 2$).

From the kinetics calculation results listed in Table 3, we can see that the kinetics triplet calculations are greatly dependable on the model selection. The activation energies calculated by Model 13 are around 225 kJ·mol^{-1}, while for Model 8, the calculation result is around 463 kJ·mol^{-1}. From Table 3, we can find that for all cases of each particle size and heating rate, the best three models are the same, i.e., first-order model (F1), Avrami–Eroféev (A3/2), and Avrami–Eroféev (A2). All three models show good linearity, larger than 0.98. However, the A3/2 and A2 models are more reasonable than the F1 model because the activation energies obtained by Avrami–Eroféev are closer to the results by isoconversional methods. Also, the experimental $f(\alpha)$ shows an increase first then decrease variation, whose variation tendency only fits the Avrami–Erofeév model. Although the dimensional diffusion model has the similar variation, its magnitude is too small to fit with experimental results.

Calculation of activation energy at each conversional extent allows the reconstruction of the pyrolysis model, which acquires pre-exponential knowledge in advance. For one fixed reaction at one known heating rate, the activation energies have a linear relation with natural logarithm of the pre-exponential factor called compensation effect, which can be expressed as $\ln A_j = a + bE_j$, where a and b are constants for one reaction, $a = \ln k_{iso}$ and $b = 1/RT_{iso}$. k_{iso} is called artificial isokinetic rate and T_{iso} is defined as artificial isokinetic temperature. The subject j means the selected model. If the model we employed in calculation is not appropriately hypothesized, then the kinetic parameter artificial isokinetic temperature may locate out of the experimental temperature.

For each model, one set of kinetic parameters can be calculated. Then all the kinetics parameters can be used for modelling compensation effects, as listed in Table 4. Results showed that all the heating rates for each particle size have good linearity, as shown in Figure 4, which allows for the prediction of the pre-exponential factor at each conversional extent.

Table 4. The values of k_{iso} and T_{iso} by model fitting methods for pyrolysis of polystyrene particles with four sizes.

Particle Size	3 K min⁻¹		5 K min⁻¹		7.5 K min⁻¹		10 K min⁻¹	
	k_{iso}	T_{iso}	k_{iso}	T_{iso}	k_{iso}	T_{iso}	k_{iso}	T_{iso}
5 μm	0.001446	704.10	0.002360	711.74	0.003338	719.24	0.004233	722.76
10 μm	0.001407	697.77	0.002181	705.57	0.003241	712.49	0.004162	716.62
15 μm	0.001447	698.59	0.002301	706.27	0.003332	713.09	0.004265	717.20
50 μm	0.001482	701.24	0.00233	708.72	0.003393	715.68	0.004423	719.17

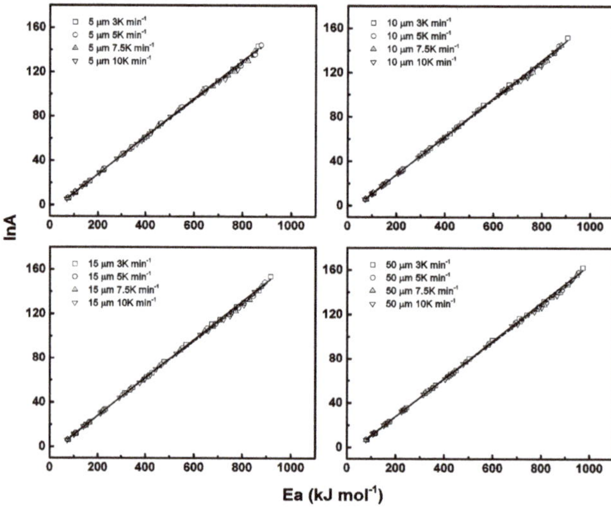

Figure 4. The isokinetic relationships (ln$A$ vs. E_a) obtained during degradation process using Coats–Redfern method for different particle sizes and heating rates.

6.4. Numerical Reconstruction

In Section 6.2, the activation energies at each conversional extent were obtained by isoconversional methods. Then, nineteen models were checked by the Coats–Redfern method to obtain a reasonable model describing polystyrene particle pyrolysis for cases of four different particle sizes. Avrami–Erofeév models (both A3/2 and A2) showed high linearity to the fitting with experimental profiles. Based on kinetic triplet results by different models, compensation effects could be employed to create numerical connection between activation energies and the pre-exponential factors, by which the pre-exponential factor at each conversional extent can also be clear. Based on the obtained pre-exponential factor on conversional extent, the calculated reaction model function can be obtained and compared with the theoretical reaction model function to examine the validity of the reaction model.

For all nineteen models, only the Avrami–Eroféev model can fit with experimental data during all conversional ranges; however, the results are still unsatisfactory to fit all heating rates well. This is because the most universally employed model in thermal kinetics is not applicable for reactions in/on media that are solid or porous structured [36]. So, when the pyrolysis kinetics are being described and refitted accurately, one accommodation function should be introduced to modify the model based on its known function. The real reaction model can be calculated by the arithmetic products of two functions, one is the accommodation function which can be expressed by α^m, and the other is a classical reaction model. The new kinetics model after modification can be expressed by

$$f(\alpha) = n\alpha^m(1 - \alpha)[-\ln(1 - \alpha)]^{1 - 1/n} \tag{15}$$

Figure 5 shows the comparisons of experimental $f(\alpha)$ points during all conversional ranges with theoretical profiles based on Equation (15) for four particle sizes. Results show that the experimental and theoretical data can match reasonably well during all conversional extents with two parameters m and n to describe the reaction model.

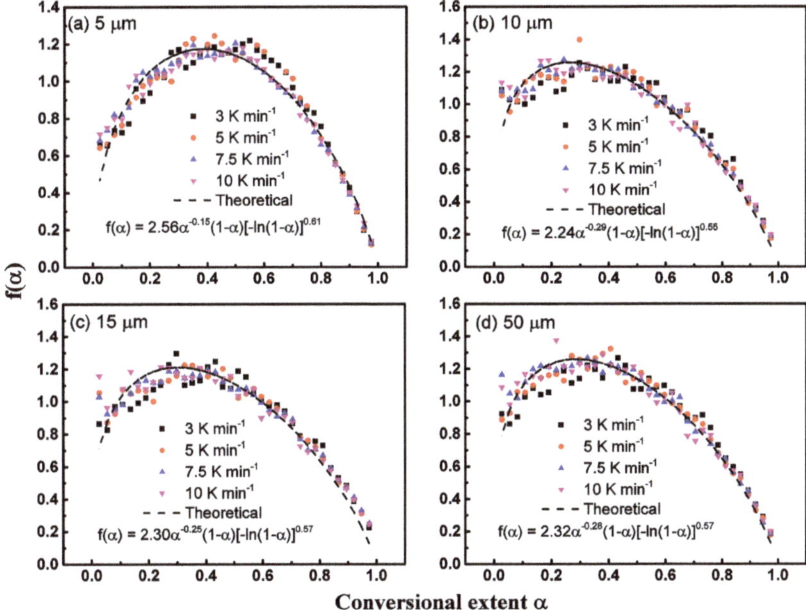

Figure 5. The experimental kinetics function $f(\alpha)$ reconstructed from isoconversional kinetic method of polystyrene pyrolysis for 3, 5, 7.5, and 10 K·min^{-1} heating rates. The dash line means the reconstructed profile of modified Avrami–Eroféev reaction model.

By further processing experimental data of each heating rate, sixteen sets of m and n parameters are obtained. We find that there is a roughly linear relationship between all m and n, which can be described by $m = 0.39n - 1.15$ with $R^2 = 0.92$. Then, the pyrolysis model function can be rewritten by

$$f(\alpha) = n\alpha^{0.39n - 1.15}(1 - \alpha)[-\ln(1 - \alpha)]^{1 - 1/n} \tag{16}$$

As shown in Figure 6, four sample size experimental data were put together for model reconstruction since the reconstruction model lines in Figure 5 show similar variations. Results showed that for all four sample sizes, the reaction model can be described as $f(\alpha) = 2.02\alpha^{-0.27}(1 - \alpha)[-\ln(1 - \alpha)]^{0.50}$. It can be concluded that the pyrolysis model, $f(\alpha)$, cannot be influenced by sample particle size because the geometric dimension cannot change the chemical reaction principles. Although the reaction model function $f(\alpha)$ cannot be influenced by particle size, the activation energies and reaction rate can be influenced greatly because the specific surface area can influence the heat transfer and evaporation rate of the particle surface.

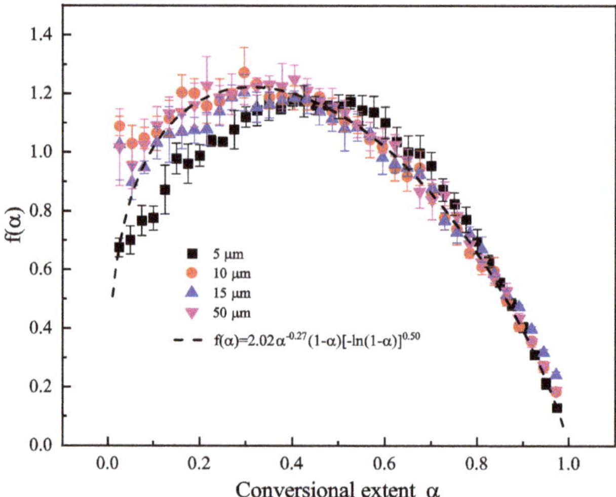

Figure 6. The experimental kinetics function $f(\alpha)$ reconstructed for 5, 10, 15, and 50 μm sample size.

It should be noted that in previous literatures about polymer pyrolysis model identification, it is far from enough that only linearity coefficients are obtained, by which the models are ranked. For each model will have its one linearity coefficient, and there must exist one model with the highest fitness; which however, does not mean that this model can describe the pyrolysis process well, especially when fitting with experimental data. Figures 5 and 6 shows that the reconstructed model can describe the experimental well after modification, though the format of the final model shows difference with traditional nineteen models. We can also call the final reaction model an apparent model, which can be regarded as the combination of several step reaction models.

6.5. Step-Reaction Separation by Distributed Activation Energy Method

By traditional kinetics methods, we can only see that the activation energies are different for different sample size, while we cannot distinguish which step reaction makes the difference on pyrolysis kinetics. So, in this section, distributed activation energy method was employed to separate the step reaction from overall pyrolysis reaction, by which we can see the weight of step reaction on activation energy for different particle sizes. Details about the mechanism of DAEM have been introduced in Section 4, and the solution of DAEM equations was based on programming MATLAB to obtain the kinetics parameters. To improve the accuracy of kinetic results, experimental data of α and dα/dt was employed to fit by DAEM model at the same time, which was judged by getting the minimum value of squared sum residuals (SSR), which can be expressed by

$$SSR = \sum_{n=1}^{e} \sum_{m=1}^{f} \left\{ \left[\alpha_{num}(T_k) - \alpha_{exp}(T_k) \right]^2 + \left[\left(\frac{d\alpha}{dT_k} \right)_{num} - \left(\frac{d\alpha}{dT_k} \right)_{exp} \right]^2 \right\} \tag{17}$$

where e and f mean all heating rates and selected experimental data points. The subscripts num and exp mean the numerical DAEM model and experimental data, respectively.

For PS pyrolysis in nitrogen, the pyrolysis mechanism has been explored a lot. It is generally acknowledged that the pyrolysis process can be divided into two steps. The first step is the pyrolysis of the main PS structure with a generation of large volatile molecules, during which the structure will show a large mass loss. The second step is the generation of single molecule styrene mainly from the large molecule and a little bit from the residual body. During the DAEM calculation, we hypothesize that PS pyrolysis process includes two reaction steps. Equations (9)–(14) were solved based on genetic algorithm (GA) in MATLAB. GA is an advanced algorithm based on Darwin's evolution theory, searching the best fitness in solving a high-dimensional optimization problem. For each new generation, GA will generate a certain amount of individuals randomly and simultaneously, among which each individual will be employed to fit with experimental data with fitness obtained. The individual with best fitness will be adopted as a parent to produce next generation. During producing, each generation process, selection, interaction, cross, and variation are all considered. Finally, one individual with best fitness is identified as the final parameters.

The aforementioned two-pseudo-component pyrolysis mechanism was employed during DAEM, and the searching ranges for four parameters, natural logarithm of pre-exponential factor, standard derivation of Gaussian distribution, activation energy, and weight factor were 5–60, 0–15 kJ mol^{-1}, 100–380 kJ mol^{-1}, and 0–1, respectively. In each heating rate, 100 points with uniform intervals were selected from the original data during the 600–900 K temperature range. Table 5 shows the DAEM calculation parameters with best fitness for two component reactions hypothesis. Figure 7 shows the activation energy distributions for both step reactions. From Table 5 and Figure 7, we can find that the activation energy distributions of 5 μm is more concentrated than the other three particle sizes especially for the second step reaction at 260–290 kJ mol^{-1}, which means the 5 μm particle is much easier to pyrolyze compared with other particles, and the first step reaction group is more concentrated. The centered activation energy increases with particle size increasing in both reaction processes, which is in accordance with the results by the isoconversional method. Obviously, the particle size effects on the second reaction are more obvious than the first step reaction.

Table 5. Distributed activation energy method (DAEM) fitness for different particle size with different heating rates.

Component	Parameter	5 μm	10 μm	15 μm	50 μm
	$\ln A_1$	38.7179	40.8492	42.6370	42.4292
Component 1		0.0063	2.2146	2.3269	2.4410
	$E_{0,1}$	262.9934	272.7594	283.5884	283.0256
	n_1	0.9004	0.8426	0.8339	0.8613
	$\ln A_2$	16.3841	18.9673	21.2544	21.8706
Component 2		4.7951	9.1801	9.1128	8.9726
	$E_{0,2}$	145.3102	155.9962	169.6331	176.2200
	n_2	0.1153	0.1614	0.1744	0.1430

Figure 8 shows the experimental α and $d\alpha/dt$, DAEM fitting α and $d\alpha/dt$, and step reaction distributions. We can find that the experimental data and DAEM fitting can match each other reasonably well for all sixteen cases. And the mass loss by the first reaction occupies most of the reaction.

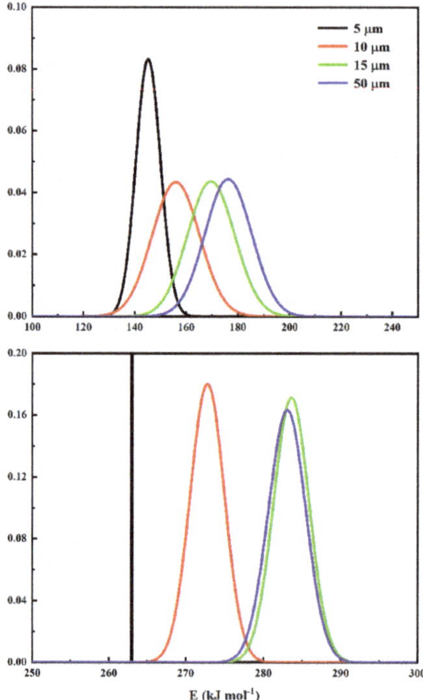

Figure 7. Activation energy distribution in distributed activation energy distribution method with Gaussian distribution for four particle sizes.

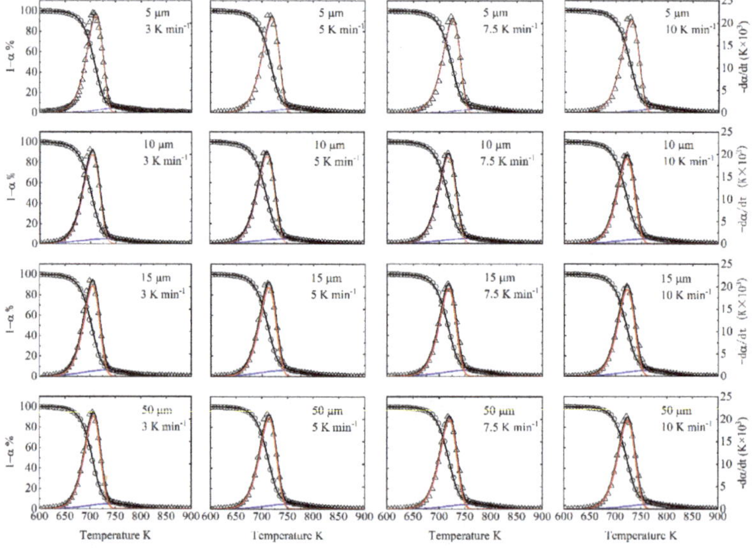

Figure 8. Comparison between DAEM calculation (solid lines, including α and $d\alpha/dt$ for overall reaction and step reactions) and experimental data (points, including α and $d\alpha/dt$) for different particle size with different heating rates.

To quantitatively show the fitness between calculation and experimental, here we use Equation (18) to evaluate the fitness, and the higher result means better fitness, here we employ the weight coefficient as 0.5, Equation (18) can be expressed as

$$Fit_{v1} = 1 - \sqrt{\sum_{m=1}^{b} \left[\left(\frac{d\alpha}{dT}\right)_{num} - \left(\frac{d\alpha}{dT}\right)_{exp}\right]^2 / f} \left/ \left[\left(\frac{d\alpha}{dT}\right)_{exp}\right]_{max}\right. \tag{18a}$$

$$Fit_{v2} = 1 - \sqrt{\sum_{m=1}^{b} \left(\alpha_{num} - \alpha_{exp}\right)^2 / f} \left/ \left(\alpha_{exp}\right)_{max}\right. \tag{18b}$$

$$Fit_v = [\kappa Fit_{v1} + (1 - \kappa)Fit_{v2}] \times 100\% \tag{18c}$$

Table 6 shows the fitness results for different heating rates during DAEM fitting. We can see all fitness are larger than 98.5%, which proves the good performance of DAEM in TGA and DTG curve prediction.

Table 6. DAEM fitness for different particle size with different heating rates.

	Fitness			
Particle Size	**3 K min^{-1}**	**5 K min^{-1}**	**7.5 K min^{-1}**	**10 K min^{-1}**
5 μm	98.06	98.01	98.20	98.29
10 μm	98.39	98.77	98.77	98.78
15 μm	98.13	98.74	98.75	98.71
50 μm	98.52	98.72	98.70	98.79

6.6. Sensitivity Analysis of DAEM Parameters

After calculating the DAEM parameters of different particles, we also need to carry out the sensitivity analysis to judge which parameter is more important and sensitive. The method to check its sensitivity is to change the target parameter by a small value and remain the rest parameters unchanged. The variation range of parameter is very small, here we employ the range ±0.1. lg(ssr) to quantitatively judge the parameter sensitivity, where ssr is the SSR with changed parameter divided by the optimal SSR value. Parameters lnA_1, σ, $E_{0,1}$, n_1, and lnA_2, σ, $E_{1,2}$, n_2, are numbered as 1–8 as shown in y-axis of Figure 9, where x-axis means the relative changed value of parameters ranging from 0.9 to 1.1, y-axis means the order of eight parameters, and the color value in Figure 9 means lg(ssr). The blue color presents that the parameter is insensitive and accurate, while the red color means sensitive to the value change. So, during calculation, we should check the accuracy of these sensitive parameter to make sure its accuracy. Obviously, the pre-exponential factor and activation energy we obtained by DAEM methods are insensitive compared with weight factor and distribution factor, which means the result is reasonably dependable. Here in Figure 9, the data is 50 μm PS pyrolysis DAEM parameters. The parameters of other particle sizes show the same weight with 50 μm particle, so here we won't discuss other particle size cases anymore.

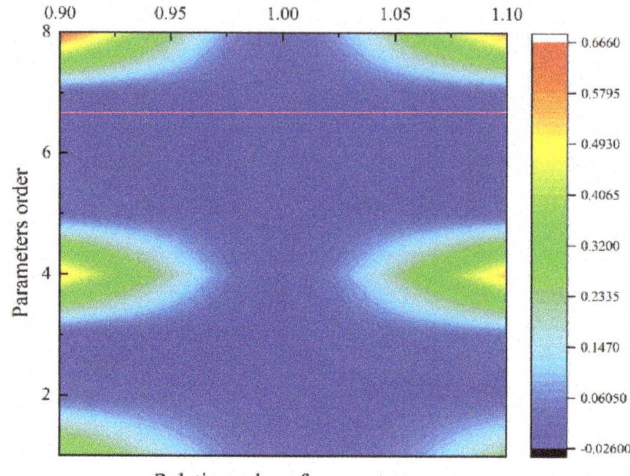

Figure 9. Sensitivity of eight DAEM parameters for 50 μm PS pyrolysis. This figure shows 50 μm PS particle case, and the sensitivity of other particles shows the same weight distribution.

7. Conclusion

Here we explore the particle size effects on pyrolysis of polystyrene from aspects of pyrolysis behavior, kinetics, reaction model, reconstruction, and validation. The final reaction model can provide scientific guidance to polymer pyrolysis modeling [22–27]. In this study, to explore the particle size effects on pyrolysis behavior, polystyrene particles with four different sizes, 5, 10, 15, and 50 μm, were selected to conduct a series of TG experiments. Isoconversional methods were employed to calculate kinetic parameters during all conversional extents. Results show that the temperature of the DTG curve peak will decrease first, then increase with particle size for the same heating rate, which may be caused by the competition of compactness and specific surface area effects. During the main pyrolysis stage, with the increase of conversional extent, activation energies decrease slowly and linearly for all four size samples. With the increase of particle size, the activation energies will increase for the same conversional extent, which means that the reaction of the largest particle is the hardest to trigger. The Avrami–Eroféev model was identified by the Coats–Redfern method as the controlling model during the polystyrene pyrolysis process. Considering the accommodation function of the reaction model, Avrami–Eroféev model was modified as $f(\alpha) = 2.02\alpha^{-0.27}(1-\alpha)[-\ln(1-\alpha)]^{0.50}$, by which the polystyrene pyrolysis process can be well explained. To find the weight of each step reaction, the DAEM model was employed to separate the step reaction from overall reaction. Results showed that both step reactions can be largely influenced by particle size, especially for the second step. For the five μm particle, the activation energy distributions in both step reactions are more concentrated and forward, and its reaction is more uniform.

Author Contributions: Conceptualization, L.J. and J.-H.S.; methodology, L.J., Q.X., O.D. and J.-H.S.; software, L.J.; validation, L.J., O.D.; formal analysis, L.J., X.-R.Y.; investigation, L.J.; resources, L.J., X.G.; data curation, L.J., J.-H.S.; writing—original draft preparation, L.J., O.D., Q.X., J.-H.S.; writing—review and editing, L.J., O.D., Q.X., J.-H.S.; visualization, L.J., M.K.K.; supervision, Q.X., J.-H.S.; project administration, L.J.; funding acquisition, L.J., Q.X., O.D., and M.K.K. All authors have read and agreed to the published version of the manuscript.

Funding: The authors would like to thank the National Natural Science Foundation of China (NSFC, Grant 51806208), the joint project of NSFC and STINT (51911530151, China side, and CH2018-7733, Sweden side), Slovenian Research Agency for project BI-CN/18-20-016, and Slovenia-China joint research fund 12-22. Oisik Das expresses his gratitude to Bio4Energy. Manja Kitek Kuzman thanks Slovenian Research Agency for financial support within the scope of the program P4-0015, as well as BI-CN/18-20-016.

Conflicts of Interest: The authors declare no conflicts of interest.

References

1. Zhang, D.; Jiang, L.; Lu, S.; Cao, C.; Zhang, H.-P. Particle size effects on thermal kinetics and pyrolysis mechanisms of energetic 5-amino-1h-tetrazole. *Fuel* **2018**, *217*, 553–560. [CrossRef]
2. Ding, Y.; Fukumoto, K.; Ezekoye, O.A.; Lu, S.; Wang, C.; Li, C. Experimental and numerical simulation of multi-component combustion of typical charring material. *Combustion Flame* **2020**, *211*, 417–429. [CrossRef]
3. Ding, Y.; Zhang, W.; Yu, L.; Lu, K. The accuracy and efficiency of GA and PSO optimization schemes on estimating reaction kinetic parameters of biomass pyrolysis. *Energy* **2019**, *176*, 582–588. [CrossRef]
4. Jiao, L.; Xu, G.; Wang, Q.; Xu, Q.; Sun, J. Kinetics and volatile products of thermal degradation of building insulation materials. *Thermochim. Acta* **2012**, *547*, 120–125. [CrossRef]
5. Jiao, L.; Sun, J. A thermal degradation study of insulation materials extruded polystyrene. *Procedia Eng.* **2014**, *71*, 622–628. [CrossRef]
6. Cheng, J.; Pan, Y.; Yao, J.; Wang, X.; Pan, F.; Jiang, J. Mechanisms and kinetics studies on the thermal decomposition of micron Poly (methyl methacrylate) and polystyrene. *J. Loss Prev. Process Ind.* **2016**, *40*, 139–146. [CrossRef]
7. Shen, J.; Wang, X.S.; Garcia-Perez, M.; Mourant, D.; Rhodes, M.J.; Li, C.Z. Effects of particle size on the fast pyrolysis of oil mallee woody biomass. *Fuel* **2009**, *88*, 1810–1817. [CrossRef]
8. Marcilla, A.; García, A.N.; Pastor, M.V.; León, M.; Sánchez, A.; Gómez, D. Thermal decomposition of the different particles size fractions of almond shells and olive stones. Thermal behaviour changes due to the milling processes. *Thermochim. Acta* **2013**, *564*, 24–33. [CrossRef]
9. Di Blasi, C. Kinetic and heat transfer control in the slow and flash pyrolysis of solids. *Ind. Eng. Chem. Res.* **1996**, *35*, 37–46. [CrossRef]
10. Hanson, S.; Patrick, J.W.; Walker, A. The effect of coal particle size on pyrolysis and steam gasification. *Fuel* **2002**, *81*, 531–537. [CrossRef]
11. Yu, D.; Xu, M.; Sui, J.; Liu, X.; Yu, Y.; Cao, Q. Effect of coal particle size on the proximate composition and combustion properties. *Thermochim. Acta* **2005**, *439*, 103–109. [CrossRef]
12. Kissinger, H.E. Reaction kinetics in differential thermal analysis. *Anal. Chem.* **1957**, *29*, 1702–1706. [CrossRef]
13. Akahira, T.; Sunose, T. Method of determining activation deterioration constant of electrical insulating materials. *Res. Rep. Chiba. Inst. Technol.* **1971**, *16*, 22–31.
14. Flynn, J.H.; Wall, L.A. A quick, direct method for the determination of activation energy from thermogravimetric data. *J. Polym. Sci. Pol. Lett.* **1966**, *4*, 323–328. [CrossRef]
15. Ozawa, T. A new method of analyzing thermogravimetric data. *Bull. Chem. Soc. Jap.* **1965**, *38*, 1881–1886. [CrossRef]
16. Tang, W.; Liu, Y.; Zhang, H.; Wang, C. New approximate formula for Arrhenius temperature integral. *Thermochim. Acta* **2003**, *408*, 39–43. [CrossRef]
17. Tang, W.; Chen, D.; Wang, C. Kinetic study on the thermal dehydration of $CaCO_3 \cdot H_2O$ by the master plots method. *Aiche J.* **2006**, *52*, 2211–2216.
18. Vyazovkin, S.; Dollimore, D. Linear and nonlinear procedures in isoconversional computations of the activation energy of nonisothermal reactions in solids. *J. Chem. Inf. Comput. Sci.* **1996**, *36*, 42–45. [CrossRef]
19. Vyazovkin, S. Advanced isoconversional method. *J. Anal. Calorim.* **1997**, *49*, 1493–1499. [CrossRef]
20. Vyazovkin, S. Evaluation of activation energy of thermally stimulated solid-state reactions under arbitrary variation of temperature. *J. Comput. Chem.* **1997**, *18*, 393–402. [CrossRef]
21. Farjas, J.; Roura, P. Isoconversional analysis of solid state transformations. *J. Anal. Calorim.* **2011**, *105*, 757–766. [CrossRef]
22. Li, K.Y.; Huang, X.; Fleischmann, C.; Rein, G.; Ji, J. Pyrolysis of medium-density fiberboard: Optimized search for kinetics scheme and parameters via a genetic algorithm driven by Kissinger's method. *Energy Fuels* **2014**, *28*, 6130–6139. [CrossRef]
23. Li, K.; Pau, D.S.W.; Wang, J.; Ji, J. Modelling pyrolysis of charring materials: Determining flame heat flux using bench-scale experiments of medium density fibreboard (MDF). *Chem. Eng. Sci.* **2015**, *123*, 39–48. [CrossRef]
24. Li, K.Y.; Pau, D.S.W.; Hou, Y.N.; Ji, J. Modeling pyrolysis of charring materials: Determining kinetic properties and heat of pyrolysis of medium density fiberboard. *Ind. Eng. Chem. Res.* **2013**, *53*, 141–149. [CrossRef]

25. Gao, X.; Jiang, L.; Xu, Q.; Wu, W.-Q.; Mensah, R.A. Thermal kinetics and reactive mechanism of cellulose nitrate decomposition by traditional multi kinetics and modeling calculation under isothermal and non-isothermal conditions. *Ind. Crop. Prod.* **2020**, *145*, 112085. [CrossRef]

26. Li, K.; Pau, D.S.; Zhang, H. Pyrolysis of polyurethane foam: Optimized search for kinetic properties via simultaneous K–K method, genetic algorithm and elemental analysis. *Fire Mater.* **2016**, *40*, 800–817. [CrossRef]

27. Chen, R.; Xu, X.; Lu, S.; Zhang, Y.; Lo, S. Pyrolysis study of waste phenolic fibre-reinforced plastic by thermogravimetry/Fourier transform infrared/mass spectrometry analysis. *Energy Convers. Manag.* **2018**, *165*, 555–566. [CrossRef]

28. Chen, R.; Li, Q.; Zhang, Y.; Xu, X.; Zhang, D. Pyrolysis kinetics and mechanism of typical industrial non-tyre rubber wastes by peak-differentiating analysis and multi kinetics methods. *Fuel* **2019**, *235*, 1224–1237. [CrossRef]

29. Kong, S.; Huang, X.; Li, K.; Song, X. Adsorption/desorption isotherms of CH4 and C2H6 on typical shale samples. *Fuel* **2019**, *255*, 115632. [CrossRef]

30. Gao, Z.; Wan, H.; Ji, J.; Bi, Y. Experimental prediction on the performance and propagation of ceiling jets under the influence of wall confinement. *Energy* **2019**, *178*, 378–385. [CrossRef]

31. Vand, V. A theory of the irreversible electrical resistance changes of metallic films evaporated in vacuum. *Proc. Phys. Soc.* **1943**, *55*, 222. [CrossRef]

32. Pitt, G.J. The kinetic of the evolution of volatile products from coal. *Fuel* **1962**, *41*, 267–274.

33. Cao, H.Q.; Duan, Q.L.; Chai, H.; Li, X.-X.; Sun, J.-H. Experimental study of the effect of typical halides on pyrolysis of ammonium nitrate using model reconstruction. *J. Hazard. Mater.* **2020**, *384*, 121297. [CrossRef] [PubMed]

34. Li, M.; Jiang, L.; He, J.J.; Sun, J.-H. Kinetic triplet determination and modified mechanism function construction for thermo-oxidative degradation of waste polyurethane foam using conventional methods and distributed activation energy model method. *Energy* **2019**, *175*, 1–13. [CrossRef]

35. Chai, H.; Duan, Q.; Jiang, L.; Gong, L.; Chen, H.; Sun, J. Theoretical and experimental study on the effect of nitrogen content on the thermal characteristics of nitrocellulose under low heating rates. *Cellulose* **2019**, *26*, 763–776. [CrossRef]

36. Koga, N.; Malek, J. Accommodation of the actual solid-state process in the kinetic model function. Part 2. Applicability of the empirical kinetic model function to diffusion-controlled reactions. *Thermochim. Acta* **1996**, *282*, 69–80. [CrossRef]

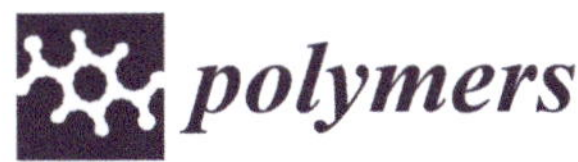 

Article

The Role of Nanoparticle Shapes and Structures in Material Characterisation of Polyvinyl Alcohol (PVA) Bionanocomposite Films

Mohanad Mousa [1,2] and Yu Dong [1,*]

[1] School of Civil and Mechanical Engineering, Curtin University, GPO Box U1987, Perth 6845, Australia; mohanadmousa616@yahoo.com

[2] Shatrah Technical Institute, Southern Technical University, Basra 61001, Iraq

* Correspondence: Y.Dong@curtin.edu.au; Tel.: +61-8-9266-9055

Received: 13 November 2019; Accepted: 15 January 2020; Published: 25 January 2020

Abstract: Three different types of nanoparticles, 1D Cloisite 30B clay nanoplatelets, 2D halloysite nanotubes (HNTs), and 3D nanobamboo charcoals (NBCs) were employed to investigate the impact of nanoparticle shapes and structures on the material performance of polyvinyl alcohol (PVA) bionanocomposite films in terms of their mechanical and thermal properties, morphological structures, and nanomechanical behaviour. The overall results revealed the superior reinforcement efficiency of NBCs to Cloisite 30B clays and HNTs, owing to their typical porous structures to actively interact with PVA matrices in the combined formation of strong mechanical and hydrogen bondings. Three-dimensional NBCs also achieved better nanoparticle dispersibility when compared with 1D Cloisite 30B clays and 2D HNTs along with higher thermal stability, which was attributed to their larger interfacial regions when characterised for the nanomechanical behaviour of corresponding bionanocomposite films. Our study offers an insightful guidance to the appropriate selection of nanoparticles as effective reinforcements and the further sophisticated design of bionanocomposite materials.

Keywords: polyvinyl alcohol (PVA); bionanocomposites; nanomechanical behaviour; thin films

---

1. Introduction

Nanoparticles in spheroidal, platelet-like, and tubular shapes as effective nanofillers have attracted materials engineers and researchers in the field of nanocomposite materials in the past few decades [1,2]. The incorporation of different nanoparticles into continuous polymer matrices has been proven to significantly alter the properties of virgin polymers, resulting in a novel-class system of polymer nanocomposites with superior properties and excellent functionalities [1,2]. In general, when embedded with a small fraction of nanoparticles being less than 10 wt %, the optical [3], mechanical [4], thermal [5], electronic [6], and antimicrobial [7] properties of polymer nanocomposites can be remarkably enhanced while maintaining some features of net polymer systems such as low density and easy processibility [2]. Such polymer nanocomposites possess a wide range of applications including medical devices, aerospace engineering, and automotive components [1]. For instance, nanocomposites reinforced with some polymeric and inorganic nanofillers such as chitosan nanoparticles and silver nanoparticles have been proven to be effective for antimicrobial treatment in dentistry [8] or for bone tissue regeneration [9]. Other studies [10,11] demonstrated that polymer nanocomposites, as exemplified by polypropylene (PP)/clay nanocomposites, have real automotive industrial potential to result in significant property improvement with only minor increasing cost if a deeper understanding of their structure–property relationship can be achieved.

The effective reinforcing mechanism is based on the fundamental concept that the chain mobility of polymeric molecules is restricted by rigid nanofillers according to the matrix–particle interfacial interactions in polymer nanocomposites [2,12]. The specific areas associated with matrix–filler interactions are known as interfacial regions with completely distinct properties from those of nanoparticles and polymer matrices individually. More importantly, the material performance of polymer nanocomposites primarily depends on the volume of interfacial regions and interfacial properties [12] in relation to critical nanofiller parameters such as nanoparticle shapes and structures.

In addition to a major concern of nanoparticle structures, nanoparticle shapes are also equally important when matrix–filler interaction is considered in polymer nanocomposites, which can be classified into three popular shapes, namely 1D platelet-like nanoparticles such as montmorillonite (MMT) clays and nanoplatelet graphene sheets, 2D tubular nanoparticles such as HNTs and carbon nanotubes (CNTs), as well as 3D spherical nanoparticles such as diamond nanoparticles and nanosilica particles and fractal-like or irregular near circular-like nanoparticles such as NBCs. In a nanocomposite system, the alteration of nanoparticle shapes means that the contact areas inevitably vary between polymer matrices and nanoparticles to effectively control the volume of their interfacial regions [12]. Most previous studies [13,14] were based on theoretical or numerical modelling approaches such as atomistic and coarse-grained molecular dynamic (MD) simulations for evaluating the matrix–filler bonding effect. Nonetheless, current computational capability and the environment may be mostly restricted to the context of single and two-particle systems by neglecting the effect of actual nanoparticle structures and shapes induced in different material processing techniques [15].

The main objective and novelty of this study lie in holistically assessing the influence of different dimensional nanoparticle shapes, structures, and contents on the effective reinforcement mechanism of PVA bionanocomposites reinforced with 1D Cloisite 30B clays, 2D HNTs, and 3D NBCs, respectively. The selection of PVA as a base polymer arises from its good biodegradability and water solubility to replace conventional petroleum-based polymers for generating much less marine plastic wastes [16]. Our study demonstrated that 3D NBCs could act as relatively new and superior carbon-based nanofillers to clay-based Cloisite 30B and HNTs for the best material performance of nanocomposites. This highlighted their great benefit to be more competitive nanoreinforcements in the manufacture of composite materials, as well as future potential to electronics, material packaging, and biomedical applications.

2. Materials and Methods

2.1. Materials

PVA (material type: MFCD00081922), as a popular water-soluble biopolymer, was purchased from Sigma Aldrich Pty. Ltd., Castle Hill, NSW, Australia with the molecular weight of 89,000–98,000 g/mol and the degree of hydrolysis of 99.0%–99.8%. Three different types of nanoparticles used in this study comprised Cloisite 30B clays, HNTs, and NBCs. In between, NBCs were directly purchased from US Research Nanomaterials, Inc. Co., Houston, TX, USA (molecular weight: 12.01 g/mol, particle density: 0.43 g/cm^3, and particle size <69.43 nm [4]). Moreover, Cloisite 30B clays with methyl, tallow, bis-2-hydroxyethyl, quaternary ammonium were supplied by Southern Clay Products, Gonzales, LA, USA while HNT powders, donated by Imerys Tableware Limited, Auckland, New Zealand, have particle dimensions of 120–140 nm in outer dimeter, 15–100 nm in inner diameter, and 0.3–1.5 μm in length [17].

2.2. Fabrication of PVA Bionanocomposite Films

All PVA bionanocomposite films reinforced with Cloisite 30B clays, HNTs, and NBCs were prepared using solution casting according to the fabrication procedure mentioned in our previous work [4]. Initially, 5 wt %/v PVA aqueous solution was prepared by mixing 10 g PVA into 190 mL deionised water under vigorous magnetic stirring at 400 rpm and 90 °C for 3 h until PVA was completely dissolved. Aqueous suspensions of all nanoparticles were achieved by mechanical mixing filler powders in deionised water with a rotor speed of 405 rpm at 40 °C for 2 h, which was followed by the ultrasonication (Model ELMA Ti–H–5, Elma Schmidbauer GmbH, Singen, Germany) at 25 kHz and 40 °C with a power intensity of 70% for 1 h. Subsequently, nanoparticle contents of 0, 3, 5 and 10 wt % were obtained by controlling PVA amounts used in each material formulation. Then, such aqueous suspensions were gradually added in a dropwise manner into PVA solutions and simultaneously subjected to mechanical mixing at 405 rpm and 40 °C for 2 h. Afterwards, their mixtures were stirred at 400 rpm and 90 °C for 1 h prior to the sonication for 30 min to achieve uniform nanoparticle dispersion. Finally, 20 mL prepared solution was poured on a glass Petri dish and allowed to dry in an air-circulating oven at 40 °C for 48 h. Subsequently, different types of PVA bionanocomposite films were stored in a silica gel-containing desiccator prior to material testing.

2.3. Characterisation Methods

In this study, nanomechanical properties of PVA bionanocomposite films were quantitatively assessed in peak force quantitative nanomechanical mapping (PFQNM) [12] via atomic force microscopy (AFM) based on a Bruker Dimension Fastscan AFM system (Bruker Corporation, Karlsruhe, Germany). A Tapping Mode Etched Silicon Probe (TESPA) was employed with the nominal spring constant of 40 N/m and tip radius of 8 nm. The image scan rate was controlled at 2 Hz with 256 × 256 digital pixel resolution. In order to remove unwanted noise, bow and tilt features from the vertical scanner (Z) and AFM topographic images were first-order flattened via the *Flatten* command with the aid of Burker Nanoscope 1.5 software (Bruker Corporation, Karlsruhe, Germany).

Fourier transform infrared (FTIR) spectrometry (PerkinElmer Spectrum 100 FTIR spectrometer, PerkinElmer, Waltham, MA, USA) was utilised to characterise the chemical bonding effects of PVA, nanoparticles as well as PVA bionanocomposites in a wavenumber range of 650–4000 cm^{-1} with a resolution of 4 cm^{-1} according to an attenuated total reflectance (ATR) method.

Additionally, X-ray diffraction (XRD) analysis was carried out using a Bruker D8 Advanced Diffractometer (Bruker Corporation, Karlsruhe, Germany). The X-ray source was Ni-filtered Cu-Kα radiation (wavelength λ = 0.1541 nm) carried out at the accelerating voltage and current of 40 kV and 40 mA, respectively. X-ray spectra were recorded in a 2θ range of 10–50° at the scan rate of 0.015°/s.

A universal testing machine (Lloyd EZ50, Lloyd Instruments Ltd., West Sussex, UK) was employed at room temperature at the crosshead speed of 10 mm/min (gauge length: 50 mm) in order to measure the tensile properties of neat PVA and PVA bionanocomposite films according to ASTM D882-02. For each material batch, six specimens were tested with the mean values and standard deviations being calculated accordingly. Moreover, the tensile toughness was also determined based on tensile energy to break (TEB) with reference to ASTM D882-02.

The fracture morphology for tensile testing specimens was evaluated with the aid of a field emission scanning electron microscope (FE-SEM, Zeiss NEON 40 EsB Cross Beam, Carl Zeiss Microscopy GmbH, Jena, Germany) at an accelerating voltage of 5 kV after being coated with platinum (layer thickness: 5 nm).

The thermal properties of PVA bionanocomposite films were examined using a combined measurement system based on thermal gravimetric analysis (TGA) and differential scanning calorimetry (DSC) (1 STAR^e system, Mettler-Toledo, Columbus, OH, USA) from 35 to 700 °C at a scan rate of 10 °C/min and a flow rate of 25 mL/min under argon atmosphere. The degree of crystallinity χ_c of PVA matrices within PVA bionanocomposites was calculated as follows:

$$\chi_C\ (\%)\ =\ \frac{\Delta H_m}{w\Delta H_m^O} \times 100\% \tag{1}$$

where ΔH_m is the measured melting enthalpy according to DSC data. ΔH_m^O = 138.6 J/g [5] is the enthalpy of fully crystalline PVA, and w is the weight fraction of PVA matrices in corresponding PVA bionanocomposites.

3. Results and Discussion

3.1. Nanoparticle Shape and Size

The morphological structures of as-received nanoparticles of Cloisite 30B clays, HNTs, and NBCs are illustrated in Figure 1. All nanoparticles powders show high irregularities in size and material morphology. However, HNTs are most likely to possess cylindrical shapes with transparent central areas running longitudinally along such cylindrical structures, as illustrated in Figure 1a,b. The outer diameters and lumen diameters of HNTs, as typical tubular nanoparticles in hollow and open-end structures, were found to be in range of 20–115 and 5–30 nm, respectively. Whereas, the lengths of HNTs vary from 50 nm to 1.5 μm. On the other hand, the morphological structures of Cloisite 30B clays were detected using an AFM tapping mode from diluted clay suspension when deposited onto the mica substrate, as revealed in Figure 1c,d. It is evident that Cloisite 30B clays possess platelet-like structures with an average particle diameter of approximately 100.75 ± 6.5 nm by measuring 925 clay particles, while their thickness varies from 1.69 to 5.9 nm, which is exemplified by a cross-sectional analysis for a typical section (A2–B2) illustrated in Figure 1d. These results suggest that clay platelet-like layered structures can consist of single platelets (thickness: ~1 nm) [18] as well as stacked layered platelets, signifying the combination of clay exfoliation and intercalation. NBC sizes were determined previously to be 69.43 nm by average [4], as shown in Figure 1e,f, which can be considered as 3D anisotropic nanoparticles as opposed to 1D platelet-like clays and 2D HNTs.

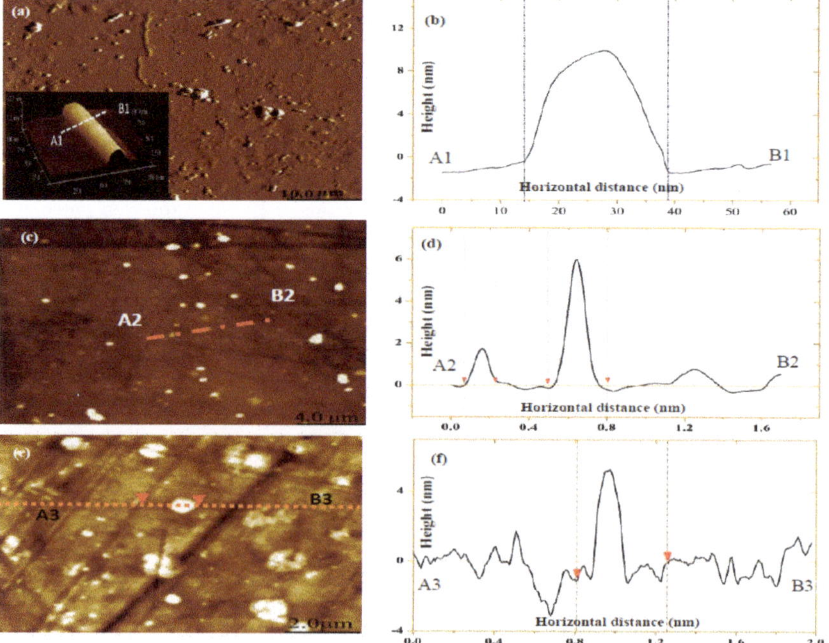

Figure 1. Atomic force microscopy (AFM) characterisation of different nanoparticles: AFM images of (**a**) HNTs, (**c**) Cloisite 30B clays, and (**e**) NBCs deposited on mica substrate in aqueous solutions, and height profiles of (**b**) HNTs, (**d**) Cloisite 30B clays, and (**f**) NBCs at typical cut sections A1–B1, A2–B2, and A3–B3, respectively.

3.2. Chemical Bonding Effect

FTIR was employed to evaluate functional groups in PVA matrices and nanoparticles as well as their associated chemical bonding effects. As-received HNTs exhibit two Al_2OH stretching bands assigned to 3691.5 and 3621 cm^{-1} in Figure 2a, resulting from OH bending in connection with two Al atoms along with other band features of inorganic aluminosilicate structures of halloysite [19]. Furthermore, FTIR peaks observed at 1004 and 906 cm^{-1} are associated with Si–O–Si and Al–OH stretchings, respectively. In comparison, as-received Cloisite 30B clays have an existing peak at 3629.6 cm^{-1} corresponding to Si–OH and Al–OH stretchings in Figure 2b. The broad band at 3405 cm^{-1} is assigned to OH groups in relation to interlayer water, while two typical bands at 2924.6 and 2853.5 cm^{-1} are related to $–CH_2$ asymmetric and symmetric stretchings, respectively [20,21]. Moreover, FTIR peaks at 1647 and 1123.3 cm^{-1} can be due to the deformation vibration of interlayer water and Si–O bending accordingly, as opposed to an assigned band at 1470 cm^{-1} arising from $–CH_2$ bending [20]. The NBC results presented in Figure 2c confirm the absence of –OH groups in their FTIR spectra, which infers much lower moisture and alcohol contents obtained in NBCs. Additionally, FTIR peaks at 2417.5 and 1567.4 m^{-1} reveal the existence of C≡H stretching [22] and C=C vibration in an aromatic system [23], respectively. On the other hand, the peak spectrum at 1696 cm^{-1} was assigned to the C=O band primarily for ionisable carboxyl groups as an indicator of surface hydrophilicity [24].

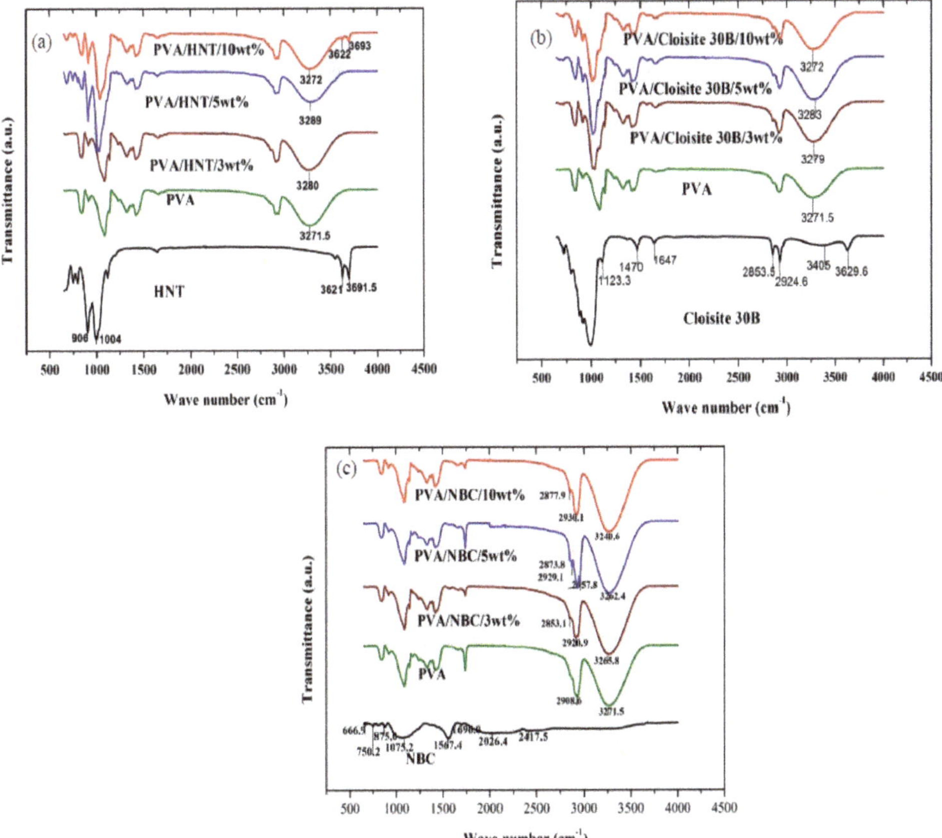

Figure 2. FTIR spectra for chemical interactions of polyvinyl alcohol (PVA) bionanocomposite films reinforced with (**a**) HNTs, (**b**) Cloisite 30B clays, and (**c**) NBCs.

In the case of bionanocomposite systems, the FTIR spectra of PVA/HNT bionancomposites and PVA/Cloisite 30B clay bionanocomposites are also illustrated in Figure 2a,b, respectively. The FTIR peak located at 3271.5 cm^{-1} associated with O–H stretching shifts to higher wavenumbers at 3280 and 3289 cm^{-1}, as well as 3279 and 3283 cm^{-1} with the inclusion of HNTs and Cloisite 30B clays at the nanoparticle contents of 3 and 5 wt %, respectively. Such a finding was attributed to the strengthening effect of hydrogen bonds between –OH groups from PVA molecules and those located on clay surfaces such as silanol groups (–SiOH), which is in good agreement with previous investigations on PVA/organomodified Cloisite Na$^+$ (OMMT) nanocomposites [21], poly (ε-caprolactone) (PCL)/Cloisite 30B clay nanocomposites [25], and PVA/chitosan (CS)/HNT nanocomposites [26]. However, when the HNT content increases up to 10 wt %, two Al$_2$OH stretchings appear for embedded HNTs in bionanocomposite films due to typical HNT agglomeration [27]. As for PVA/NBC bionancomposite films, increasing the NBC content from 0 to 10 wt % leads to the band-peak shift to a lower wavenumber at 3240.6 cm^{-1} owing to large amounts of hydroxyl groups in PVA molecules [4] as well as strong filler–matrix bonding. As a result, hydrogen bonds are generated to be intertwined at PVA/NBC interfaces with a broad O–H band. Such a variation associated with –OH stretching vibration proves the formation of hydrogen bonds, which is in good accordance in PVA/graphene nanocomposites [27] and PVA/bamboo charcoal (BC) nanocomposites [5]. The aforementioned results fail to show existing new bands in PVA films with the inclusion of both HNTs and Cloisite 30B clays. On the contrary, the addition of 3 and 5 wt % NBCs within PVA matrices in bionanocomposite films gives rise to a new band in relation to –CH$_2$-asymmetric and symmetric stretchings [4]. Such a finding can be ascribed to typical NBC porous structures enabling absorbing molecular chains of hydrophilic polymers such as PVA with the combined mechanical and chemical bondings. More consistently, the incorporation of HNTs and Cloisite 30B clays in PVA bionanocomposites shifts the hydroxyl peaks of PVA to relatively high wavenumbers compared to the addition of NBCs [4]. Such results indicate that the numbers of hydrogen bonding generated in PVA/HNT bionanocomposites and PVA/Cloisite 30B clay bionanocomposites are higher when compared with those detected in PVA/NBC bionanocomposites, owing to different chemical structures of nanofillers. Since NBCs do not possess –OH peaks, most hydrogen bonding formed in bionanocomposites can arise from hydroxyl groups of PVA molecules. Whereas, existing –OH peaks detected in HNTs and Cloisite 30B clays in PVA bionanocomposites are believed to further facilitate the formation of more hydrogen bonds within PVA matrices.

3.3. XRD Patterns

XRD analysis is a very useful material characterisation method to evaluate the crystalline structures of polymers and composites as well as to determine *d*-spacing values between clay interlayers. By monitoring the position and intensity of basal reflections from distributed silicate layers, nanocomposite structures (i.e., intercalated or exfoliated) as well as clay aggregated structures can be identified accordingly [28,29]. The XRD patterns of HNTs and corresponding nanocomposites are presented in Figure 3a. HNT patterns possess three major peaks of (001), (020)/(110), and (002) located at 2θ = 11.9°, 20°, and 24.9°, leading to *d*-spacing values of 0.74, 0.44, and 0.37 nm, respectively. The relevant peak taking place at 2θ = 24.9° is attributed to the presence of silica in the form of cristobalite and quartz [30]. After the incorporation of HNTs into PVA matrices in nanocomposite systems, the XRD characteristic peak at 2θ = 11.9° appears to be very weak at the low HNT content of 3 wt %. Similar XRD peaks have been detected at 2θ = 12.5° and 12.6° with slight peak shift at relatively high HNT contents of 5 and 10 wt %, respectively. Overall, decreasing the HNT content significantly reduces the intensity of XRD peaks for all PVA/HNT nanocomposites, which may be due to uniform HNT dispersion at the low HNT content levels. Such a phenomenon suggests that the disappearance or intensity reduction of XRD peaks at low HNT contents results from uniform HNT dispersion in a more randomly oriented manner. On the other hand, the reappearance of XRD peaks at higher HNT contents is indicative of possible HNT agglomeration.

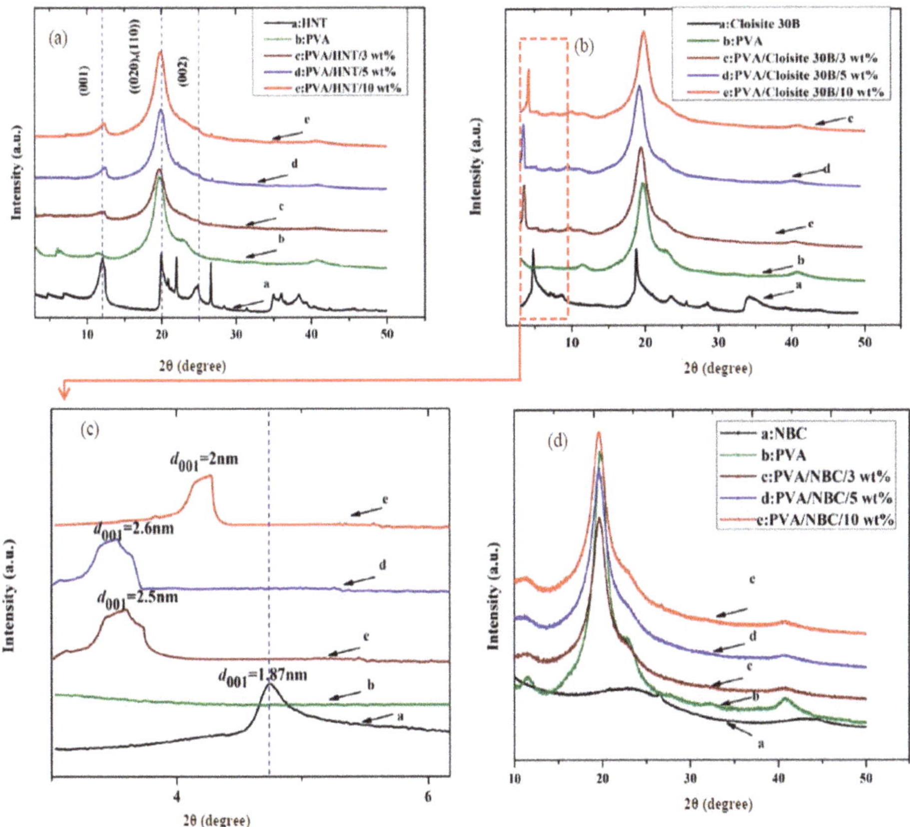

Figure 3. XRD patterns for PVA bionanocomposites reinforced with (**a**) HNTs, (**b,c**) Cloisite 30B clays with both wide and small diffraction angles, respectively, and (**d**) NBCs.

The XRD patterns of Cloisite 30B clays reveal the diffraction peak at $2\theta = 4.72°$ corresponding to the *d*-spacing value of 1.87 nm, as shown in Figure 3b,c. The (001) diffraction peak shifted to lower angles, as evidenced by the *d*-spacing values of 2.5, 2.6, and 2 nm for PVA/Cloisite 30B nanocomposites at the clay contents of 3, 5 and 10 wt %, respectively. This phenomenon clearly arises from the diffusion of polymeric chains inside clay interlayers to induce clay intercalation in agreement with PVA/Na$^+$ MMT nanocomposites [21] and PLA/Cloisite 30B nanocomposites [31]. The XRD peak for PVA alone appears at $2\theta = 19.7°$, which is associated with the total (101) crystalline atactic formation of PVA molecular chains [32] to slightly shift to lower diffraction angles when increasing the clay content in PVA bionanocomposites. The occurrence of PVA molecular chains at the (101) crystalline plane suggests that PVA matrices evolve towards crystalline structures under more constraints. A similar behaviour was also reported in PVA/clay nanocomposites [32,33], which is ascribed to the strong chemical interactions between nanofillers and polymer matrices. The aforementioned results indicate that Cloisite 30B clays are successfully intercalated and/or exfoliated by PVA molecular chains, and HNTs are homogenously dispersed at their low contents within continuous PVA matrices, which is attributed to active interactions between PVA matrices and clay nanoparticles due to strong hydrogen bonding taking place between carboxyl groups of PVA molecules and hydroxyl groups in the interlayer areas of Cloisite 30B clays or on the surface edges of HNTs [28].

In comparison, the XRD patterns of NBCs demonstrate two broad XRD peaks, as depicted in Figure 3d. The broad peaks located at $2\theta = 22.9°$ are associated with the sharp peaks of graphite assigned to the (002) diffraction plane [34]. Besides, the second broad peak detected at $2\theta = 43.6°$ characterises 2D in-plane symmetry (101) along with graphene layers. Moreover, XRD patterns of PVA/NBC bionanocomposites only show the diffraction angles from PVA, as illustrated in Figure 3d, which is consistent with the previous finding [35] in PVA/5 wt % graphene oxide (GO) nanocomposites with a clear disappearance sign of GO diffraction peaks in regular and periodic structures, leading to individually exfoliated GOs in PVA matrices.

3.4. Topographic Surface Morphology and Roughness

To assess nanofiller dispersion within PVA matrices, 3D height mapping images of PVA and PVA bionanocomposites are exhibited in Figure 4. As illustrated in Figure 4b, HNT nanoparticles are separated from one another with better HNT dispersion in PVA bionanocomposites reinforced with 3 wt % HNTs, as opposed to typical clay agglomeration and clustering issues beyond the 3 wt % HNTs shown in Figure 4c,d. An excessive amount of HNTs results in decreasing intraparticle spacing along with the higher intramolecular bonding of HNTs, leading to particle agglomeration [36]. Besides, the average root mean square (R_q) value as an indicator of the surface roughness of PVA bionanocomposites has been reported to be 2.4 ± 0.13 nm at the HNT content of 3 wt % when compared with 1.9 ± 0.17 nm for neat PVA, as shown in Figure 4a. It is suggested that the smooth surfaces of PVA/HNT bionanocomposites remain with the incorporation of HNTs at relatively low HNT contents, owing to their uniform dispersion. Such a finding is consistent with PVA/HNT hydrogels mentioned elsewhere [37]. At the low HNT content of 3 wt %, the smooth surfaces of PVA/HNT bionanocomposites can also be ascribed to the combination of strong interactions and good compatibility between HNTs and PVA matrices. Nonetheless, increasing the HNT content up to 5 and 10 wt % leads to much higher surface roughness (i.e., R_q = 4.54 ± 0.18 and 13.1 ± 0.23 nm, respectively) due to the presence of prevalent HNT aggregates [38].

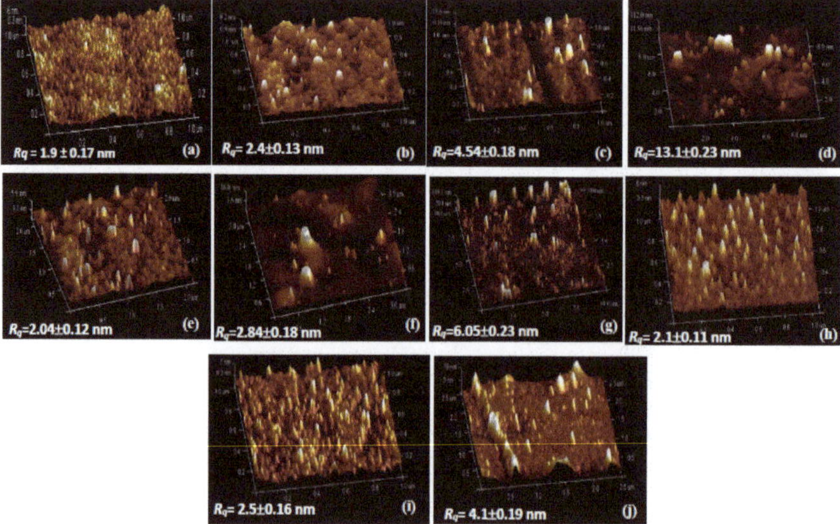

Figure 4. 3D AFM height mapping images of (**a**) PVA and PVA bionanocomposites reinforced with (**b**) 3 wt % HNTs, (**c**) 5 wt % HNTs, (**d**) 10 wt % HNTs, (**e**) 3 wt % Cloisite 30B clays, (**f**) 5 wt % Cloisite 30B clays, (**g**) 10 wt % Cloisite 30B clays, (**h**) 3 wt % NBCs, (**i**) 5 wt % NBCs, and (**j**) 10 wt % NBCs.

On the other hand, PVA/Cloisite 30B clay bionanocomposites and PVA/NBC bionanocomposites reveal different dispersibilities as opposed to PVA/HNT bionanocomposites. When nanofiller contents are below 10 wt %, spiky nanoparticles appear to be separated from one another, resulting in the homogeneous dipsersion of Cloisite 30B clays and NBCs within PVA matrices, as shown in Figure 4e,f, as well as Figure 4h,i accordingly. In particular, as the nanofiller content increases from 3 to 5 wt %, R_q values increase moderately from 2.04 ± 0.12 to 2.84 ± 0.18 nm for PVA/Cloisite 30B clay bionanocomposites, as well as from 2.1 ± 0.11 to 2.5 ± 0.16 nm for PVA/NBC bionanocomposite in contrast with 1.9 ± 0.17 nm for neat PVA. This finding suggests that the smooth surfaces for PVA bionanocomposites are evident at the low nanofiller contents of Cloisite 30B clays and NBCs, which is in good agreement with previous studies of PVA/nanocelloluse composite films [39]. On the contrary, the inclusion of 10 wt % Cloisite 30B clays and NBCs in PVA bionanocomposites consistently gives rise to increasing Rq values up to 6.05 ± 0.23 and 4.1 ± 0.19 nm, respectively, which are also far higher than that of neat PVA at 1.9 ± 0.17 nm. Such results indicate that the presence of aggregated Cloisite 30B clays and NBCs results in a much higher surface roughness on PVA surfaces, as expected. In omparison, the R_q value of 4.1 ± 0.19 nm for PVA/NBC bionanocomposites appears to be relatively low as compared with those of other PVA nanocomposites reinforced with carbon-based fillers such as PVA/reduced graphene oxide (rGO) nanocomposites with a R_q value of 4.6 ± 0.55 nm based on deposition layers [40].

Notwithstanding that the same manufacturing process condition and nanofiller contents have been utilised for preparing PVA bionanocomposite films, different nanoparticle types play an important role in changing the degree of surface roughness. Overall, with increasing the nanofiller content, the surface roughness of PVA bionanocomposites in this study is enhanced to a different extent, as evidenced by increasing the maximum relative change of surface roughness up to 589.4%, 218.4%, and 115.8%, respectively with the inclusion of HNTs, Cloisite 30B clays, and NBCs at the same filler content of 10 wt % shown in Figure 5. Such results suggest that NBCs may have better ability to be dispersed uniformly in PVA matrices as opposed to HNTs and Cloisite 30B clays due to their least increasing level in surface roughness, especially when beyond 5 wt % in filler content. Whereas, the effect of different nanoparticle shapes and sizes on the surface roughness of PVA bionancomposites becomes less pronounced at low filler contents below 3 wt %.

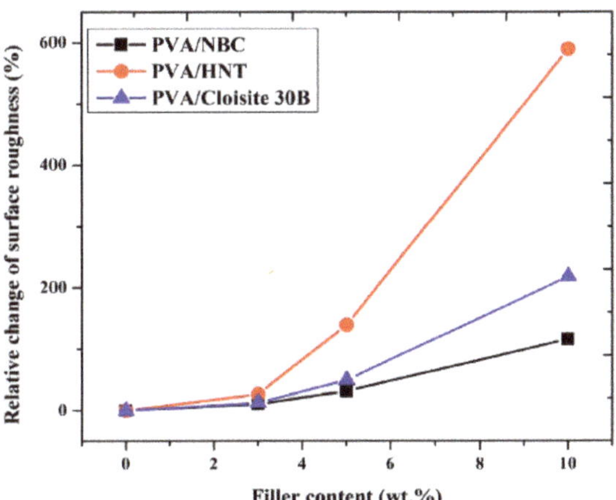

Figure 5. Relative change of surface roughness in terms of filler content in different PVA bionanocomposites.

3.5. Aspect Ratios of Embedded Nanofillers in PVA Bionanocomposites

The aspect ratio of nanofillers is regarded as one of key factors in reinforcement efficiency and mechanical performance of nanocomposites, which is generally defined as the ratio between the largest dimensions over the smallest dimension of nanofillers. According to this fundamental concept, the largest dimension of nanofillers can be represented by the lengths of tubular HNTs and platelet-like Cloisite 30B clays or the diameters of NBCs, while the smallest dimension is represented by the diameter of HNTs or thickness of Cloisite 30B clays and NBCs [28].

For instance, it is well known that when nanoclays are uniformly dispersed within polymer matrices, the formation of their exfoliated or intercalated structures leads to the improvement of the mechanical performance of nanocomposites to different extent, which is totally different from agglomerated nanoclays, resulting in the deterioration of their mechanical properties [1].

To investigate the degree of clay-exfoliated structures in detail, the height profiles of clay platelets relative to those of PVA matrices have been determined (Figure S1 in the Supporting Information). The thickness of 3 wt % Cloisite 30B clays within PVA matrices in bionanocomposites appears to be in range of 0.85–1.43 nm, suggesting typical exfoliated clay structures in dispersion. MMT clays are well known to be exfoliated when their thickness is similar to that of individual clay platelets (i.e., ~1 nm) [18]. Gaume et al. [33] and other groups [41,42] also detected intercalated and exfoliated structures of MMT clays in the thickness range of 1.3–5 nm.

The surface roughness mentioned earlier can be associated with nanofiller shapes and sizes since HNTs and Cloisite 30B clays may possess relatively high aspect ratios when compared with those of NBCs despite an existing 'nanofiller waviness' issue. HNTs and Cloisite 30B clays with high aspect ratios inevitably undergo considerable wavy nanofiller formation, thus undermining their homogeneous dispersion within polymer matrices [28]. Moreover, ultrasonication, as an effective fine nanoparticle dispersion technique in this study, also enables potentially damaging nanofiller structures particularly by applying high-power intensity or using a longer sonication time [43]. As such, the specific sizes/dimensions of nanofillers may vary to different extent, which are required to be determined for embedded HNTs, Cloisite 30B clays, and NBCs in PVA bionanocomposites to calculate their actual aspect ratios, as evidenced in Figure 6. The relevant frequency distribution of nanofiller dimensions are presented in Figures S2–S4 in Supporting Information as the reference. It is clearly revealed that the aspect ratios of nanofillers increase from 5.91 to 10.60 for HNTs in Figure 6a–c, as well as 5.75 to 8.17 for NBCs in Figure 6g–i with increasing the nanofiller content from 3 to 10 wt %. In contrast, the aspect ratios of Cloisite 30B clays decrease from 22.70, 12.38 to 13.46 when increasing the clay contents from 3, 5 to 10 wt % accordingly, despite their overall highest aspect ratios among all the nanofillers, as shown in Figure 6d–f. Such findings imply that the majority of Cloisite 30B clays tend to form exfoliated or intercalated clay structures with relatively high aspect ratios. However, the decrease in the aspect ratios of Cloisite 30B clays can be associated with more severe clay aggregation. It is very convincing that the aspect ratios of nanofillers can be greatly influenced by nanofiller shapes, and apparently 3D NBC nanoparticles have relatively low aspect ratios when compared with 1D platelet-like Cloisite 30B clays [28].

Overall, the aspect ratio may play a significant role in mechanical performance of nanocomposites when nanofiller shapes or structures are only considered within polymer matrices. However, for different types of nanofillers, several other factors such as the number of particles per unit volume, interphase modulus, interphase volume, and surface area, as well as the ratio of interphase volume per nanoparticle volume should also be taken into account for their overall material properties.

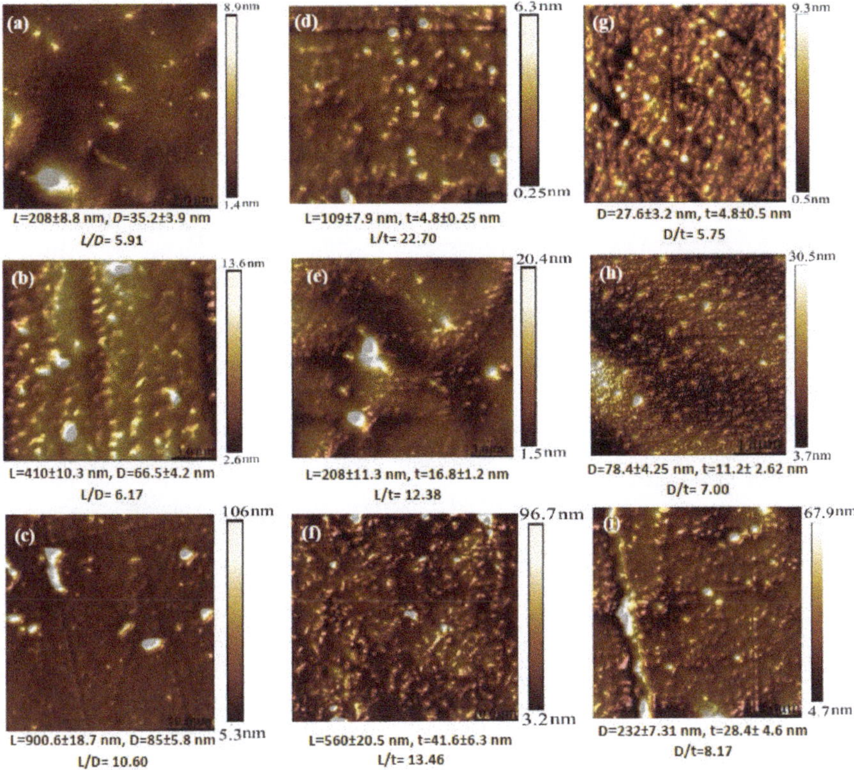

Figure 6. AFM topographic images of PVA bionanocomposites associated with aspect ratios of embedded fillers: (**a**) 3 wt % HNTs, (**b**), 5 wt % HNTs, (**c**) 10 wt % HNTs, (**d**) 3 wt % Cloisite 30B clays (**e**) 5 wt % Cloisite 30B clays, (**f**) 10 wt % Cloisite 30B clays, (**g**) 3 wt % NBCs (**h**) 5 wt % NBCs, and (**i**) 10 wt % NBCs.

3.6. Mechanical Properties

Figure 7 displayed the mechanical properties of PVA bionanocomposites reinforced with HNTs, Cloisite 30B clays, and NBCs at different nanofiller contents. Overall, the tensile moduli of such bionanocomposites increase significantly in a monotonic manner with the increasing nanofiller content, as shown in Figure 7a. The addition of only 3 wt % HNTs, Cloisite 30B clays, and NBCs result in increasing the tensile moduli by 40%, 52%, and 70.67% as opposed to that of near PVA at 2.08 GPa, which is in good accordance with the previous results obtained in PVA/starch/ glycerol (GL)/HNT nanocomposites [44] and PVA/chitosan/HNT nanocomposites [45]. More remarkably, the maximum increases in tensile modulus are achieved by 61.5%, 84.1%, and 123% with the addition of 10 wt % HNTs, Cloisite 30B clays, and NBCs, respectively when compared with that of neat PVA. This phenomenon usually takes place for most polymers filled with more rigid inorganic nanoparticles, as the reinforcements lead to much stiffer nanocomposite materials [46]. In particular, NBCs induce more reinforcement efficiency as nanofillers when compared with Cloisite 30B clays and HNTs. This can be clearly seen from the overall relatively high tensile moduli of PVA/NBC bionanocomposites, arising from much closer interactions between PVA matrices and NBCs via the formation of mechanical and hydrogen bonds owing to highly porous structures of NBCs. On the other hand, a different trend for the tensile strengths of PVA/NBC bionanocomposites is clearly revealed from those of PVA/Cloisite 30B clay bionanocomposites and PVA/HNT bionanocomposites, as shown in Figure 7b. The tensile strength of PVA/HNT nanocomposites is improved by 23% with the addition of 3 wt % HNTs relative to that of neat PVA at 70.32 MPa. Nonetheless, a drastic strength-decreasing tendency takes place with the strength reductions of 3.2% and 13.9% when embedded with 5 and 10 wt % HNTs accordingly. Such results indicate that the enhancement of tensile strengths for PVA/HNT nanocomposites depends on effective stress transfer from PVA matrices to HNTs, resulting from homogeneous HNT dispersion within PVA matrices. On the contrary, increasing the HNT content inevitably causes noticeable particle aggregation with more stress concentration sites around HNT agglomerates as a result of potential crack initiation to deteriorate the mechanical performance of bionanocomposites. With repect to PVA/Cloisite 30B clay bionanocomposites, their tensile strengths are increased by 18.8% and 28.4% with the incorporation of 3 and 5 wt % Cloisite 30B clays, respectively. This finding is ascribed to more uniform clay dispersion as well as the formation of stronger matrix–filler network structures, resulting from increasing hydrogen bonding between these constituents due to larger clay surface areas [47,48]. When the Cloisite 30B clay content increases up to 10 wt %, the tensile strength of PVA/Cloisite 30B clay bionanocomposites is decreased by 5.16% as opposed to that of near PVA. This phenomenon suggests that the aggregation of nanofillers at high clay content levels can undermine the tensile strengths of bionanocomposites. On the contrary, the tensile strengths of PVA/NBC bionanocomposites have the initial improvement up to 147.94 MPa (by a maximum level of 110.4%) when the NBC content increases from 0 to 3 wt %. Beyond 3 wt % NBCs, the tensile strengths of such bionanocomposites tend to decline until they reach the lowest strength levels of 96.34 MPa at the NBC content of 10 wt %. However, such lowest strength levels are still better than that of neat PVA. Overall, both tensile moduli and tensile strengths of PVA/NBC bionanocomposites are consistently superior to those of PVA/HNT bionanocomposites and PVA/Cloisite 30B bionanocomposites to confirm the most effective reinforcement efficiency of NBCs among all three different nanofillers.

The elongation at break and tensile toughness of PVA/Cloisite 30B clay bionanocomposites and PVA/NBC bionanocomposites were continuously decreased, especially beyond the nanofiller content of 3 wt %, as shown in Figure 7c,d. The maximum decreasing levels by approximately 59.5% and 58% in elongation at break have been detected for PVA bionanocomposites reinforced with 10 wt % Cloisite 30B clays and NBCs, respectively. This finding can be associated with the stiffening effect from filler reinforcements of NBCs and Cloisite 30B clays to restrict the mobility of PVA molecular chains, thus resulting in the overall flexibility reduction in bionanocomposite films [49]. As for PVA/HNT bionanocomposites, elongation at break and tensile toughness are increased by 12.7% and 16.9% with the incoproration of 3 wt % HNTs. Beyond that, they both remarkably diminish until maximum reductions of 50% and 45.3% take place at the HNT content of 10 wt %, respectively, as opposed to those of PVA. The former finding can be explained by good particle–matrix interactions with more uniform particle dispersion at low HNT contents. Whereas, the latter result can be associated with typical particle agglomeration at high HNT contents up to 10 wt % with the disapearance of the 'nano effect' of HNTs, since most HNT aggregates become less favourable microfillers with poor particle dispersion. As such, those HNT aggregates act as typical defects with high stress concentration prone to crack initiation towards mechanical failure, thus leading to poor material toughness [36].

In this study, the incorporation of three different nanofillers (i.e., HNTs, Cloisite 30B clays, and NBCs) has successfully enhanced the mechanical properties of PVA bionanocomposite films. According to our results, the highest increasing level among PVA/HNT bionanocomposites and PVA/NBC bionanocomposites can be achieved at the filler content of 3 wt %, as opposed to the optimum content of 5 wt % for PVA/Cloisite 30B clay bionanocomposites. Nonetheless, such an increasing rate achieved in PVA bionanocomposites using three types of nanofillers appears to be quite different, which is associated with various nanofiller features in terms of their structures and geometries, as well as the degree of compatibility between nanofillers and polymer matrices. With respect to nanofiller shape, it is well known that NBCs are regarded as 3D nanofillers as opposed to 2D nanofillers for HNTs and 1D nanofillers for Cloisite 30B clays. Different nanofiller shapes thereby influence the overall interfacial areas between fillers and polymer matrices, which plays a key role in the improvement of tensile strengths of nanocomposites with different levels of filler–matrix interactions. The second aspect is related to the structures, particularly the location of hydroxyl groups for nanofiller structures and amounts of hydroxyl groups within nanofillers. In the case of NBCs, hydroxyl groups are located inside their pores, which tend to more closely interact with PVA from a 3D point of view. As for HNTs, the majority of hydroxyl groups are constrained in inner tubes between layers, which makes the matrix–HNT interaction limited to the inner tubes of HNTs only. Moreover, in the case of Cloisite 30B clays, hydroxyl groups are located between layered structures, which means that the interactions between polymer matrices and platelet-like clays are limited to small highly constrained interlayer areas. The highest tensile moduli and tensile strength of bionanocomposite films have been achieved with the incorporation of NBCs relative to that of PVA. Several reasons can explain the above-mentioned results in relation to the mechanical properties of bionanocomposites. First, 3D nanofiller shape of NBCs can be generated at low nanofiller contents and in small particle sizes with relatively large interfacial areas, as compared with 2D HNTs and 1D Cloisite 30B clays. Liu and Brinson [2] investigated the effect of nanofiller geometry on the reinforcing efficiency of nanocomposites, which shows that at a low nanofiller content with the random nanofiller orientation, the transverse modulus of nanoparticle-based nanocomposites significantly exceeded those of nanotube-based nanocomposites, as well as nanoplatelet-based nanocomposites. Schadler et al. [50] reported that in the case of a nanocomposite system, with the incorporation of nanoparticles and nanotubes having a nanofiller diameter of 10 nm at the volume fraction of 10 vol%, the volume fraction of interfacial polymers was about 30% in the case of nanoparticle-based nanocomposites as opposed to only 10% for nanotube-based nanocomposites. The second reason in relation to the high mechanical performance of PVA/NBC bionanocomposites can be ascribed to the chemical structures of nanofillers in terms of the amounts and locations of hydroxyl groups in order to control the nanofiller dispersion within bionanocomposites, thus reflecting upon the bonding between polymer matrices and nanofillers. Pakzad et al. [51] indicated that the number and nature of hydrogen bonds had a substantial effect on the mechanical properties of nanocomposites. In the case of PVA/3 wt % NBC bionanocomposites, NBCs have highly porous structures with a large amount of hydroxyl groups located inside these pores when NBCs are uniformly dispersed. As confirmed by the FTIR and XRD results, polymeric chains enter these pores and form both hydrogen and mechanical bondings. Such two bonding types can be particularly recognised for NBCs as compared to Cloisite 30B clays and HNTs, thus significantly reflecting upon the enhanced mechanical properties of nanocomposites [4,5]. In case of PVA/5 wt % Cloisite 30B clay bionanocomposites in comparison to PVA/5 wt % HNT bionanocomposites, the strong adhesion of clays in polymer matrices associated with uniform clay dispersion leads to the strong interfacial bonding between nanoclays and polymer matrices, which thus significantly contributes to the improvement of mechanical properties of bionanocomposites.

The dispersion state of nanofillers can also influence the mechanical properties of PVA bionanocomposites. As mentioned earlier, NBCs have a better dispersion state than Cloisite 30B clays and HNTs. The incorporation of NBCs within PVA matrices yields smoother bionanocomposite films with higher tensile strength when compared with those of PVA/HNT bionanocomposites and PVA/Cloisite 30B clay bionanocomposites. The better dispersion state of NBCs improves their intercation with PVA matrices, thus leading to the higher tensile strengths of PVA/NBC bionanocomposites. On the contrary, increasing the nanofiller content appears to induce higher surface roughness as well as lower tensile strengths of nanocomposites, which indicates that nanofiller agglomeration apparently has detrimental effect on the improvement of tensile strength. This is particularly the case for PVA/HNT bionanocomposites due to the poor HNT dispersion state. On the contrary, PVA/3 wt % HNT bionanocomposites yield much higher elongation at break and fracture toughness as opposed to those of neat PVA, which are different from PVA/3 wt % NBC bionanocomposites and PVA/5 wt % Cloisite 30B clay bionanocomposites with corresponding lower values. Such results can be clearly explained by two major reasons. The first reason is ascribed to the number of nanoparticles depending on the volume and volume fraction of nanoparticles in bionanocomposites. At the same volume fraction, the number of NBCs is significantly larger than those of tubular HNTs or platelet-like Cloisite 30B clays. As such, this finding results in increasing the number of available reinforcements for improving the matrix rigidity and then decreasing the fracture toughness [43,52]. The second reason is related to the mechanism of fracture toughness, including the pre-crack effect for the fracture of nanocomposites. In general, crack deflection and crack pinning are most well-known mechanisms resulting in an increase in fracture energy [53], and consequently an increase in fracture toughness of nanocomposites. In both mechanisms mentioned earlier, the crack growth path can increase as long as those cracks reach nanofiller regions and the reinforcement shape highly affects the amount of crack deviation from their initial path. Since HNTs have larger lateral dimensions in comparison with NBCs, the cracks tend to pass over longer distances in PVA bionanocomposites reinforced with HNTs. Moreover, crack bridging is also well recognised as a fracture mechanism in nanocomposites reinforced with nanoparticles with a high aspect ratio [43]. An ideal situation in this mechanism occurs when nanotube fillers are still embedded in matrices while aligned in a perpendicular direction to crack faces. Consequently, PVA/HNT bionanocomposites achieve less reduction in fraction toughness when compared with the PVA/Cloisite 30B bionanocomposites and PVA/NBC bionanocomposites in Figure 7d.

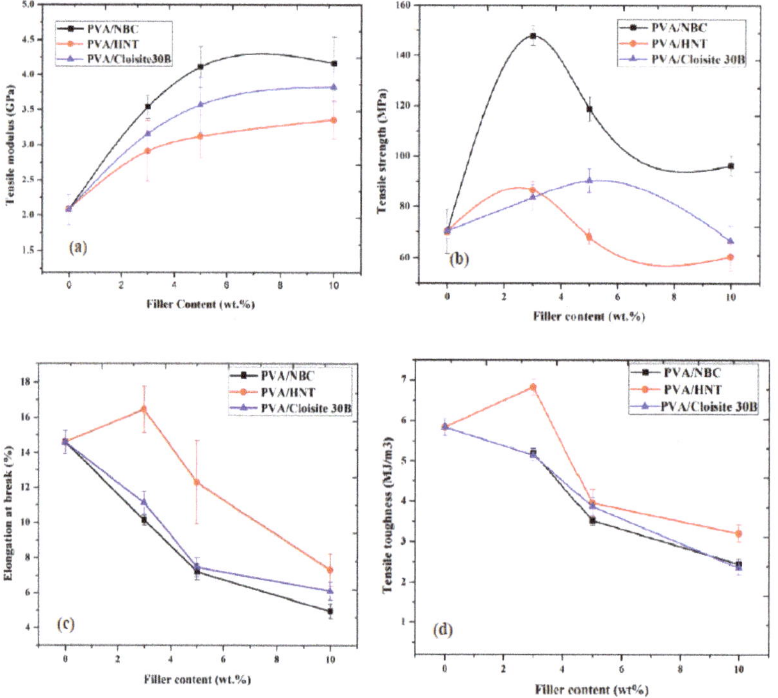

Figure 7. Mechanical properties of PVA bionanocomposites at different filler contents: (**a**) tensile modulus. (**b**) tensile strength, (**c**) elongation at break, and (**d**) tensile toughness.

3.7. Fracture Morphology

Figure 8 shows typical SEM micrographs of cross-sectional fracture surfaces for PVA, PVA/HNT bionanocomposites, PVA/Cloisite 30B clay bionanocomposites, and PVA/NBC bionanocomposites. It can be clearly seen in Figure 8b,e that PVA bionanocomposites reinforced with 3 wt % of HNTs and Cloisite 30B clays reveal much rougher fractured surfaces when compared with that of neat PVA films, as illustrated in Figure 8a. Moreover, 3 wt % HNTs or Cloisite 30B clays are distributed uniformly within PVA matrices. The good dispersion of both nanoparticles and the strong interaction between clay particles and polymer matrices clearly contribute to the reinforcing effect, as reflected by the increase in both tensile strength and elastic modulus. Nevertheless, in both PVA/HNT bionanocomposite and PVA/Cloisite 30B clay bionanocomposite systems, uniform multi-layered structures have not been achieved similar to those detected in PVA/3 wt % NBC bionanocomposites, as illustrated in Figure 8h. Such results are indicative of high NBC dispersability as compared with those of HNTs and Cloisite 30B clays, resulting in the highest mechanical performance. Meanwhile, at the HNT content of 5 wt %, particle–particle interactions are more favourable than their particle–matrix counterparts, as evidenced by more filler agglomeration in the presence of debonding and microvoid effects depicted in Figure 8c. Such defects in nanocomposite systems give rise to the decreasing tensile strengths of PVA/HNT bionanocomposites. However, as for PVA/5 wt % Cloisite 30B clay bionanocomposites, the clay dispersion appears to be still relatively uniform with the presence of small particle agglomeration shown in Figure 8f. With increasing the nanofiller contents of HNTs and Cloisite 30B clays from 5 to 10 wt %, the fracture surfaces of bionanocomposites films are altered from ductile characteristic to more brittle behaviour, as illustrated in Figure 8d,g, respectively. Similar phenomena are also found in PVA/NBC bionanocomposites according to Figure 8j. As is well known, decreasing surface roughness reveals that the failure mode of PVA bionanocomposite films can be quite different by changing from ductile to brittle fracture [54], which is consistent with the reduced mechanical properties of bionanocomposite films in this study.

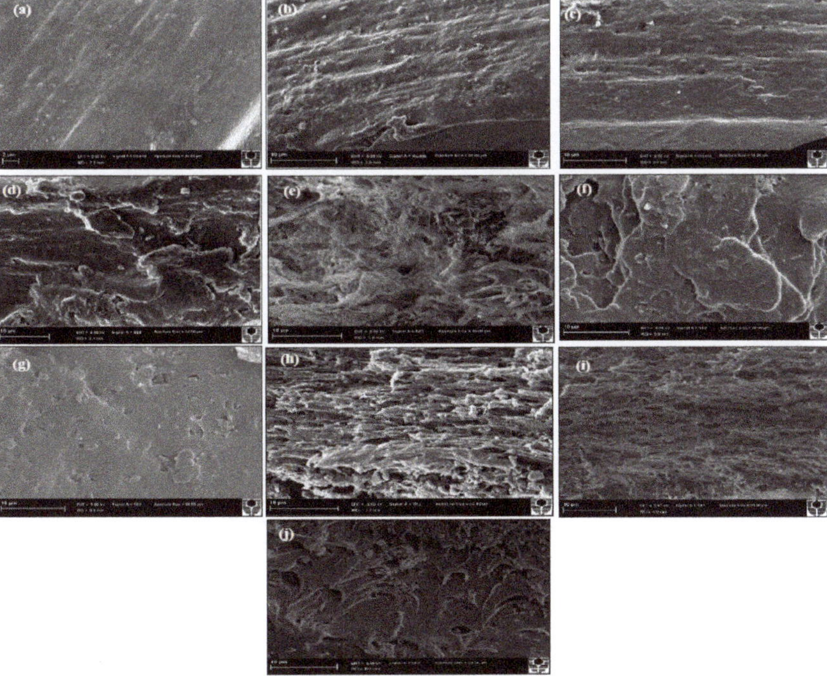

Figure 8. SEM micrographs of tensile fracture surfaces: (**a**) PVA, (**b**) PVA/3 wt % HNT bionanocomposites, (**c**) PVA/5 wt % HNT bionanocomposites, (**d**) PVA/10 wt % HNT bionanocomposites, (**e**) PVA/3 wt % Cloisite 30B clay bionanocomposites, (**f**) PVA/5 wt % Cloisite 30B bionanocomposites, (**g**) PVA/10 wt % Cloisite 30B clay bionanocomposites, (**h**) PVA/3 wt % NBC bionanocomposites, (**i**) PVA/5 wt % NBC bionanocomposites and (**j**) PVA/10 wt % NBC bionanocomposites. Note that Figure 8a shows the SEM micrograph with a scale bar of 2 μm while the rest of micrographs are labelled with a scale bar of 10 μm.

3.8. Thermal Properties

PVA is a water-soluble semicrystalline polymer, in which high physical interchain and intrachain interactions exist because of the typical hydrogen bonding between hydroxyl groups. The inclusion of nanoclays with hydroxyl groups can alter the intramolecular and intermolecular interactions of PVA molecular chains. This may affect both the crystallisation behaviour and physical structures of PVA. Similar observations can be found in previous studies dealing with PVA/HNT bionanocomposites [46,55].

Figure 9 shows the DSC results of PVA/HNT bionanocomposites, PVA/Cloisite 30B clay bionanocomposites, and PVA/NBC bionanocomposites. The summarised data of these thermal characteristics are reported in Table S1 in Supporting Information. For PVA/HNT bionanocomposites, it is clearly seen that the glass transition temperature T_g of PVA becomes unchanged with the addition of HNTs in bionanocomposite films, implying that HNTs do not play an important role in inhibiting the chain mobility of PVA molecules. Qiu and Netravali [46] also reported a similar result in T_g with the incorporation of HNTs into PVA. Such a finding might be related to the reduction in the entanglements and interactions of PLA polymeric chains with HNT inclusions. The relatively unchanged T_g in PVA/HNT bionanocomposites, as compared to that of neat PVA, may arise from different nanofiller geometries. The diameters of HNTs are at a nanoscaled level as opposed to submicron- or microsized tubular lengths that considerably exceed the typical gyration radii of polymeric chains [4]. As a result, HNTs cannot be completely wrapped by PVA molecular chains leading to many voids surrounding HNT particles. On the contrary, the high T_g values for all PVA/Cloisite 30B clay bionanocomposites are evident, as opposed to that of neat PVA. With the incorporation of 3, 5 and 10 wt % Cloisite 30B clays, the T_g values of such PVA bionanocomposites are moderately enhanced up to 67.5, 70.2 and 71.8 °C, respectively when compared with that of neat PVA at 65.19 °C. This phenomenon can be attributed to the confinement of polymeric chains by intercalated clay structures to prevent their segmental motions [1], which has also been recorded in PVA/MMT nanocomposites [56,57], PVA/bentonite nanocomposites [58], as well as PVA/starch/MMT nanocomposites [59]. In the case of PVA/NBC bionanocomposites, the T_g increases monotonically up to 75.06 °C with increasing the NBC content from 0 to 10 wt % accordingly. The incoporation of rigid NBC particles can restrict the chain mobility of PVA matrices so that higher T_g values are required for the phase transformation of nanocomposites from a glassy state to a rubbery state. This finding is well known for many types of nanofillers such as nanoclays, GOs, CNTs, HNTs, etc. [1]. Overall, the T_g values of PVA bionanocomposite films with the incorporation of NBCs and Cloisite 30B clays are much higher than those of PVA/HNT bionanocomposites. Such results indicate that NBCs and Cloisite 30B clays can restrict PVA chains more efficiently, as evidenced by the enhanced mechanical properties of corresponding bionanocomposite films. According to previous studies [60,61], the phenomenon of increasing T_g is primarily associated with the reduction in polymeric chain mobility by incorporating inorganic nanofillers. The incorporation of nanoparticles into polymer matrices can change the distribution of chain segments, which is most likely due to a change in the chain packing density in the vicinity of nanofiller surfaces. It should be noted that filler geometry may play a critical role to influence T_g. NBCs and Cloisite 30B clays have different nanofiller shapes to render the absorption of polymeric chains with entangled structures on their surfaces when nanofiller diameters are comparable to the gyration radii of polymeric chains. As such, it leads to increasing the packing density for polymeric chains and restricting their chain mobility as a result of higher T_g values. However, the incorporation of HNTs into PVA matrices has a minor impact on increasing T_g instead, which is consistent with the previous work [55]. Although the diameter of HNTs is on the nanometer scale, their length turns to be submicron- or microsized, which becomes considerably higher than the typical gyration radii of polymeric chains. As a consequence, it is very difficult for polymeric chains to cover entire HNT structures. Moreover, the presence of microvoids along HNT lengths could offer free sites for the segments of polymeric chains, resulting in an insignificant increase of T_g [55].

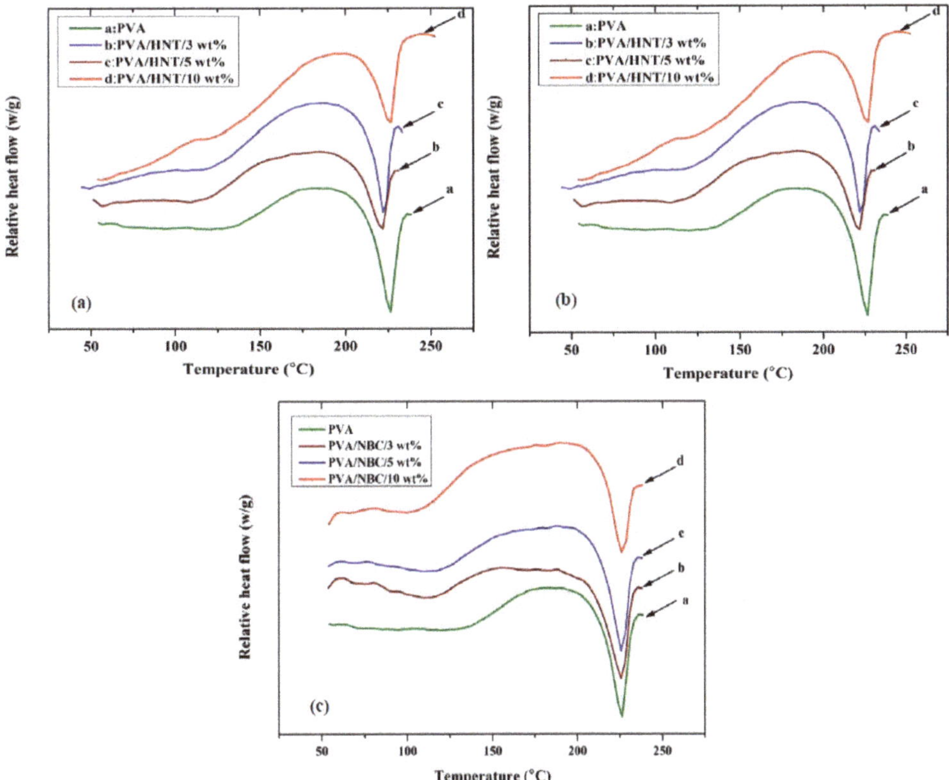

Figure 9. Differential scanning calorimetry (DSC) thermograms of PVA bionanocomposites reinforced with (**a**) HNTs, (**b**) Cloisite 30B clays, and (**c**) NBCs. The curves are shifted vertically for clarity.

The degree of crystallinity (χ_c) of PVA slightly increases from 36.65% for neat PVA to 38.2% and 37.2%, 40% for corresponding bionanocomposites with the incorporation of 5 wt % HNTs and Cloisite 30B clays and 10 wt % of NBCs, respectively. This suggests that such nanofillers have minor effect on the crystalline phases of PVA matrices in bionanocomposites. On the other hand, the melting temperature T_m of PVA bionanocomposites virtually has no change with the addition of Cloisite 30B clays and NBCs, as evidenced by the given T_m ranges of 220.44–221.62 °C and 221–225 °C, respectively when compared with that of neat PVA at 222.91 °C. However, PVA/HNT bionanocomposites possess a moderate increase in T_m up to 226.67 °C with the inclusion of 10 wt % HNTs. A similar phenomenon has also been noticed in PHBV/HNT nanocomposites [28] with their T_m values being increased from 169 to 173 °C when incoprorated with 5 wt % HNTs. Based on their XRD results, thicker and more oriented HNT/PHBV structures could be formed, leading to higher melting temperatures.

The thermal decomposition behaviours of PVA/HNT bionanocomposites, PVA/Cloisite 30B clay bionanocomposites, and PVA/NBC bionanocomposites have been evaluated using thermogravimetric analysis (TGA) with the corresponding results being presented in Figures 10 and 11, as well as Table S1 in Supporting Information. The relevant results for both systems reveal the existence of three major degradation steps according to previous studies [21]. Initially, the first degradation takes place at 107 °C owing to the breakage of hydrogen bonds, impurities, and monomers of vinyl alcohol. Then, the second degradation occurring at 274 °C involves a dehydration reaction on PVA molecular chains, the degradation of main backbones, as well as the decomposition of organic clays. This process is accompanied by a drastic mass change caused by the removal of organic compounds such as

CO$_2$ and the long molecular chains of alkyl derivatives. Finally, the third degradation step appears at a temperature level below 429 °C with more complexity including the further degradation of polyene residues to yield the carbon and hydrocarbon. The incorporation of HNTs, Cloisite 30B clays, and NBCs can increase the thermal stability of PVA by reducing the weight loss and increasing the decomposition temperatures, as presented in Figures 10 and 11. As for PVA/HNT bionanocomposites, the decomposition temperature at 5% weight loss $T_{5\%}$ increases from 200.2 °C for PVA to 265.3, 268.1, and 270.2 °C for PVA bionanocomposites reinforced with 3, 5 and 10 wt % HNTs, respectively. Such a finding suggests that HNTs work as an effective barrier material to heat and mass transfer. Moreover, the intrinsic hollow tubular structures of HNTs can produce the traps for volatile particles, thus improving the thermal stability by delaying the mass transfer during a decomposition process. Moreover, as clearly seen from the derivative thermogravimetry (DTG) curves in Figure 11a, the maximum decomposition temperature T_d of PVA shifts to a higher temperature level, which means that the dehydration process is hindered, resulting from strong interactions between PVA matrices and HNTs, as well as an important role of HNTs as good barrier materials to increase the thermal resistance of PVA bionanocomposites. Furthermore, the second DTG peaks of PVA/HNT bionanocomposites reinforced with 5 and 10 wt % HNTs are much wider than that of neat PVA in the presence of main and side peaks as compared to the single DTG peak for PVA at the same step. This signifies that a single peak for PVA can be attributed to the eliminated reaction while the side and main peaks of PVA/HNT bionanocomposites correspond to the eliminated reaction as well as the overlap of continual elimination and chain-scission reaction with the requirement of more energy to accrue at high temperatures [60].

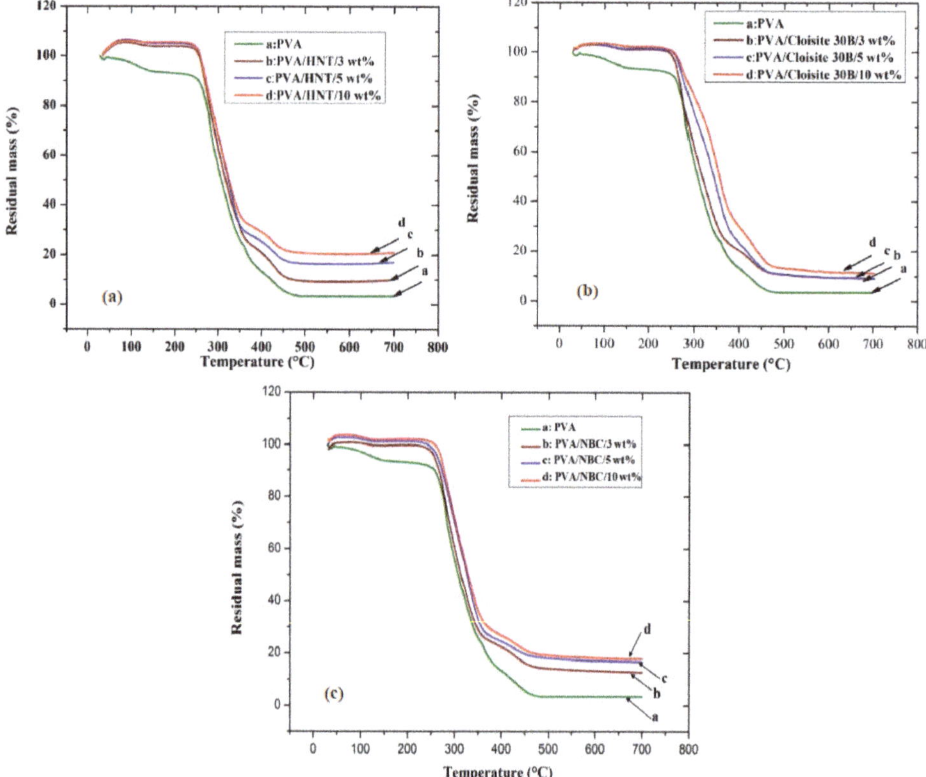

Figure 10. TGA curves for PVA bionanocomposites reinforced with (**a**) HNTs, (**b**) Cloisite 30B clays, and (**c**) NBCs.

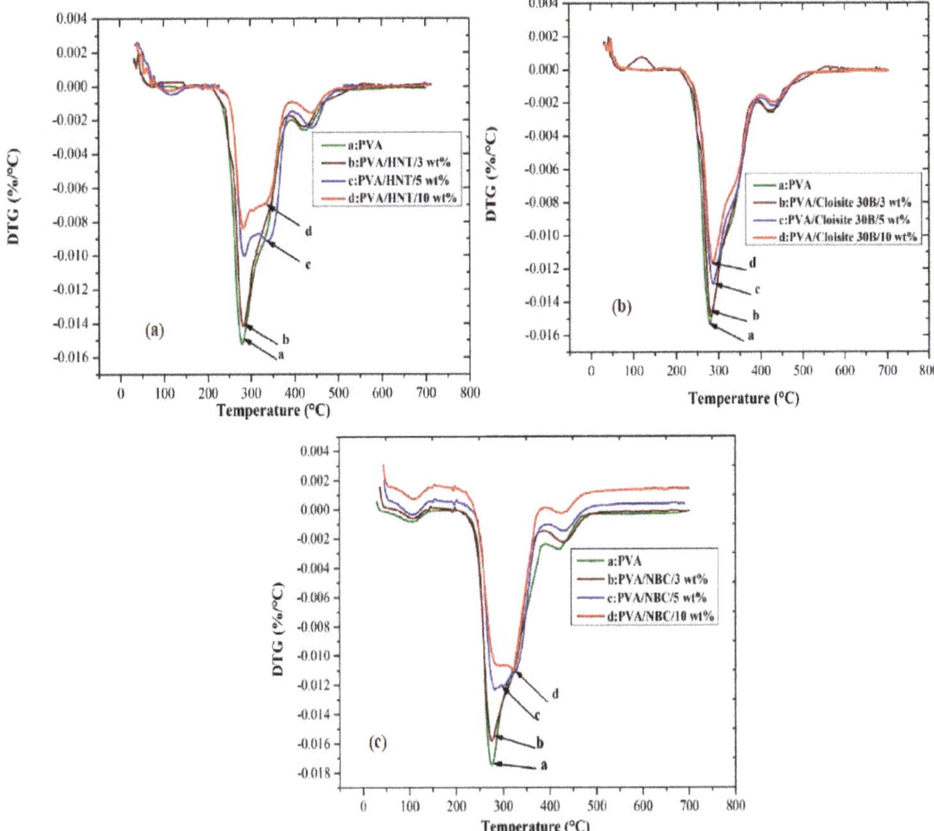

Figure 11. DTG curves for PVA bionanocomposites reinforced with (**a**) HNTs, (**b**) Cloisite 30B clays, and (**c**) NBCs.

On the other hand, in the case of PVA/Cloisite 30B clay bionanocomposites, with increasing the clay content, the thermal stability of bionanocomposites improves as compared to that of PVA, which is evidently demonstrated from consistently high $T_{5\%}$, the decomposition temperature at 80% weight loss $T_{80\%}$, and T_d values shown in Figure 10b. For instance, the $T_{80\%}$ of PVA increases from 363.5 to 407 °C with the inclusion of 5 wt % Cloisite 30B clays. Such a result is in good agreement with previous studies of PVA/MMT nanocomposites [21,62]. Moreover, the shift in the decomposition temperatures T_d for PVA/Cloisite 30B bionanocomposites depicted in Figure 11b suggests the hindrance of a dehydration process. Such a finding in thermal stability is associated with the presence of nanolayers acting as the barriers to maximise the heat insulation and minimise the permeability of volatile degradation products in the materials. This increase is also attributed to the decrease in oxygen permeability related to good clay dispersion in PVA matrices.

Moreover, the thermal stabilities of PVA/NBC bionanocomposites are improved significantly relative to that of PVA, as evidenced by consistently higher $T_{5\%}$, $T_{80\%}$, and T_d values, as shown in Figure 10c. The degree of thermal stability of bionanocomposites is even more pronounced when incorporated with NBCs in relation to $T_{5\%}$ and $T_{80\%}$. The $T_{5\%}$ of PVA/3 wt % NBC bionanocomposites was determined to be 256.3 °C, and it increases to 262.95 °C with the incorporation of 5 wt % NBCs, which is relatively similar to that of PVA/5 wt % Cloisite 30B bionanocomposites at 261.1 °C. On the other hand, the $T_{5\%}$ of PVA/3 wt % HNT bionanocomposites appears to be determined at 265.29 °C, which is significantly higher as compared with PVA bionanocomposites reinforced with NBCs and Cloisite 30B clays. On the contrary, the $T_{80\%}$ of PVA/3 wt % NBC bionanocomposites has been found to be 390.67 °C and reaches 440.28 °C with the inclusion of 5 wt % NBCs, which is significantly higher than those of PVA bionanocomposites reinforced with HNTs and Cloisite 30B clays. This result infers that the maximum thermal stability is achieved in the presence of NBCs as compared with HNTs and Cloisite 30B clays. As opposed to other nanofillers, the better NBC dispersion within PVA matrices takes place along with the higher barrier towards the thermal degradation. Therefore, such a barrier effect can counterbalance the degradation drawback with the further improvement of thermal stability [5].

The shift of maximum decomposition temperatures for the first and second degradation steps T_d and $T_d{'}$ in Figure 11c also means the obstruction of the dehydration process, which can result from the interaction between the hydroxyl groups of PVA and the hydroxyl groups on NBCs, as confirmed from our previous FTIR results. Furthermore, the mass loss process occurring in the second DTG peaks suggests that the thermal decomposition of PVA bionancomposites requires more reaction activation energy with the higher reaction order [63]. This finding may be attributed to the existence of NBCs working as effective barrier materials to limit the exothermicity of pyrolysis reaction with the better thermal resistance of PVA bionanocomposites. The wider DTG peaks of PVA/NBC bionanocomposites beyond 3 wt % NBCs at the second decomposition step demonstrates a similar trend to those of the corresponding PVA/HNT bionanocomposites along with the same dual-peak effect, as mentioned elsewhere [63].

4. Conclusions

PVA bionanocomposites reinforced with Cloisite 30B clays, HNTs, and NBCs have been successfully prepared and characterised. The following conclusions can be drawn.

The properties of PVA/HNT bionanocomposites are remarkably affected when embedding HNTs, which primarily depend on the HNT content. Nanofiller dispersion can lead to various morphological structures resulting in enhanced mechanical properties at different levels for such bionanocomposites. In particular, the incorporation of 3 wt % HNTs has improved mechanical properties, while increasing the HNT content beyond that causes the decreases in tensile strength, elongation at break, and tensile toughness of PVA/HNT bionanocomposites, which is possibly associated with the typical filler–matrix debonding effect. Moreover, the thermal properties of PVA/HNT bionanocomposites in terms of the degree of crystallinity, melting temperature, and thermal stability are enhanced with increasing the HNT content as opposed to those of neat PVA films.

The morphological structures of PVA/Cloisite 30B clay bionanocomposites demonstrate uniform clay dispersion within PVA matrices in combined clay exfoliated and intercalated structures, which are in agreement with those obtained from XRD results. The tensile strengths and Young's moduli of PVA/Cloisite 30B clay bionanocomposite films are increased considerably with increasing the clay content up to 5 wt % amid decreasing elongation at break and fracture toughness. The thermal properties of PVA/Cloisite 30B clay bionanocomposites are improved compared to those of neat PVA films due to the strong hydrogen-bonding interactions between PVA matrices and nanofillers.

The effect of different nanofiller shapes and structures on the properties of PVA/NBC bionanocomposite films reveals that the maximum tensile strength and tensile modulus can be achieved with the incorporation of NBCs. This can be related to the large amount of interphase resulting from a high degree of filler dispersion in the case of PVA/NBC bionanocomposites relative to those reinforced with HNTs and Cloisite 30B clays. Moreover, the thermal stability of PVA/NBC bionanocomposites is remarkably enhanced with the inclusion of NBCs in contrast to those incorporated with HNTs and Cloisite 30B clays. This can be ascribed to the uniform dispersion of NBCs to generate more efficient interfacial regions as opposed to other nanofillers.

Supplementary Materials: The following are available online at http://www.mdpi.com/2073-4360/12/2/264/s1, Figure S1: Characterisations of PVA/ 3wt% Cloisite 30B clay bionanocomposites: (a) height mapping image and (b) height profiles on cut-line sections A_{S1}-B_{S1} and A_{S2}-B_{S2}, Figure S2: Frequency distributions of dimensions of

HNTs embedded within PVA/HNT bionanocomposites at different HNT contents: (a) and (b) for HNT length and diameter (3 wt% HNTs), (c) and (d) for HNT length and diameter (5 wt% HNTs), as well as (e) and (f) for HNT length and diameter (10 wt% HNTs), Figure S3: Frequency distributions of dimensions of Cloisite 30B clays embedded within PVA/Cloisite 30B clay bionanocomposites at different clay contents: (a) and (b) for clay length and thickness (3 wt% Cloisite 30B clays), (c) and (d) for clay length and thickness (5 wt% Cloisite 30B clays), as well as (e) and (f) for length and thickness (10 wt% Cloisite 30B clays), Figure S4: Frequency distributions of dimensions of NBCs embedded within PVA/NBC bionanocomposites at different NBC contents: (a) and (b) for NBC thickness and diameter (3 wt% NBCs), (c) and (d) for NBC thickness and diameter (5 wt% NBCs), as well as (e) and (f) for NBC thickness and diameter (10 wt% NBCs), Table S1: Thermal properties of PVA bionanocomposite films.

Author Contributions: M.M. did the experimental work and the characterisation of material samples, M.M. and Y.D. analysed the data, jointly prepared, and approved the final research manuscript. All authors have read and agreed to the published version of the manuscript.

Funding: This research is financially supported by the Higher Committee for Education Development (HCED) in Iraq through a scholarship awarded to M.M to study at Curtin University, Australia.

Conflicts of Interest: The authors declare no conflict of interest.

References

1. Mousa, M.H.; Dong, Y.; Davies, I.J. Recent advances in bionanocomposites: Preparation, properties, and applications. *Int. J. Polym. Mater. Polym. Biomater.* **2016**, *65*, 225–254. [CrossRef]

2. Liu, H.; Brinson, L.C. Reinforcing efficiency of nanoparticles: A simple comparison for polymer nanocomposites. *Compos. Sci. Technol.* **2008**, *68*, 1502–1512. [CrossRef]

3. Dung, M.X.; Choi, J.-K.; Jeong, H.-D. Newly synthesized silicon quantum dot–polystyrene nanocomposite having thermally robust positive charge trapping. *ACS Appl. Mater. Interfaces* **2013**, *5*, 2400–2409. [CrossRef] [PubMed]

4. Mousa, M.; Dong, Y. Strong poly (vinyl alcohol) (PVA)/bamboo charcoal (BC) nanocomposite films with particle size effect. *ACS Sustain. Chem. Eng.* **2017**, *6*, 467–479. [CrossRef]

5. Mousa, M.; Dong, Y.; Davies, I.J. Eco-friendly polyvinyl alcohol (PVA)/bamboo charcoal (BC) nanocomposites with superior mechanical and thermal properties. *Adv. Compos. Mater.* **2018**, *27*, 499–509. [CrossRef]

6. Maliakal, A.; Katz, H.; Cotts, P.M.; Subramoney, S.; Mirau, P. Inorganic oxide core, polymer shell nanocomposite as a high K gate dielectric for flexible electronics applications. *JACS* **2005**, *127*, 14655–14662. [CrossRef]

7. Bogdanović, U.; Vodnik, V.; Mitrić, M.; Dimitrijević, S.; Skapin, S.D.; Zunic, V.; Budimir, M.; Stoiljkovic, M. Nanomaterial with high antimicrobial efficacy-copper/polyaniline nanocomposite. *ACS Appl. Mater. Interfaces* **2015**, *7*, 1955–1966. [CrossRef]

8. Makvandi, P.; Gu, J.T.; Zare, E.N.; Ashtari, B.; Moeini, A.; Tay, F.R.; Niu, L. Polymeric and inorganic nanoscopical antimicrobial fillers in dentistry. *Acta Biomater.* **2020**, *101*, 69–101. [CrossRef]

9. Makvandi, P.; Ali, G.W.; Sala, F.D.; Abdel-Fattah, W.I.; Borzacchiello, A. Hyaluronic acid/corn silk extract based injectable nanocomposite: A biomimetic antibacterial scaffold for bone tissue regeneration. *Mater. Sci. Eng. C* **2020**, *107*, 110195. [CrossRef]

10. Garcés, J.M.; Moll, D.J.; Bicerano, J.; Fibiger, R.; McLeod, D.G. Polymeric nanocomposites for automotive applications. *Adv. Mater.* **2000**, *12*, 1835–1839. [CrossRef]

11. Naskar, A.K.; Keum, J.K.; Boeman, R.G. Polymer matrix nanocomposites for automotive structural components. *Nat. Nanotechnol.* **2016**, *11*, 1026–1030. [CrossRef] [PubMed]

12. Mousa, M.; Dong, Y. Novel three-dimensional interphase characterisation of polymer nanocomposites using nanoscaled topography. *Nanotechnology* **2018**, *29*, 385701. [CrossRef] [PubMed]

13. Komarov, P.; Mikhailov, I.; Chiu, Y.T.; Chen, S.M.; Khalatur, P. Molecular dynamics study of interface structure in composites comprising surface-modified SiO_2 nanoparticles and a polyimide matrix. *Macromol. Theory Simul.* **2013**, *22*, 187–197. [CrossRef]

14. Meng, D.; Kumar, S.K.; Lane, J.M.D.; Grest, G.S. Effective interactions between grafted nanoparticles in a polymer matrix. *Soft Matter* **2012**, *8*, 5002–5010. [CrossRef]

15. Ganesan, V.; Jayaraman, A. Theory and simulation studies of effective interactions, phase behavior and morphology in polymer nanocomposites. *Soft Matter* **2014**, *10*, 13–38. [CrossRef]

16. Wierckx, N.; Narancic, T.; Eberlein, C.; Wei, R.; Drzyzga, O.; Magnin, A.; Ballerstedt, H.; Kenny, S.T.; Pollet, E.; Avérous, L.; et al. Plastic biodegradation: Challenges and opportunities. In *Consequences of Microbial Interactions with Hydrocarbons, Oils, and Lipids: Biodegradation and Bioremediation*; Steffan, R., Ed.; Springer: Cham, Switzerland, 2019; pp. 1–29.

17. Haroosh, H.J.M. Investigating Novel Biopolymeric Nanostructures for Drug Delivery. Ph.D. Thesis, Curtin University, Perth, Australia, 2014.

18. Ploehn, H.J.; Liu, C. Quantitative analysis of montmorillonite platelet size by atomic force microscopy. *Ind. Eng. Chem. Res.* **2006**, *45*, 7025–7034. [CrossRef]

19. Aloui, H.; Khwaldia, K.; Hamdi, M.; Fortunati, E.; Kenny, J.M.; Buonocore, G.G.; Lavorgna, M. Synergistic effect of halloysite and cellulose nanocrystals on the functional properties of PVA based nanocomposites. *ACS Sustain. Chem. Eng.* **2016**, *4*, 794–800. [CrossRef]

20. Ramadan, A.R.; Esawi, A.M.; Gawad, A.A. Effect of ball milling on the structure of Na^+-montmorillonite and organo-montmorillonite (Cloisite 30B). *Appl. Clay Sci.* **2010**, *47*, 196–202. [CrossRef]

21. Mallakpour, S.; Shahangi, V. Bio-modification of Cloisite Na^+ with chiral L-leucine and preparation of new poly (vinyl alcohol)/organo-nanoclay bionanocomposite films. *Synth. React. Inorg. Met. Org. Nano Met. Chem.* **2013**, *43*, 966–971. [CrossRef]

22. Fu, D.; Zhang, Y.; Lv, F.; Chu, P.K.; Shang, J. Removal of organic materials from TNT red water by bamboo charcoal adsorption. *Chem. Eng. J.* **2012**, *193*, 39–49. [CrossRef]

23. She, B.; Tao, X.; Huang, T.; Lu, G.; Zhou, Z.; Guo, C.; Dang, Z. Effects of nano bamboo charcoal on PAHs-degrading strain *Sphingomonas* sp. GY2B. *Ecotoxicol. Environ. Saf.* **2016**, *125*, 35–42. [CrossRef] [PubMed]

24. Lorenzoni, M.; Evangelio, L.; Verhaeghe, S.; Nicolet, C.; Navarro, C.; Pérez-Murano, F. Assessing the local nanomechanical properties of self-assembled block copolymer thin films by peak force tapping. *Langmuir* **2015**, *31*, 11630–11638. [CrossRef] [PubMed]

25. Babu, S.S.; Mathew, S.; Kalarikkal, N.; Thomas, S. Antimicrobial, antibiofilm, and microbial barrier properties of poly (ε-caprolactone)/cloisite 30B thin films. *Biotech* **2016**, *6*, 249. [CrossRef] [PubMed]

26. Huang, D.; Wang, W.; Kang, Y.; Wang, A. A chitosan/poly (vinyl alcohol) nanocomposite film reinforced with natural halloysite nanotubes. *Polym. Compos.* **2012**, *33*, 1693–1699. [CrossRef]

27. Abdullah, Z.W.; Dong, Y.; Davies, I.J.; Barbhuiya, S. PVA, PVA blends, and their nanocomposites for biodegradable packaging application. *Polym. Plast. Technol. Eng.* **2017**, *56*, 1307–1344. [CrossRef]

28. Carli, L.N.; Crespo, J.S.; Mauler, R.S. PHBV nanocomposites based on organomodified montmorillonite and halloysite: The effect of clay type on the morphology and thermal and mechanical properties. *Compos. Part A Appl. Sci. Manuf.* **2011**, *42*, 1601–1608. [CrossRef]

29. Ray, S.S.; Okamoto, M. Polymer/layered silicate nanocomposites: A review from preparation to processing. *Prog. Polym. Sci.* **2003**, *28*, 1539–1641.

30. Swapna, V.; Suresh, K.; Saranya, V.; Rahana, M.; Stephen, R. Thermal properties of poly (vinyl alcohol)(PVA)/halloysite nanotubes reinforced nanocomposites. *Inter. J. Plast. Technol.* **2015**, *19*, 124–136.

31. Rhim, J.-W.; Hong, S.-I.; Ha, C.-S. Tensile, water vapor barrier and antimicrobial properties of PLA/nanoclay composite films. *LWT-Food Sci. Technol.* **2009**, *42*, 612–617. [CrossRef]

32. Strawhecker, K.; Manias, E. Structure and properties of poly (vinyl alcohol)/Na^+ montmorillonite nanocomposites. *Chem. Mater.* **2000**, *12*, 2943–2949. [CrossRef]

33. Gaume, J.; Taviot-Gueho, C.; Cros, S.; Rivaton, A.; Therias, S.; Gardette, J.-L. Optimization of PVA clay nanocomposite for ultra-barrier multilayer encapsulation of organic solar cells. *Sol. Energy Mater. Sol. Cells* **2012**, *99*, 240–249. [CrossRef]

34. Li, R.S.; Li, X.; Deng, Q.; Li, D. Three kinds of charcoal powder reinforced ultra-high molecular weight polyethylene composites with excellent mechanical and electrical properties. *Mater. Des.* **2015**, *85*, 54–59. [CrossRef]

35. Liang, J.; Huang, Y.; Zhang, L.; Wang, Y.; Ma, Y.; Guo, T.; Chen, Y. Molecular-level dispersion of graphene into poly (vinyl alcohol) and effective reinforcement of their nanocomposites. *Adv. Funct. Mater.* **2009**, *19*, 2297–2302. [CrossRef]

36. Cheng, Z.L.; Qin, X.X.; Liu, Z.; Qin, D.Z. Electrospinning preparation and mechanical properties of PVA/HNTs composite nanofibers. *Polym. Adv. Technol.* **2017**, *28*, 768–774. [CrossRef]

37. Azmi, S.; Razak, S.I.A.; Abdul Kadir, M.R.; Iqbal, N.; Hassan, R.; Nayan, N.H.M.; Abdul Wahab, A.H.; Shaharuddin, S. Reinforcement of poly (vinyl alcohol) hydrogel with halloysite nanotubes as potential biomedical materials. *Soft Mater* **2017**, *15*, 45–54. [CrossRef]

38. Ormanci-Acar, T.; Celebi, F.; Keskin, B.; Mutlu-Salmanli, O.; Agtas, M.; Turken, T.; Tufani, A.; Imer, D.Y.; Ince, G.O.; Demir, T.U.; et al. Fabrication and characterization of temperature and pH resistant thin film nanocomposite membranes embedded with halloysite nanotubes for dye rejection. *Desalination* **2018**, *429*, 20–32. [CrossRef]

39. Hakalahti, M.; Mautner, A.; Johansson, L.-S.; Hanninen, T.; Setala, H.; Kontturi, E.; Bismarck, A.; Tammelin, T. Direct interfacial modification of nanocellulose films for thermoresponsive membrane templates. *ACS Appl. Mater. Interfaces* **2016**, *8*, 2923–2927. [CrossRef]

40. Humood, M.; Qin, S.; Song, Y.; Polychronopoulou, K.; Zhang, Y.; Grunlan, J.C.; Polycarpou, A.A. Influence of graphene reduction and polymer cross-linking on improving the interfacial properties of multilayer thin films. *ACS Appl. Mater. Interfaces* **2016**, *9*, 1107–1118. [CrossRef]

41. Carrado, K.A.; Thiyagarajan, P.; Elder, D.L. Polyvinyl alcohol-clay complexes formed by direct synthesis. *Clays Clay Miner.* **1996**, *44*, 506–514. [CrossRef]

42. Theng, B.K.G. Clay–polymer interactions: Summary and perspectives. *Clays Clay Miner.* **1982**, *30*, 1–10. [CrossRef]

43. Alishahi, E.; Shadlou, S.; Doagou, R.S.; Ayatollahi, M.R. Effects of carbon nanoreinforcements of different shapes on the mechanical properties of epoxy-based nanocomposites. *Macromol. Mater. Eng.* **2013**, *298*, 670–678. [CrossRef]

44. Abdullah, Z.W.; Dong, Y. Preparation and characterisation of poly (vinyl) alcohol (PVA)/starch (ST)/halloysite nanotube (HNT) nanocomposite films as renewable materials. *J. Mater. Sci.* **2018**, *53*, 3455–3469. [CrossRef]

45. Khoo, W.; Ismail, H.; Ariffin, A. Tensile, swelling, and oxidative degradation properties of crosslinked polyvinyl alcohol/chitosan/halloysite nanotube composites. *Inter. J. Polym. Mater. Polym. Biomater.* **2013**, *62*, 390–396. [CrossRef]

46. Qiu, K.; Netravali, A.N. Halloysite nanotube reinforced biodegradable nanocomposites using noncrosslinked and malonic acid crosslinked polyvinyl alcohol. *Polym. Compos.* **2013**, *34*, 799–809. [CrossRef]

47. Sapalidis, A.A.; Katsaros, F.K.; Kanellopoulos, N.K. PVA/montmorillonite nanocomposites: Development and properties. In *Nanocomposites and Polymers with Analytical Methods*; Cuppoletti, J., Ed.; InTechOpen: London, UK, 2011; pp. 29–50.

48. Asad, M.; Saba, N.; Asiri, A.M.; Jawaid, M.; Indarti, E.; Wanrosli, W. Preparation and characterization of nanocomposite films from oil palm pulp nanocellulose/poly (Vinyl alcohol) by casting method. *Carbohydr. Polym.* **2018**, *191*, 103–111. [CrossRef]

49. Raheel, M.; Yao, K.; Gong, J.; Chen, X.-C.; Liu, D.-T.; Lin, Y.C.; Cui, D.M.; Siddiq, M.; Tang, T. Poly (vinyl alcohol)/GO-MMT nanocomposites: Preparation, structure and properties. *Chin. J. Polym. Sci.* **2015**, *33*, 329–338. [CrossRef]

50. Schadler, L.; Brinson, L.; Sawyer, W. Polymer nanocomposites: A small part of the story. *JOM* **2007**, *59*, 53–60. [CrossRef]

51. Pakzad, A.; Simonsen, J.; Yassar, R.S. Gradient of nanomechanical properties in the interphase of cellulose nanocrystal composites. *Compos. Sci. Technol.* **2012**, *72*, 314–319. [CrossRef]

52. Santos, K.; Castel, C.D.; Liberman, S.; Oviedo, M.; Mauler, R. Polyolefin-based nanocomposite: The effects of processing aids. *J. Appl. Polym. Sci.* **2011**, *119*, 1567–1575. [CrossRef]

53. Wetzel, B.; Rosso, P.; Haupert, F.; Friedrich, K. Epoxy nanocomposites–fracture and toughening mechanisms. *Eng. Fract. Mech.* **2006**, *73*, 2375–2398. [CrossRef]

54. Li, Y.; Yang, T.; Yu, T.; Zheng, L.; Liao, K. Synergistic effect of hybrid carbon nantube–graphene oxide as a nanofiller in enhancing the mechanical properties of PVA composites. *J. Mater. Chem.* **2011**, *21*, 10844–10851. [CrossRef]

55. Liu, M.; Guo, B.; Du, M.; Jia, D. Drying induced aggregation of halloysite nanotubes in polyvinyl alcohol/halloysite nanotubes solution and its effect on properties of composite film. *Appl. Phys. A* **2007**, *88*, 391–395. [CrossRef]

56. Yu, Y.-H.; Lin, C.Y.; Yeh, J.M.; Lin, W.H. Preparation and properties of poly(vinyl alcohol)-clay nanocomposite materials. *Polymer* **2003**, *44*, 3553–3560. [CrossRef]

57. Sirousazar, M.; Kokabi, M.; Hassan, Z.M.; Bahramian, A.R. Polyvinyl alcohol/Na-montmorillonite nanocomposite hydrogels prepared by freezing-thawing method: Structural, mechanical, thermal, and swelling properties. *J. Macromol. Sci Part B* **2012**, *51*, 1335–1350. [CrossRef]

58. Jose, T.; George, S.C.; Maria, H.J.; Wilson, R.; Thomas, S. Effect of bentonite clay on the mechanical, thermal, and pervaporation performance of the poly(vinyl alcohol) nanocomposite membranes. *Ind. Eng. Chem. Res.* **2014**, *53*, 16820–16831. [CrossRef]

59. Ali, S.S.; Tang, X.; Alavi, S.; Faubion, J. Structure and physical properties of starch/poly vinyl alcohol/sodium montmorillonite nanocomposite films. *J. Agric. Food Chem.* **2011**, *59*, 12384–12395. [CrossRef] [PubMed]

60. Nakane, K.; Yamashita, T.; Iwakura, K.; Suzuki, F. Properties and structure of poly (vinyl alcohol)/silica composites. *J. Appl. Polym. Sci.* **1999**, *74*, 133–138. [CrossRef]

61. Mbhele, Z.; Salemane, M.; Van Sittert, C.; Nedeljković, J.; Djoković, V.; Luyt, A. Fabrication and characterization of silver–polyvinyl alcohol nanocomposites. *Chem. Mater.* **2003**, *15*, 5019–5024. [CrossRef]

62. Mondal, D.; Mollick, M.M.R.; Bhowmick, B.; Maity, D.; Bain, M.K.; Rana, D.; Mukhopadhyay, A.; Dana, K.; Chattopadhyay, D. Effect of poly (vinyl pyrrolidone) on the morphology and physical properties of poly (vinyl alcohol)/sodium montmorillonite nanocomposite films. *Prog. Nat. Sci. Mater. Int.* **2013**, *23*, 579–587. [CrossRef]

63. Peng, Z.; Kong, L.X. A thermal degradation mechanism of polyvinyl alcohol/silica nanocomposites. *Polym. Degrad. Stab.* **2007**, *92*, 1061–1071. [CrossRef]

Article

Application of Adaptive Neuro-Fuzzy Inference System in Flammability Parameter Prediction

Rhoda Afriyie Mensah [1], Jie Xiao [1], Oisik Das [2], Lin Jiang [1,*], Qiang Xu [1,*] and Mohammed Okoe Alhassan [1]

[1] School of Mechanical Engineering, Nanjing University of Science and Technology, Nanjing 210094, China; ramensah@ymail.com (R.A.M.); xiaojieblue@163.com (J.X.); mohaokoe2020@yahoo.com (M.O.A.)
[2] The Division of Material Science, Department of Engineering Sciences and Mathematics, Luleå University of Technology, Luleå 97187, Sweden; oisik.das@ltu.se
* Correspondence: ljiang@njust.edu.cn (L.J.); xuqiang@njust.edu.cn (Q.X.); Tel.: +86-125-84317023 (Q.X.)

Received: 12 December 2019; Accepted: 1 January 2020; Published: 5 January 2020

Abstract: The fire behavior of materials is usually modeled on the basis of fire physics and material composition. However, significant strides have been made recently in applying soft computing methods such as artificial intelligence in flammability studies. In this paper, multiple linear regression (MLR) was employed to test the degree of non-linearities in flammability parameter modeling by assessing the linear relationship between sample mass, heating rate, heat release capacity (HRC) and total heat release (THR). Adaptive neuro-fuzzy inference system (ANFIS) was then adopted to predict the HRC and THR of the extruded polystyrene measured from microscale combustion calorimetry experiments. The ANFIS models presented excellent predictions, showing very low mean training and testing errors as well as reasonable agreements between experimental and predicted datasets. Hence, it can be inferred that ANFIS can handle the non-linearities in flammability modeling, making it apt as a modeling technique for accurate and effective flammability assessments.

Keywords: flammability; heat release rate; microscale combustion calorimetry; multiple linear regression; adaptive neuro-fuzzy inference system

---

1. Introduction

Material flammability analysis is carried out using flammability characteristics obtained from fire experiments. Fire experiments are grouped into small-scale, bench-scale and full-scale experiments depending on the size of sample required. In reality, full-scale fire experiments provide the best estimates of material flammability since they imitate actual fire scenarios. However, due to the high cost, operational and technical challenges, small-scale experiments have been adopted for flammability evaluations. To validate the accuracy of small-scale experiments, correlation analysis conducted between the different scales of experiments shows feasible relationships in the test results [1–4].

Microscale combustion calorimetry (MCC) is a small-scale fire experiment operated on the principles of oxygen consumption calorimetry. MCC has received significant attention in recent years and remains the most commonly cited fire experiment for polymer flammability assessments due to the wealth of data measured from it [5,6]. Amongst the data is the heat release capacity (HRC), a flammability parameter and material property measured exclusively from the 'Method A' [6] of MCC. HRC integrates thermal stability and combustion properties, hence rendering it the best predictor of a materials response to fire. HRC is defined as the ratio of the maximum value of the specific heat release rate to the average heating rate of a sample [5–9]. The total heat released (THR) is a measure from the total heat released when a sample undergoes a complete combustion in an oxygen atmosphere. THR is also a flammability property calculated from the results measured from the Method A procedure [7,9,10]. Several chemistry-based models, such as inverse modeling, quantitative structure-property relationship

(QSPR), quantitative structure-activity relationship (QSAR) and additive molar group contribution methods, have been developed over the years to estimate these important fire safety parameters (HRC and THR). Although effective, some disadvantages, like large prediction errors, have been identified in the traditional modeling approaches and need to be addressed.

Until now, correlation analysis and statistical prediction models have been appropriate analytical tools in material flammability [11]. The significance of the prediction models lies in the employment of robust techniques and strategies for the accurate estimation of flammability parameters, whereas correlation analysis assesses the relationship between the predicted or measured parameters. Statistical analysis is mostly empirical, devoid of the chemical compositions and physical structure of the material under consideration [12]. It is imperative to note that fire experiments are demanding, expensive and time consuming. Similarly, traditional flammability parameter predictions involve sophisticated fire modeling and calibration, requiring a great deal of expertise and computing power. It is, therefore, quite convenient for researchers to opt for these system theoretical models as opposed to conceptual models. Statistical models in recent times largely embrace the artificial neural network (ANN) due to its ability to capture complex nonlinearities in a system when compared to linear regression methods. ANN mimics the operation of the human brain by processing information available to the input layers to achieve a desirable output.

Generalized regression and the ANN's feed-forward back propagation methods have been applied in flammability studies to predict peak heat release rate, heat release capacity, total heat released, etc., with a high level of accuracy [13,14]. Deviation of the predicted results from the actual or experimental results was seen to be low when compared with classical methods including quantity structure activity/property relationships. Alternative methods, such as the group method of data handling neural network, also enhanced the predictability of the aforementioned methods, as reported by Mensah et al. [15]. It is well-known that the overall performance of these models highly depends on physical variables such as the network architecture, transfer and activation functions. Despite their simple implementation, the iterative process employed to obtain optimal variables for the prediction makes them quite cumbersome.

Adaptive neuro-fuzzy inference system (ANFIS) was developed in the 1990's on the principles of the Takagi-Sugeno fuzzy inference system [16]. This method has been applied in several research areas with an excellent degree of accuracy; however, it has not been applied in in flammability studies. It is a hybrid analytical method, i.e., it combines the merits of the neural network and theories of fuzzy logic systems in its operation [17]. Each concept plays an important role in achieving the required result. While neural networks control the representation of information and the physical architecture, fuzzy logic systems imitate human reasoning and increase the model's ability to manage uncertainty within the system [17]. ANFIS basically learns the features of a given data and alters the system parameters to suit the required error criterion of the system in order to generate an output. It utilizes less computing power with shorter training times and, therefore, serves as a suitable method for the prediction of flammability parameters [18,19].

In the current study, an MCC experiment was conducted with extruded polystyrene samples at heating rates ranging from 0.1 to 3.5 K s^{-1}. This research explored the applicability of ANFIS in the prediction of HRC and THR derived from the experiment. The degree of accuracy was determined by comparing the root mean squared error (RMSE) criterion, the coefficient of correlation and the coefficient of determination. A comparative analysis was carried out with multiple linear regression and the feed-forward back propagation neural network to show the efficacy, accuracy and superiority of ANFIS. The modeling results obtained from this research will help validate the robustness of ANFIS and its continual usage in future flammability assessments.

2. Experimental Methods

2.1. Material

The flammability of pure extruded polystyrene (XPS) obtained from Zhengbang New Building Material Co. Ltd. located at Zaozhuang, China was studied. The samples were cut from large boards of XPS into milligram sizes for the experiments. The Mettler AX-205 Analytical Semi Micro Balance Delta Range from Hamilton company in Reno, NV, US. The instrument has a readability of 0.01 mg and a weighing range of 81 g was used to weigh the samples. The material properties are listed in Table 1 [5,11].

Table 1. Properties of XPS.

Property	Value
Thermal conductivity/Wm^{-1} K^{-1}	0.1316
Thermal diffusivity/m^2 s^{-1}	0.4201
Specific heat capacity/kJ g^{-1} K^{-1}	1.34
LOI %	19.3
Density, ρ/kg m^{-3}	52.6
Density of molten material, ρ/kg m^{-3}	828

2.2. Microscale Combustion Calorimetry (MCC)

The MCC experiment took place at the VTT Technical research center of Finland in MCC-2 equipment from Govmark (Farmingdale, NY, US) Limited. According to standards in ASTM D7309-13 [20], the experimental procedure applied was in line with Method A. Milligram samples taken from extruded polystyrene boards were weighed and prepared for the MCC experiment. Samples of mass ranging from 1 to 4 mg were heated at a temperature of 75 to 600 °C in a pyrolyzer under heating rates ranging from 0.1 to 3.5 K s^{-1}. The volatile pyrolysis products were removed from the pyrolyzer by nitrogen gas and were oxidized with excess oxygen at 900 °C in a tubular combustion furnace. Oxygen consumption calorimetry was applied for calculating the heat release rate from the volumetric flow rate and the oxygen concentration of the gases that flowed out of the combustor [6,13,14,20]. The samples were tested in three replicates and an average of the measured results was recorded. The samples were labelled as xps_1_0.1 representing the first sample tested under 0.1 K s^{-1}, and so on. The heat release temperature, time to heat release and heat release rate were measured and recorded. HRC was obtained by dividing the specific heat release rate by the corresponding heating rate. Additionally, THR was calculated from the area under the specific heat release rate against time plots at a given heating rate.

2.3. Adaptive Neuro-Fuzzy Inference System (ANFIS)

The artificial neural network has a unique quality of learning the input and output datasets for the system and reproducing accurate values to match the data. Fuzzy logic, on the other hand, has the capability of interpreting, organizing, representing and also adding an element of reasoning to an applied data. A Fuzzy Inference System (FIS) is made up of four distinct components, namely a fuzzifier, fuzzy rules, inference engine and a de-fuzzifier [17]. With a given input dataset, the output of an FIS is determined by building the fuzzy rules, fuzzifying the inputs with the membership functions, developing a rule strength and finding its consequences. The consequences are then put together to obtain an output distribution, which is then further de-fuzzified. There are two types of FIS, the Mamdani and Sugeno types. The Mamdani type FIS requires the use of fuzzy rules to link fuzzy set to outputs, which are de-fuzzified to produce scalar variables. The Sugeno FIS is quite similar to the Mamdani type, however, no output distribution or output membership function is included in the system. Instead, to obtain the output, the inputs are multiplied by a constant and the results are added [21].

A combination of ANN and FIS, therefore, employs the architecture of ANN with its learning ability and integrates fuzzy reasoning to add logic and the prior knowledge effect. With this method, ANN accurately learns the membership functions of a fuzzy logic system in order to build the input data of the model, which is organized as fuzzy IF-THEN rules by the FIS. This hybridization is carried out to ensure the optimization of the parameters used in developing the FIS with an application of a learning algorithm for input-output mapping. The architecture of a typical ANFIS structure that has two input variables with five layers is presented in Figure 1 [21]. It must be noted that the squares and circles in Figure 1 represent adaptive nodes and fixed nodes, respectively.

The first layer has four adaptive nodes showing the premise parameters A_1, A_2, B_1, B_2. The fuzzy IF-THEN rules for Figure 1 are described below.

Rule 1: If x is A_1 and y is B_1, then

$$f_1 = p_1 x + q_1 y + r_1. \tag{1}$$

Rule 2: If x is A_2 and y is B_2, then

$$f_2 = p_2 x + q_2 y + r_2 \tag{2}$$

where p_i, q_i, r_i are the consequent parameters and $i = 1,\ 2$. The first layer has two inputs x and y representing the heating rate and sample mass, respectively, with one output (either HRC or THR). The values in each input variable are changed to a membership value using the assigned membership functions. The membership function usually applied for ANFIS is the generalized bell function. The output of Layer 1, which is also the fuzzy membership value, is denoted as O_i, representing the value for any ith node in layer j. The operations in the adaptive nodes are shown in Equations (1) and (2).

$$O_{1,i} = \mu A_i(x),\ i = 1,\ 2 \tag{3}$$

$$O_{1,i} = \mu B_{i-2}(y),\ i = 3,\ 4 \tag{4}$$

From Equations (1) and (2), $O_{1,i}$ represents the membership function (generalized bell function, triangular function or Gaussian function) of the fuzzy set A_1, A_2, B_1, B_2, which also shows the connection between the input set x and y and the fuzzy set. The variables A_i and B_{i-2} are all parameters in the ith node of layer j.

$$\mu_{A_i}(x) = \left\{ 1 + \left[\frac{(x - c_i)}{a_i^2} \right]^{b_i} \right\}^{-1} \tag{5}$$

The membership functions can be expressed in mathematical forms as Equations (4)–(7).

$$\begin{aligned}
\text{Triangular}:\ \mu_x(a) &= \frac{(a-x)}{(y-x)}, x \le a \le y \\
&= \frac{(z-a)}{(z-y)}, y \le a \le z \\
&= 0
\end{aligned} \tag{6}$$

$$\text{Gaussian}:\ \mu_x(a) = \frac{1}{1 + \left(\frac{a-z}{x} \right)^2} \tag{7}$$

Bell shaped:

$$\mu_{Xi}(a) = \frac{1}{1 + \left(\frac{a-z_i}{x_i} \right)^{2x_i}}, i = 1, 2, \dots \tag{8}$$

$$\mu_{Yi}(b) = \frac{1}{1 + \left(\frac{b-z_j}{x_j} \right)^{2y_j}}, j = 1, 2, \dots \tag{9}$$

Altering the consequent parameters of the membership function will subsequently produce a different membership function and ensures the flexibility in defining membership functions.

Layer 2 contains fixed nodes that operate on multiplication rules. In Layer 2, the product of the various input signals is obtained to generate rule-firing strengths. This operation is presented in Equation (8).

$$O_{2,i} = \omega_i = \mu_{A_i}(x) \times \mu_{B_i}(x), i = 1, 2 \tag{10}$$

Normalization of the firing strengths attained in the second layer takes place in Layer 3. The ratio of the firing strength of the ith rule to the sum of rules in the model is assessed at this point. The mathematical expression of the normalization process is shown in Equation (9) [22–25].

$$O_{3,i} = \bar{\omega}_i = \frac{\omega_i}{\omega_1 + \omega_2}, i = 1, 2 \tag{11}$$

The rules for the outputs are computed in the fourth layer. The consequent parameters are adjusted until an optimal value is obtained with minimal errors. This layer is made up of adaptive nodes, which helps in calculating the total output of the developed model. The output of Layer 3, $\bar{\omega}_i$, is multiplied by a parameter set $\{a_i, b_i, c_i\}$ to get the output of Layer 4 [26].

$$O_{4,i} = \bar{\omega}_i f_i = \bar{\omega}_i (p_i x + q_i y + r_i) \tag{12}$$

Lastly, the various outputs in Layer 4 are added up to obtain the final output of the ANFIS model. Layer 5 has one fixed node with a summation function operation [17].

$$O_{5,i} = \sum_i \bar{\omega}_i f_i \tag{13}$$

The neuro-fuzzy app designer in Matlab provides a very simple platform for ANFIS predictions. After loading the training and test data, the app trains the data to shape the membership functions and generates fuzzy rules for the calculation of the output data. The language recognition and reasoning aspect is handled by the fuzzy logic part of the app.

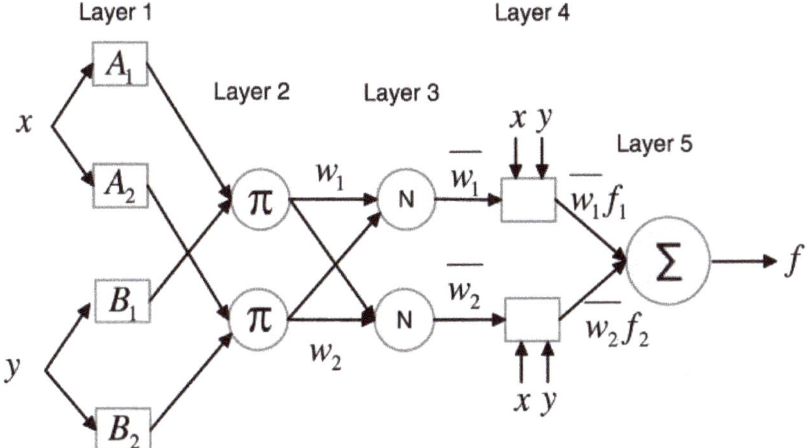

Figure 1. Structure of Adaptive Neuro Fuzzy Inference System.

2.4. Multiple Linear Regression (MLR)

Regression analysis is used to evaluate the cause-effect relationship among variables in a given dataset with the aim of developing prediction equations. Multiple linear regression is a statistical method applied to describe how several explanatory variables define a corresponding dependent variable. MLR basically models the linear relationship between dependent and independent variables [27]. MLR fits a linear equation of the form shown in Equation (12) to the observed data. The coefficients in the fitted equation show the effect of or the changes in the dependent variable when the independent variables change by one unit, while the constant attached (ε) shows the value of the dependent variable if all the other variables are zero.

$$y_i = \beta_0 + \beta_1 x_{i1} + \beta_2 x_{i2} + \ldots + \beta_p x_{ip} + \varepsilon \tag{14}$$

where for any $i = n$ observations, y_i is the dependent variable (HRC and THR), x_i represents the independent variable (sample mass and heating rate) and the y-intercepts are denoted by β_0 and β_p, representing the slope coefficients of x_i. Lastly, the error obtained during the modeling process is represented by ε. The degree of linearity is evaluated using the coefficient of determination, while the error term accounts for the variation or the difference between the predicted and actual variables. To ascertain the suitability of conducting MLR on a specific dataset, various tests such as the linearity, normality, missing value test and extreme value test are conducted [28,29].

2.5. Model Implementation

ANFIS prediction technique was applied to estimate the heat release capacity and total heat released of extruded polystyrene samples. The Sugeno method was used since it is known to display faster convergence and better accuracy than the Mamdani method [15]. A trial and error method was used to select the optimal membership function for the model. The membership function that presented the least root mean squared error was chosen. The other variables such as the optimization method (hybrid or back propagation), method of generating FIS (sub-clustering or grid partition), the number of membership functions within a hidden layer, the types of composition function and interference were selected based on the minimum error approach. The MCC experimental data, divided into training and testing sets, were used as input data in the neuro-fuzzy designer app built in MATLAB (R2018a). The computer used for the training has the following specifications: 64-bit operating system and a 4 GB memory with an i3-4005U CPU @ 1.70 GHz processor. The suitable structure for the model was selected depending on the data size and application. The necessary parameters were selected and the model was trained to evaluate the learning ability and determine the structural parameters (consequence and premise) with an optimization algorithm. The hybrid optimization algorithm is an integration of the gradient descent and the least squares method [21]. The outputs of the various nodes are forward propagated until it reaches the fourth layer. The consequent parameters in this section are determined by the least-squares method. The errors attained are back propagated, and the premise parameters are altered and adjusted using the gradient descent algorithm. The error factor in ANFIS is defined as presented in Equation (13).

$$E = \sum_{k=1}^{n} \left(f_k - f_k' \right)^2 \tag{15}$$

Basically, the hybrid method employs different algorithms for each of the training parts, hence, eliminating the local minima convergence and increasing the performance of the model. The overall performance was assessed using the test patterns in the Neuro-Fuzzy Designer app [21].

3. Results and Discussion

3.1. MCC Experimental Results

The specific heat release rate of XPS measured during the MCC experiment is plotted against temperature in Figure 2. The figure affirms the relationship between HRR, heating rate and temperature at peak HRR (pTemp), which is that the heat release rate and the corresponding heat release temperature increases with the increasing heating rate. Figures 3 and 4 also show the variation of HRC with respect to sample mass and heating rate. On average, 1.5 mg samples had the highest HRC values compared to the other masses. More distinct lines at lower heating rates are shown in Figure 4, which are the HRC values versus the inverse of heating rate [11,14,30–32].

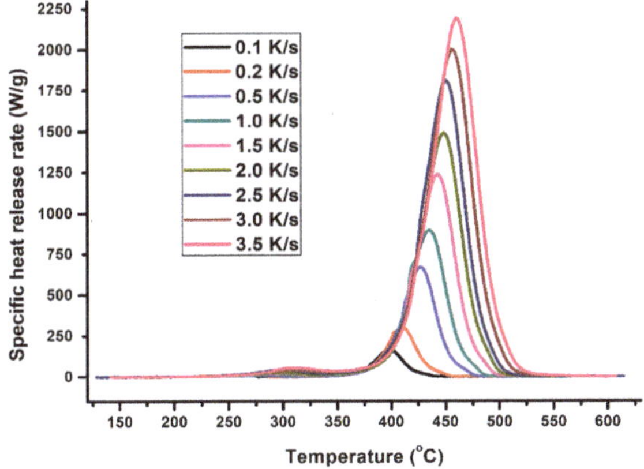

Figure 2. Plot of specific heat release rate versus temperature.

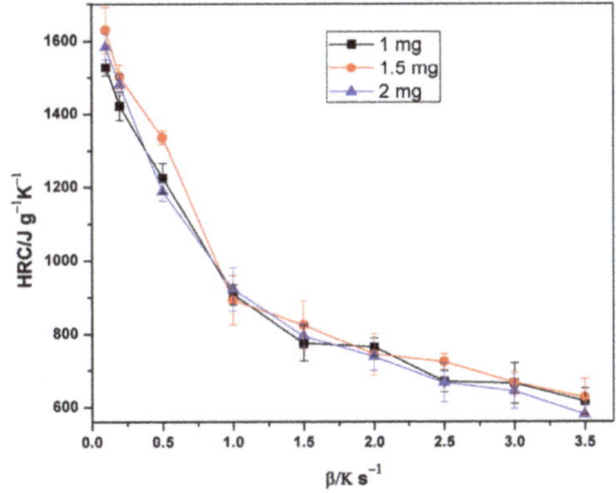

Figure 3. Plot of Heat Release Capacity versus heating rate for different sample masses.

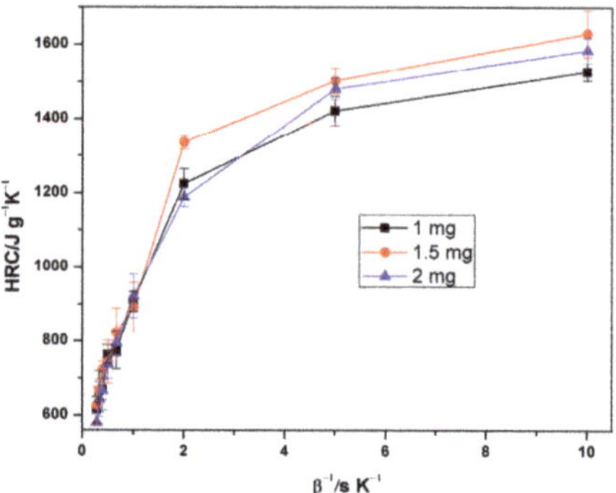

Figure 4. Plot of HRC versus inverse of heating rate for different sample masses.

3.2. Statistical Analysis

To determine the regression equation using multiple regression analysis, HRC and THR were selected as the independent variables with heating rate and sample mass being the dependent variables. The descriptive statistics of the input data are listed in Table 2. The analysis of variance showing the influence of HRC and THR on heating rate and sample mass in this regression analysis is presented in Tables 3 and 4. It is clearly seen that the dependent variables have a greater significance in the estimation of HRC than THR. The test statistic of HRC has an F-value of 53.85, which is larger than the critical value $F_{0.05,\ 2,25} = 3.385$. This analysis signifies that there is a significant statistical difference in the means of the variables. However, the F-value for THR is quite smaller than the critical value; hence, the null hypothesis for equal population means cannot be rejected.

Table 2. Descriptive statistics of experimental data.

	N	Mean	SD	Sum	Min	Max
HRC/J g^{-1} K^{-1}	28	966.64571	349.69697	27,066.08	580.4	1630
THR/kJ g^{-1}	28	32.10357	1.29257	898.9	28.6	34.6
Heating rate	28	1.56786	1.1757	43.9	0.1	3.5
Mass	28	1.49607	0.41362	41.89	0.93	2.11

Table 3. Analysis of variance for HRC.

	DF	Sum of Squares	Mean Square	F Value	Prob > F
Model	2	2.68×10^{6}	1.34×10^{6}	53.85	8.68×10^{-10}
Error	25	622,061.52	24,882.46		
Total	27	3.31×10^{6}			

Table 4. Analysis of variance for Total Heat Release.

	DF	Sum of Squares	Mean Square	F Value	Prob > F
Model	2	3.20	1.60	0.95	0.39
Error	25	41.91	1.68		
Total	27	45.11			

In Table 5, the multiple linear regression model summarized for HRC and THR are presented. It can be seen that the adjusted R-square for HRC is higher than THR. HRC has a linear relationship with sample mass and heating rate, while THR is almost constant throughout the range of heating rates applied. It should be noted that to get a very accurate prediction of these flammability parameters, especially for THR, a method that can handle non-linear modeling could be used. Hence, the next section applies ANFIS networks in the prediction of HRC and THR.

Table 5. Summary of regression analysis.

HRC/J g^{-1} K^{-1}			THR/kJ g^{-1}		
Variable	**Value**	**Std. Error**	**Variable**	**Value**	**Std. Error**
Constant	1392.82	120.32	Constant	31.42	0.99
Heating rate	−267.94	25.82	Heating rate	0.29	0.22
Sample mass	−4.07	73.4	Sample mass	0.16	0.61
Adjusted R^2		0.8	Adjusted R^2		0.033

3.3. ANFIS Network Prediction Results

The present study employed ANFIS networks to model the relationship between sample mass, heating rate, heat release capacity and total heat release rate measured from the MCC experiment. To develop the ANFIS model, the hold-out data splitting technique was adopted. Twenty-four randomly selected data-points out of the 28 experimental data were used for training, while the remaining 4 represented the test data for the model. For improved accuracy, the test data covered the entire range of the available dataset. Table 6 shows the datasets used for developing the models [14].

Table 6. Training and testing datasets.

	β/K s^{-1}	Mass/m	THR/kJ g^{-1}	HRC/J g^{-1} K^{-1}
	0.1	1.00	29.3 ± 0.9	1528 ± 23.5
	0.1	1.98	31.6 ± 0.7	1585 ± 33.3
	0.2	2.02	30.9 ± 0.3	1481.5 ± 28.5
	0.5	0.93	32.9 ± 0.5	1224.2 ± 39
	0.5	1.38	34.5 ± 1.8	1336.0 ± 18.2
	1.0	0.99	31.1 ± 0.7	907.1 ± 40.8
	1.0	1.52	31.9 ± 0.5	892.3 ± 67.3
	1.0	2.03	32.7 ± 0.3	922.6 ± 59.3
	1.5	1.02	33.2 ± 0.8	774.7 ± 27.5
	1.5	1.99	32.1 ± 0.9	795.0 ± 33.4
Training Set	1.5	1.45	32.2 ± 0.5	824.3 ± 65.1
	2.0	0.99	33.3 ± 1.3	763.8 ± 49.3
	2.0	1.48	34.6 ± 0.7	744.2 ± 57.8
	2.0	1.99	32.1 ± 0.7	737.4 ± 37.4
	2.5	1.07	32.7 ± 0.5	669.8 ± 28.6
	2.5	1.53	32.5 ± 0.8	723.6 ± 21.3
	2.5	2.11	33.0 ± 0.6	667.28 ± 53.2
	3.0	0.97	32.6 ± 0.5	664.6 ± 55.2
	3.0	1.49	32.6 ± 5.5	666.6 ± 30.2
	3.0	2.08	32.8 ± 2.5	643.7 ± 47.2
	3.5	1.02	32.3 ± 1.6	615.1 ± 35.8
	3.5	1.41	31.1 ± 0.5	626.3 ± 50.9
	3.5	1.94	32.5 ± 0.9	580.4 ± 26.7
	0.1	1.46	32.4 ± 0.07	1630.0 ± 62.4
	0.2	1.06	28.6 ± 3.0	1422.0 ± 28.3
Testing Test	0.2	1.52	31.0 ± 1.5	1503.0 ± 33.8
	0.5	1.96	31.5 ± 2.8	1188.8 ± 25.7

The membership function for the model was selected by trial and error, and the hybrid learning algorithm was adopted for the training process. The model structure for HRC and THR, as illustrated in Figure 5, consists of two inputs, three membership functions for each input and one output.

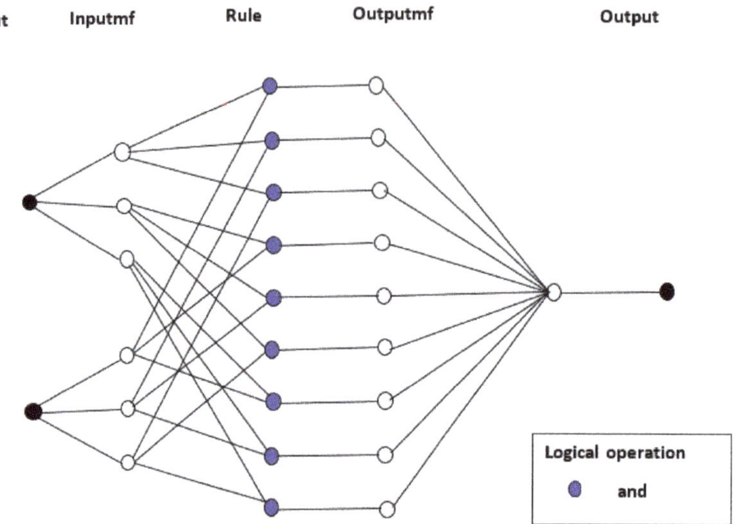

Figure 5. ANFIS model structure.

Three logical operators—and, or and not—are adopted in ANFIS applications. However, depending on the fuzzy logic rules extracted, any of the operators can be used to suit the structure of input data. In this research, only the 'and' logical operator was utilized.

The neurons in Layer 3 consist of fuzzy rules, the conditions of each rule and the consequences. The fuzzy IF-THEN rules generated for the membership functions from the input data of the developed models are detailed from 1–10. These conditional statements describe how the outputs were formulated according to the three membership functions applied.

1. If (input1 is in1mf1) and (input2 is in2mf1), then (output is out1mf1) (1).
2. If (input1 is in1mf1) and (input2 is in2mf2), then (output is out1mf2) (1).
3. If (input1 is in1mf1) and (input2 is in2mf3), then (output is out1mf3) (1).
4. If (input1 is in1mf2) and (input2 is in2mf1), then (output is out1mf4) (1).
5. If (input1 is in1mf2) and (input2 is in2mf2), then (output is out1mf5) (1).
6. If (input1 is in1mf2) and (input2 is in2mf3), then (output is out1mf6) (1).
7. If (input1 is in1mf3) and (input2 is in2mf1), then (output is out1mf7) (1).
8. If (input1 is in1mf3) and (input2 is in2mf2), then (output is out1mf8) (1).
9. If (input1 is in1mf3) and (input2 is in2mf3), then (output is out1mf9) (1).

The models were trained using 100 iterations. The modeling parameters for the developed ANFIS network after the training process are as listed in Table 7.

Table 7. Specifications for ANFIS model.

Variable	HRC/J g^{-1} K^{-1}	THR/kJ g^{-1}
	Value	Value
Number of nodes	53	35
Number of linear parameters	24	9
Number of nonlinear parameters	32	12
Total number of parameters	56	21
Number of training data pairs	24	24
Number of checking data pairs	0	0
Number of fuzzy rules	9	9

Plots of experimental data against predicted data from the HRC ANFIS model during training and testing are illustrated in Figures 6–8. From the simulation, the minimal training Root Mean Squared Error (RMSE) was 0.0224, while the average testing error obtained was 0.625. It is quite clearly seen that the predicted data show a close proximity to the experimental data. A surface plot demonstrating the relationship between the predicted HRC, sample mass and heating rate is presented in Figure 8. The shape of the curve is similar to the one illustrated in Figure 3; hence, the plots from the ANFIS model show that the model has a high predictive ability.

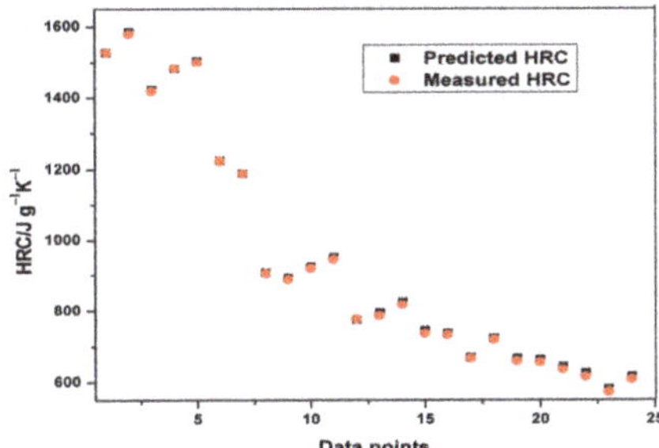

Figure 6. Plot of predicted and measured HRC for training.

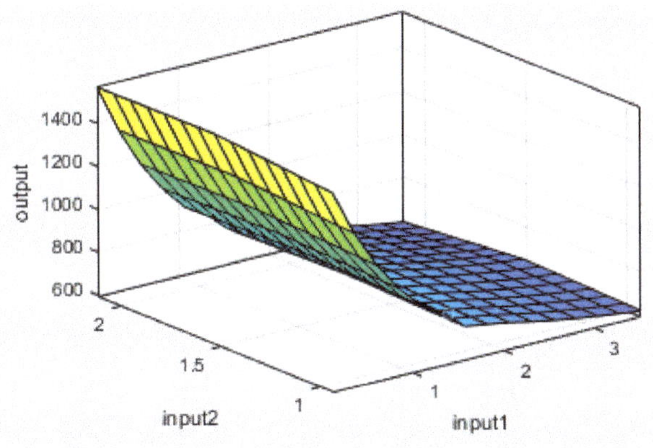

Figure 7. Surface plot of output (HRC), input2 (sample mass) and input1 (heating rate).

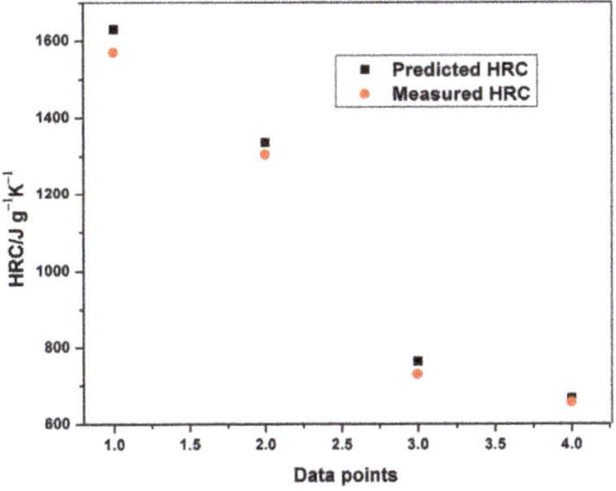

Figure 8. Plot of predicted and measured HRC for testing.

Similarly, plots of experimental and predicted THR datasets were obtained from the Neuro-Fuzzy Designer app. The minimum average training and testing RMSE for THR were 0.00781 and 0.9395, respectively.

Table 8 indicates the performance of ANFIS models in estimating THR and HRC from the MCC experiment. The basic attributes considered are the adjusted R-squared and the root mean squared errors. It is quite obvious from Table 8 that the RMSEs in all the predictions are less than one, indicating an excellent performance. Although, the training of THR outperformed the other models in terms of prediction errors, no obvious differences can be observed. The learning ability of the developed THR model was more accurate than the generalization one, as shown in the training and testing plots (Figures 9–11). Furthermore, the training of the THR model was better than the HRC model, while the test results of HRC outperformed THR. In general, the predictions were in good agreement and fitted the experimental data accurately. Considering the R^2 values obtained, one notable conclusion can be made: the model predicted HRC better than THR since both training and testing of HRC had the best results. This is due to the fact that HRC has a direct and significant statistical relationship with the input parameters, whereas THR is almost constant at any given heating rate and sample mass, thus presenting an uneven statistical distribution. It should also be noted that the test results are an indication of the excellent ability of the developed models to predict data beyond the limits of the training range.

Table 8. Performance of ANFIS models.

Statistical Indicator	HRC		THR	
	Training	**Testing**	**Training**	**Testing**
R^2	0.99994	0.99904	0.99315	0.9148
RMSE	0.0224	0.625	0.00781	0.9395

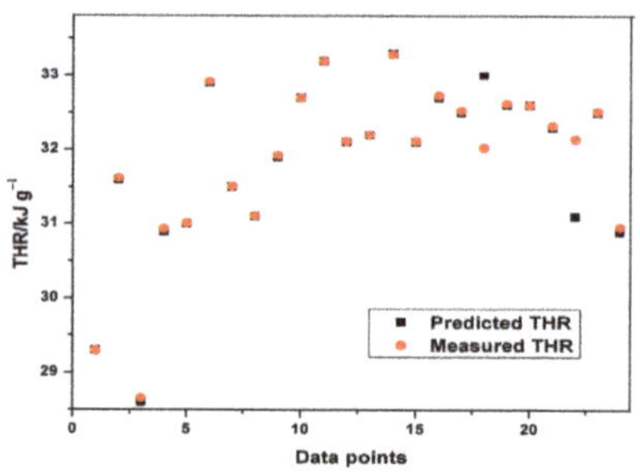

Figure 9. Plot of predicted and measured THR for training.

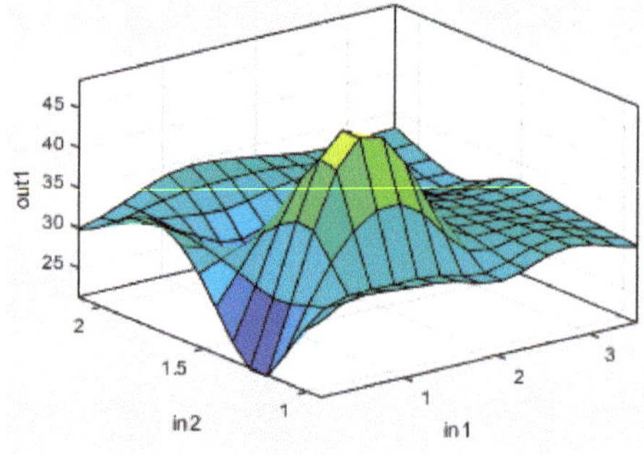

Figure 10. Surface plot of out1 (THR), in2 (sample mass) and in1 (heating rate).

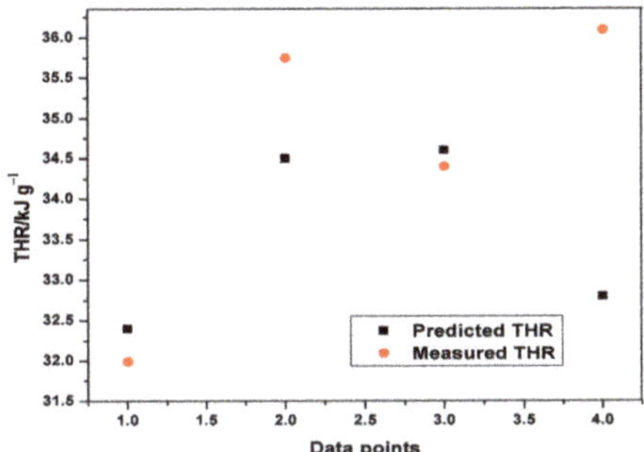

Figure 11. Plot of predicted and measured THR for testing.

The average training and testing errors in the present study have been compared with the results obtained from prediction of HRC and THR with the feed-forward back propagation neural network (FFBPNN) by Mensah et al. [14]. Table 9 shows the RMSE obtained from both the models. Similarly, Figures 12–15 give a visual representation of the variations in the predicted data from the ANFIS and FFBPNN models.

Table 9. Error comparison from ANFIS and Feed Forward Back Propagation Neural Network models.

Model	HRC			THR		
	Training	Testing	Training Time (s)	Training	Testing	Training Time (s)
ANFIS	0.0224	0.625	7.8	0.00781	0.9395	7.35
FFBPNN	0.382	0.980	13.3	0.457	1.048	12.26

Although the results from both ANFIS and FFBPNN models are seemingly good, the comparison in Table 9 indicates the presence of significant differences in the attainable prediction errors as well as the training time. From the table, the ANFIS models attained very low average errors in all cases (both training and testing). The high errors presented in the ANN models could be attributed to the limited amount of data used for the simulation. The results further affirm the superiority and accuracy in the application of ANFIS over ANN.

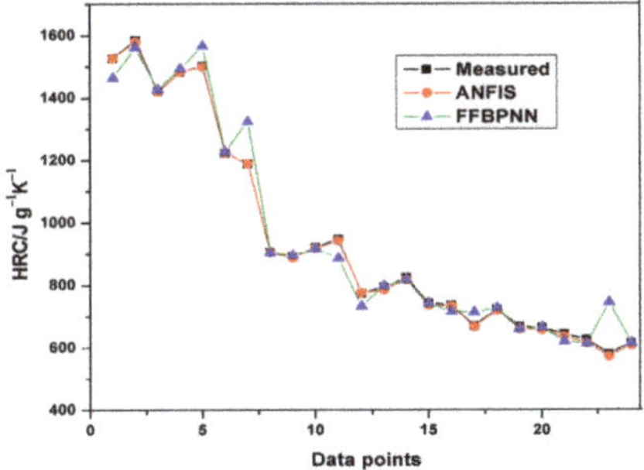

Figure 12. Comparison of HRC training results.

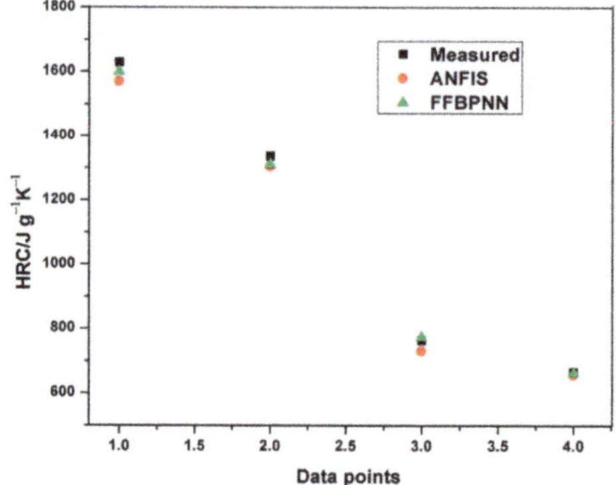

Figure 13. Comparison of HRC testing results.

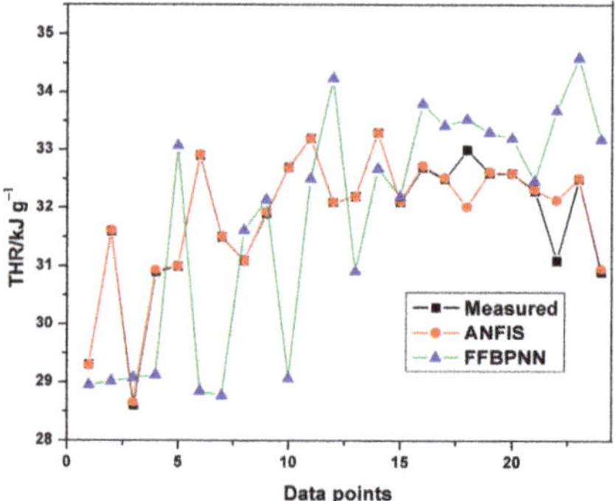

Figure 14. Comparison of THR training results.

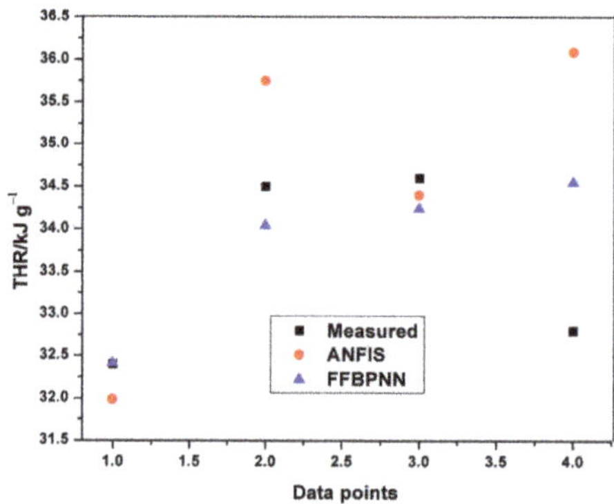

Figure 15. Comparison of THR testing results.

The combination of fuzzy reasoning and artificial neural networks optimizes and improves the learning and generalization capabilities of models. The ability of the system to tackle non-linearities in datasets is also greatly enhanced. This improvement can be observed in the application of ANFIS in flammability studies covered in the present study. The insignificant RMSE values obtained show that ANFIS is suitable for predicting HRC and THR from MCC experiments. With sufficient training, testing data and the right selection of input parameters, this modeling method can be accurately extended to a double scale analysis, such as the prediction of cone calorimeter test data from MCC test results.

4. Conclusions

The adaptive neuro-fuzzy inference system is an artificial intelligence-based computing predictive technique that combines fuzzy inference and the artificial neural network. The method has been applied in various research areas for predicting an output from various input variables. An attempt has been made in the present study to predict HRC and THR measured from the Method A procedure of the MCC experiment. This was done after realizing the degree of non-linearities in flammability parameter modeling using multiple linear regression. While developing the ANFIS models, sample mass and heating rate were used as input variables. The training and testing datasets consisted of 24 and 4 data points, respectively. The research showed that ANFIS has a great potential in flammability simulations and assessments and can, therefore, be used accurately and reliably in flammability studies.

Author Contributions: Conceptualization, R.A.M. and M.O.A.; methodology, R.A.M., M.O.A. and J.X.; software, R.A.M.; validation, O.D., L.J. and Q.X.; formal analysis, R.A.M.; investigation, R.A.M.; resources, O.D., L.J. and Q.X.; data curation, O.D., L.J. and Q.X.; writing—original draft preparation, R.A.M.; writing—review and editing, O.D., L.J., R.A.M.; visualization, O.D., L.J., R.A.M.; supervision, L.J.; project administration, L.J.; funding acquisition, O.D., L.J. and Q.X. All authors have read and agreed to the published version of the manuscript.

Acknowledgments: The authors would like to thank the National Natural Science Foundation of China (NSFC, Grant 51776098), the joint project of NSFC and STINT (51911530151, China side, and CH2018-7733, Sweden side), and NSFC (Grant 51706219). Oisik Das expresses his gratitude to *Bio4Energy*.

Conflicts of Interest: The authors declare no conflict of interest.

References

1. Lyon, R.E.; Richard, W. *A Microscale Combustion Calorimeter*; Federal Aviation Administration Washington DC Office of Aviation Research: Washington, DC, USA, 2002; No. DOT/FAA/AR-01/117.
2. Hostikka, S.; Matala, A. Pyrolysis Model for Predicting the Heat Release Rate of Birch Wood. *Combust. Sci. Technol.* **2017**, *189*, 1373–1393. [CrossRef]
3. Lyon, R.E.; Walters, R.N.; Stoliarov, S.I. Screening flame retardants for plastics using microscale combustion calorimetry. *Polym. Eng. Sci.* **2007**, *47*, 1501–1510. [CrossRef]
4. Schartel, B.K.; Pawlowski, H.; Richard, E.L. Pyrolysis combustion flow calorimeter: A tool to assess flame retarded PC/ABS materials. *Thermochim. Acta* **2007**, *462*, 1–14. [CrossRef]
5. Xu, Q.; Jin, C.; Majlingova, A.; Restas, A. Discuss the heat release capacity of polymer derived from microscale combustion calorimeter. *J. Therm. Anal. Calorim.* **2018**, *133*, 649–657. [CrossRef]
6. *Standard Test Method for Determining Flammability Characteristics of Plastics and Other Solid Materials Using Microscale Combustion Calorimetry*; ASTM D7309; American Society for Testing and Materials: West Conshohocken, PA, USA, 2013.
7. Keshavarz, M.H.; Dashtizadeh, A.; Motamedoshariati, H.; Soury, H. A simple model for reliable prediction of the specific heat release capacity of polymers as an important characteristic of their flammability. *J. Therm. Anal. Calorim.* **2017**, *128*, 417–426. [CrossRef]
8. Yang, C.Q.; He, Q. Textile heat release properties measured by microscale combustion calorimetry: Experimental repeatability. *Fire Mater.* **2012**, *36*, 127–137. [CrossRef]
9. Lyon, R.E.; Walters, R.N. Heat release capacity. In Proceedings of the Fire and Materials Conference, San Francisco, CA, USA, 22 January 2001; pp. 285–300.
10. Lyon, R.E.; Walters, R.N. Thermal analysis of polymer flammability. *Bridg. Centuries Sampe's Mater. Process. Technol.* **2000**, *24*, 1721–1729.
11. Mensah, R.A.; Xu, Q.; Asante-Okyere, S.; Jin, C.; Bentum-Micah, G. Correlation analysis of cone calorimetry and microscale combustion calorimetry experiments. *J. Therm. Anal. Calorim.* **2018**, *136*, 589–599. [CrossRef]
12. Yilmaz, I.; Kaynar, O. Multiple regression, ANN (RBF, MLP) and ANFIS models for prediction of swell potential of clayey soils. *Expert Syst. Appl.* **2011**, *38*, 5958–5966. [CrossRef]
13. Asante-Okyere, S.; Xu, Q.; Mensah, R.A.; Jin, C.; Ziggah, Y.Y. Generalized regression and feed forward back propagation neural networks in modelling flammability characteristics of polymethyl methacrylate (PMMA). *Thermochim. Acta* **2018**, *667*, 79–92. [CrossRef]

14. Mensah, R.A.; Jiang, L.; Xu, Q.; Asante-Okyere, S.; Jin, C. Comparative evaluation of the predictability of neural network methods on the flammability characteristics of extruded polystyrene from microscale combustion calorimetry. *J. Therm. Anal. Calorim.* **2019**, *138*, 3055–3064. [CrossRef]

15. Jang, J.S. ANFIS: Adaptive-network-based fuzzy inference system. *IEEE Trans. Syst. Man Cybern.* **1993**, *23*, 665–685. [CrossRef]

16. Sharma, M. Artificial neural network fuzzy inference system (ANFIS) for brain tumor detection. *arXiv* **2012**, arXiv:1212.0059.

17. Atuahene, S.; Bao, Y.; Ziggah, Y.; Gyan, P.; Li, F. Short-Term Electric Power Forecasting Using Dual-Stage Hierarchical Wavelet-Particle Swarm Optimization-Adaptive Neuro-Fuzzy Inference System PSO-ANFIS Approach Based on Climate Change. *Energies* **2018**, *11*, 2822. [CrossRef]

18. Sihag, P.; Tiwari, N.K.; Ranjan, S. Prediction of unsaturated hydraulic conductivity using adaptive neuro-fuzzy inference system (ANFIS). *ISH J. Hydraul. Eng.* **2019**, *25*, 132–142. [CrossRef]

19. Emiroğlu, M.; Beycioğlu, A.; Yildiz, S. ANFIS and statistical based approach to prediction the peak pressure load of concrete pipes including glass fiber. *Expert Syst. Appl.* **2012**, *39*, 2877–2883. [CrossRef]

20. Lyon, R.E.; Walters, R.N. Pyrolysis combustion flow calorimetry. *J. Anal. Appl. Pyrolysis* **2004**, *71*, 27–46. [CrossRef]

21. Hadi, A.A.; Wang, S. A Novel Approach for Microgrid Protection Based upon Combined ANFIS and Hilbert Space-Based Power Setting. *Energies* **2016**, *9*, 1042. [CrossRef]

22. Lee, C.C. Fuzzy logic in control systems: Fuzzy logic controller. *IEEE Trans. Syst. Man Cybern.* **1990**, *20*, 404–418. [CrossRef]

23. Kisi, O.; Haktanir, T.; Ardiclioglu, M.; Ozturk, O.; Yalcin, E.; Uludag, S. Adaptive neuro-fuzzy computing technique for suspended sediment estimation. *Adv. Eng. Softw.* **2009**, *40*, 438–444. [CrossRef]

24. Zarandi, M.H.F.; Türksen, I.B.; Sobhani, J.; Ramezanianpour, A.A. Fuzzy polynomial neural networks for approximation of the compressive strength of concrete. *Appl. Soft Comput.* **2008**, *8*, 488–498. [CrossRef]

25. Takagi, T.; Sugeno, M. Fuzzy identification of systems and its applications to modeling and control. *IEEE Trans. Syst. Man Cybern.* **1985**, *1*, 116–132. [CrossRef]

26. Al-Sulaiman, M.A. Applying of an adaptive neuro fuzzy inference system for prediction of unsaturated soil hydraulic conductivity. *Biosci. Biotechnol. Res. Asia* **2015**, *12*, 2261–2272. [CrossRef]

27. Pal, M.; Bharati, P. Introduction to Correlation and Linear Regression Analysis. In *Applications of Regression Techniques*; Springer: Singapore, 2019; pp. 1–18.

28. Liu, D.; Xu, Z.; Fan, C. Predictive analysis of fire frequency based on daily temperatures. *Nat. Hazards* **2019**, *97*, 1175–1189. [CrossRef]

29. Zhang, Y.; Shen, L.; Ren, Y.; Wang, J.; Liu, Z.; Yan, H. How fire safety management attended during the urbanization process in China? *J. Clean. Prod.* **2019**, *236*, 117686. [CrossRef]

30. An, W.; Jiang, L.; Sun, J.; Liew, K.M. Correlation analysis of sample thickness, heat flux, and cone calorimetry test data of polystyrene foam. *J. Therm. Anal. Calorim.* **2015**, *119*, 229–238. [CrossRef]

31. Fan, C.L.; Zhang, S.; Jiao, Z.; Yang, M.; Li, M.; Liu, X. Smoke spread characteristics inside a tunnel with natural ventilation under a strong environmental wind. *Tunn. Undergr. Space Technol.* **2018**, *82*, 99–110. [CrossRef]

32. Gao, X.; Jiang, L.; Xu, Q. Experimental and theoretical study on thermal kinetics and reactive mechanism of nitrocellulose pyrolysis by traditional multi kinetics and modeling reconstruction. *J. Hazard. Mater.* **2019**, 121645, in press. [CrossRef]

 *polymers*

Article

Development of Coffee Biochar Filler for the Production of Electrical Conductive Reinforced Plastic

Mauro Giorcelli * and Mattia Bartoli

Politecnico di Torino, C.so Duca degli Abruzzi 24, 10129 Torino, Italy; mattia.bartoli@polito.it
* Correspondence: mauro.giorcelli@polito.it; Tel.: +39-0110-904-327

Received: 15 October 2019; Accepted: 18 November 2019; Published: 21 November 2019

Abstract: In this work we focused our attention on an innovative use of food residual biomasses. In particular, we produced biochar from coffee waste and used it as filler in epoxy resin composites with the aim to increase their electrical properties. Electrical conductivity was studied for the biochar and biochar-based composite in function of pressure applied. The results obtained were compared with carbon black and carbon black composites. We demonstrated that, even if the coffee biochar had less conductivity compared with carbon black in powder form, it created composites with better conductivity in comparison with carbon black composites. In addition, composite mechanical properties were tested and they generally improved with respect to neat epoxy resin.

Keywords: electrical properties; mechanical properties; recycling; epoxy resin

1. Introduction

Anthropogenic waste stream management is one of the main unresolved problems of industrialized societies [1,2]. In the food waste sector, coffee residuals could be considered not only a waste material but a resource. Recently Christoph Sänger [3] reported that worldwide coffee production was 159.7 million of bags in crop year 2017/18 (about 9.6 MTons), with a mean of 5 kg/capita per year in traditional markets (Germany, Italy, France, USA and Japan) and an increasing consumption in emerging markets (South Korea, Russia, Turkey and China). The coffee waste stream becomes a relevant problem not only after consumption but also during the wet processing of coffee beans when 1 ton of fresh berries results in only about 400 kg of wet waste pulp. Several solutions have been proposed to solve the problem of waste coffee biochar, such as the production of biogas [4] and flavours [5], use as filler in ceramics [6] or as absorbent for the removal of basic dyes from aqueous solutions [7]. Coffee wastes have been also used as feedstock for pyrolytic conversion producing hydrogen-rich gas [8] and fuel-quality biochar [9]. Biochar has been used not only as solid fuel but also as high performance material [10,11], as a flame retardant additive [12,13], for electrochemical [14] and energy storage applications [15] and for production of composites [16–19].

Traditionally, in the realm of carbon fillers in polymer composites, carbon black (CB) plays the main role especially in the automotive field with an estimated consumption of 8.1 MTon/year according to data released by the International CB association [20]. CB has been used for producing conductive composites [21] but, as recently reported by Quosai et al. [22], coffee-based biochar also shows remarkably conductive properties. Furthermore, coffee biochar production has an indisputable advantage if compared with CB. Coffee biochar production uses a food waste stream while oil-based feedstock is required for CB production. This decreases the environmental impact of the production process [23–25].

Among different polymers, in this work we focused our attention on epoxy resins doped with these two carbon fillers. As is well known, epoxy resin is a thermoset polymer widely applied in the field of coatings [26], adhesives [27], casting [28], potting [29], composites [30], laminates [31] and encapsulation of semiconductor devices [32]. Epoxy resins are used intensively because of their peculiar properties such as high strength, good stiffness, good thermal stability and excellent heat, moisture and chemical resistance [33,34]. Another, unneglectable advantage of epoxy resin is the possibility of being dispersed into the cross-linked polymeric matrix additives, such as micro-encapsulated amines [35,36], that could be realised after material failure promoting the self-healing process of the epoxy composite [37].

In the field of composites materials, production of conductive reinforced plastic materials has attracted an increasing interest in the last few decades [38,39]. Large-scale application fields deserve particular attention. For example conductive epoxy resin has a large-scale application in the field of coatings and adhesives [40]. In these large-scale applications, filler cost is a crucial issue. Epoxy resins have been used as a polymeric host for plenty of carbonaceous materials for the production of conductive reinforced materials [41–44], but the cost of carbon filler has to be take in account. High-cost carbon fillers such as carbon nanotubes and graphene are problematic for large-scale applications. These carbon fillers induce an increment of its electrical and mechanical properties in the host polymer matrix [45–48] but are not a suitable choice for industrial scale production. This is mainly due to the high-cost, up to 300 k\$/kg [49], and the problem of low productivity of the plants is well known [50]. Thus, low cost carbon fillers which are not derived from fossil fuels, such as CB, are a topic of relevant interest.

In this study, we investigated the use of biochar derived from pyrolytic conversion of the coffee waste stream, such as low cost carbon fillers derived by recycling materials. Results were compared with CB-based composites. Mechanical properties were also investigated for full composite characterization.

2. Materials and Methods

2.1. Carbonaceous Materials Preparation and Characterization

Exhausted coffee powder was selected as a real case study. It was collected from Bar Katia (Turin, Italy) supplied by Vergnano (Arabica mixture). Coffee was collected and dried at 105 °C for 72 h. Coffee samples (100 g) were pyrolyzed using a vertical furnace and a quartz reactor, heating rate of 15 °C/min and kept at the final temperature (400, 600, 800 and 1000 °C) for 30 min in an argon atmosphere. Samples were named as C400, C600, C800 and C1000 respectively. Biochar was grinded using a mechanical mixer (Savatec BB90E) for 10 min in order to decrease the particle size. Commercial CB (VULCAN® 9 N115) was used to compare with coffee biochar.

Ash contents of coffee and carbon-based materials (biochars and CB) were evaluated using a static furnace set at 550 or 800 °C respectively for 6 h.

All samples were investigated from morphological point of view using a field emission scanning electrical microscopy (FE-SEM, Zeis SupraTM 40, Oberkochen, Germany). The microscope was equipped with an energy dispersive X-ray detector (EDX, Oxford Inca Energy 450, Oberkochen, Germany) that was used to explore the carbon composition of biochars.

Particle size distribution of carbon fillers was evaluated using a laser granulometry (Fritsch Analysette 22, Idar-Oberstein, Germany) after a dispersion in ethanol and sonication in an ultrasonic bath for 10 min.

Coffee, biochars and CB were analysed through FT-IR (Nicolet 5700, Thermoscientific, Waltham, US) on attenuated total reflectance (ATR) mode (Smartorbit, Thermoscientific) in the range from 500 to 4000 cm^{-1}.

Biochars and CB were analysed through Raman spectroscopy using Renishaw® Ramanscope InVia (H43662 model, Gloucestershire, UK).

2.2. Composites Preparation

Biochar, derived from coffee, and commercial CB containing epoxy composites were produced using a two component bis-phenol A (BPA) diglycidyl resin (CORES epoxy resin, LPL). Carbonaceous filler (15 wt. %) were dispersed into epoxy monomer using a tip ultrasonicator apparatus (Sonics Vibra-cell) for 15 min. After the addition of the curing agent, the mixture was ultrasonicated for another 2 min and left into the moulds for 16 h at room temperature. A final thermal curing was performed using a ventilated oven (I.S.C.O. Srl "The scientific manufacturer") at 70 °C for 6 h.

2.3. Electrical Characterization

The measurement set-up was derived from Gabhi et al. [51] and is sketched in Figure 1a for fillers and Figure 1b for composites. The instrument was composed of two solid copper cylinders, 30 mm in diameter and 5 cm in length, encapsulated in a hollow Plexiglas cylinder with a nominal inner diameter of 30 mm in the case of filler electrical characterization. In this configuration, the inner diameter was slightly higher so that it was possible to force the copper rods inside the Plexiglas cavity and the upper rod could slide inside the cylinder during the measurement. This arrangement created an internal chamber between the two cylinders, where the carbon powder could be inserted. In the case of composites, the Plexiglas cylinder was removed and the sample was positioned between the aligned copper cylinders. The electrical resistance of the powders or composites was measured at increasing loads (up to 1500 bar) applied by a hydraulic press (Specac Atlas Manual Hydraulic Press 15T). Electrically insulating sheets were placed between the conductive cylinders and the load surfaces in order to ensure that the electrical signal passed through the sample. The resistance of the carbon fillers was measured using an Agilent 34401A multimeter.

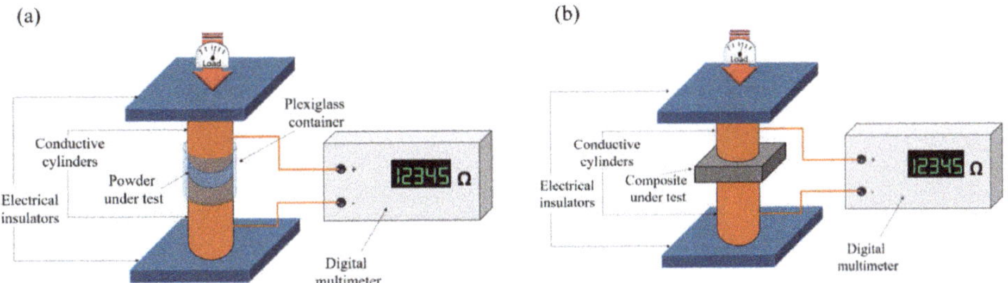

Figure 1. Sketch of measurement set-up for conductivity study of (**a**) carbon fillers and (**b**) composite.

2.4. Composites Mechanical Characterization

Carbonaceous materials containing composites were produced as dog-bone shaped according to the ASTM 638 procedure. Samples were tested using a mechanical stress test (MTS) machine (MTS Q-test10) in tensile test mode until break point. Data were analysed using a self-developed software compiled using Matlab.

2.5. Data Analysis

Statistical analysis used were based on t-tests with a significance level of 0.05 ($p < 0.05$) were carried out using Excel™ software (Microsoft Corp.) and the "data analysis" tool.

3. Results

3.1. Carbonaceous Materials Characterization

Pyrolysis of spent grounds coffee proceeded according to the mechanism reported by Setter et al. [52]. The main mechanisms that occurred during the degradative processes were those related with decomposition of the small lignin fraction [53] and the most abundant polysaccharides (i.e., cellulose and hemicellulose) [54] with the formation of bio-oils rich in anhydrosugars, furans and acetic acid with trace of aromatics [55–57].

Ash content of feedstock and carbonaceous materials were preliminary investigated and summarized in Figure 2.

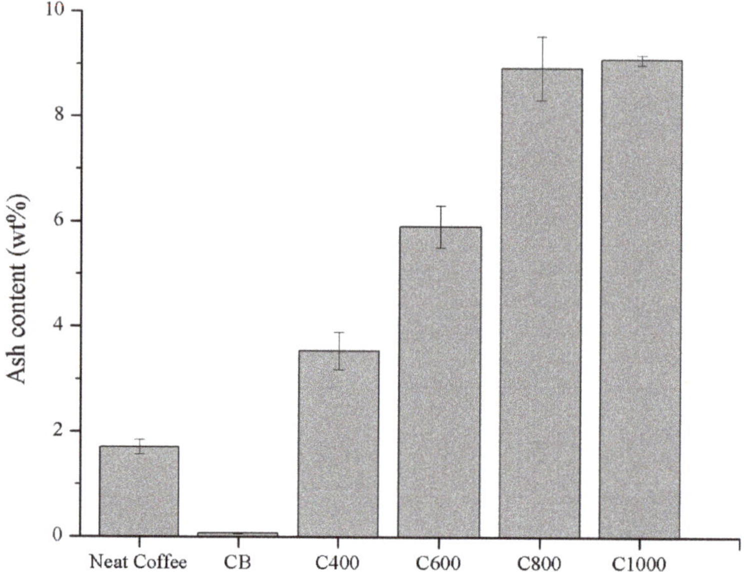

Figure 2. Ash contents of neat coffee, carbon black (CB) and coffee biochar samples heated at 400, 600, 800 and 1000 °C (C400, C600, C800 and C1000 respectively). Columns marked with different letters were significantly different ($p < 0.05$).

The ash content of neat coffee was 1.70 ± 0.14 wt. % and it increased with temperature increments reaching a value around 9 wt. % in the case of C800 (8.92 ± 0.61 wt. %) and C1000 (9.09 ± 0.09 wt. %). As expected, CB showed a very low ash content (0.07 ± 0.01 wt. %) according to Medalia et al. [58] mainly as oxides. Ash content increment at higher temperatures was imputable to advance pyrolytic degradation of the organic matrix leading to the concentration of inorganic residue [59] that did not undergo any temperature induced degradation.

The effect of pyrolytic temperature on biochar morphology was studied using FE-SEM as shown in Figure 3. Neat coffee displayed flaked collapsed structures (Figure 3a, b) that was retained by C400 after pyrolysis at 400 °C (Figure 3e,f). With the increase of temperature to 600 °C lead to the formation of porous structures with average diameters close to 30 μm separated by carbon lamellae with a thickness around 1 μm (Figure 3g,h). At 800 °C, the biochar recovered lost the structure due to the massive release of volatile organic matters during the overall pyrolytic process that induced the collapse of carbonaceous structures together with an improved grindability [60]. At 1000 °C, the increased temperature allowed the massive formation of carbon–carbon bonds that promoted the stabilization of the porous architecture with nanoscale lamellae structures. CB showed a typical highly aggregate spherule-based shape with average diameter of single particles around 50 nm.

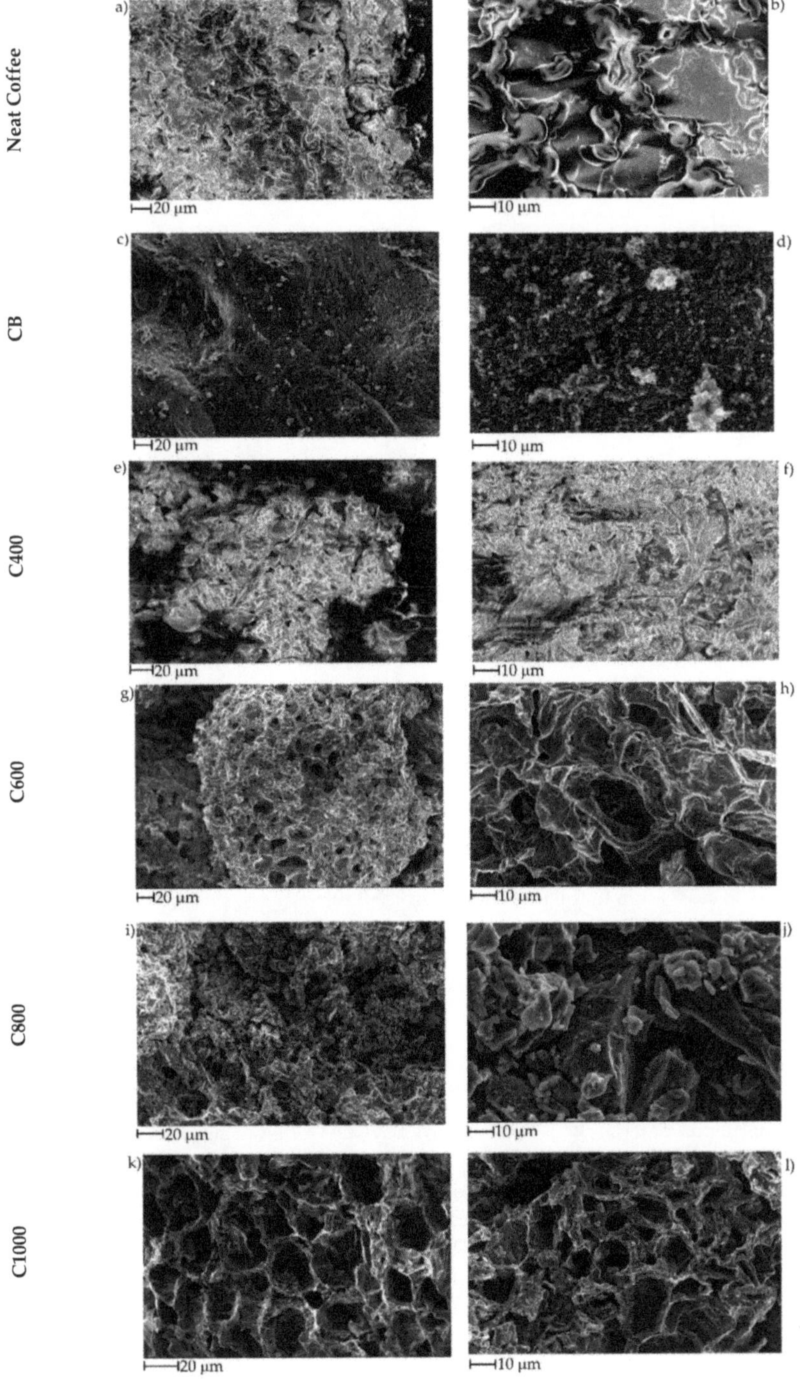

Figure 3. FE-SEM captures of (**a**,**b**) neat coffee, (**c**,**d**) CB, (**e**,**f**) C400, (**g**,**h**) C600, (**i**,**j**) C800 and (**k**,**l**) C1000.

Organic component of significant carbonaceous materials was also analysed using both FT-IR and Raman spectrometry techniques. Among carbonaceous materials, we reported neat coffee, C1000 and CB. Results are shown in Figure 4.

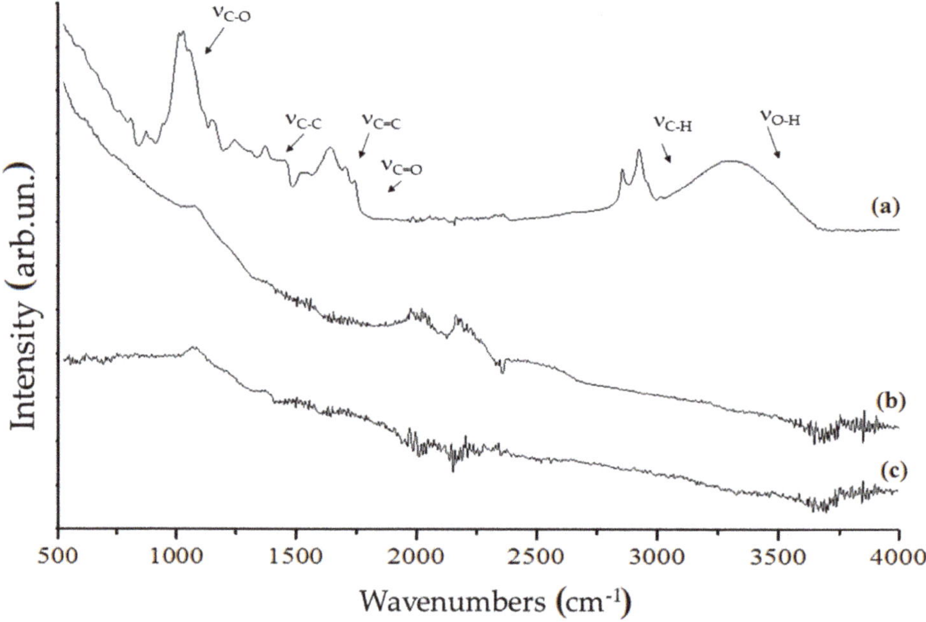

Figure 4. FT-IR spectra (ATR mode) of (**a**) neat coffee, (**b**) coffee biochar produced at 1000 °C and (**c**) CB in the range of 500–4000 cm^{-1}.

The FT-IR spectrum of neat coffee showed the broad band of ν_{O-H} (3300–3500 cm^{-1}), the bands of saturated ν_{C-H} (2850–2950 cm^{-1}), $\nu_{C=O}$ (1710–1741 cm^{-1}) due to the carboxylic functionalities, $\nu_{C=C}$ (1540–1638 cm^{-1}) due to the presence of aromatic structures, saturated and unsaturated δ_{C-H} (1370–1440 cm^{-1}), saturated ν_{C-C} (1243 cm^{-1}), ν_{C-O} (1030–1148 cm^{-1}) and out-of-plane δ_{O-H} below 700 cm^{-1}. Those bands clearly identified a lignocellulosic derived matrix with massive presence of polysaccharides and aromatics. C1000 did not show any of the characteristic bands of organic matrix but show an envelope of bands below 1800 cm^{-1} due to carbon skeletal movements. Contrary, CB showed low bands intensity below 1000 cm^{-1} due the lower variety of carbon structure embedded into particles.

Raman spectra normalized on G peak are shown in Figure 5. Coffee biochars had the typical profiles of amorphous materials [61] in contrast to CB which was more graphitic. The graphitic structure for CB could be observed by the deep gorge between D and G peaks and their shaped structure. An increase of I_D/I_G ratio was evident for biochars moving from a pyrolytic temperature of 400 to 1000 °C. This increase of I_D/I_G ratio could be ascribed to the progressively loss of residual functional groups with the increase of temperature. This observation was also supported by the decrease of fluorescence [62]. Due to the loss of less intense parts of these weak interactions, biochar underwent an appreciable disorganization together with aromatic structure formation, in particular up to 600 °C, without the completion of a proper graphitization process that occurs at higher temperature [63].

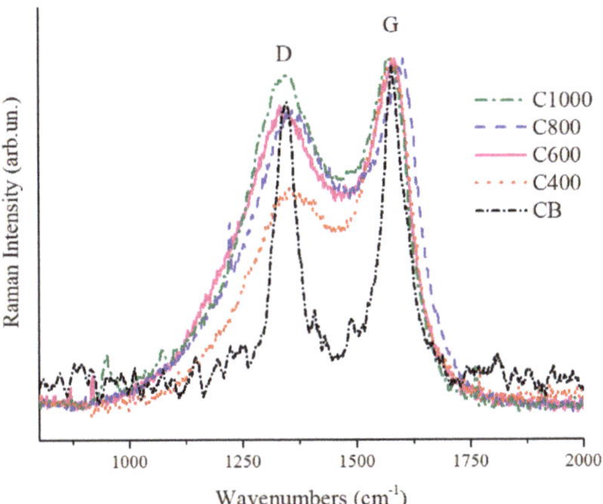

Figure 5. Magnification of Raman spectra in the range from 800 to 2000 cm^{-1} of C400, C600, C800, C1000 and CB.

The evolution of biochar structures due to temperature increment could be monitored though Raman according to Ferrari et al. [61]. Accordingly, the D peaks (Figure 5) showed wave numbers close to 1350 cm^{-1} that is typical of transition from amorphous carbon to nanocrystalline graphite. At the same time the biochar G peaks showed wavenumbers close to 1580 cm^{-1} with exception of C1000. This last showed a G peak at 1600 cm^{-1} due the high amount of nanocrystalline domains not yet rearranged in the ordered structure [64].

The above mentioned consideration was also supported by EDX analysis that showed the carbon content that significantly increased from C400 to C600–C1000 while oxygen content decreased. Carbon content of C600–C1000 were not significantly different from CB even if CB showed a more ordered structure. This support the hypothesis that the driving force of the biochar enhanced conductivity is the reorganization of nanocrystalline domains and not merely the carbon content, shown in Figure 6. Traces of Mg, P, K and Ca were also detected.

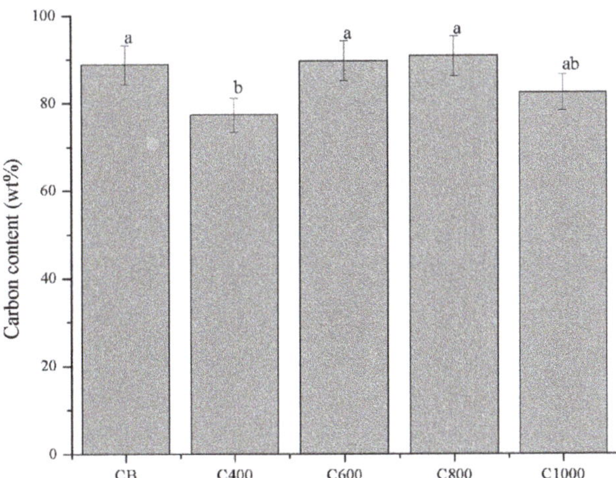

Figure 6. Carbon content of biochars and CB. Evaluated though energy dispersive X-ray (EDX) analysis. Error values were reported as 5% of the detected values according with Laskin et al. [65]. Columns marked with different letters are significantly different ($p < 0.05$).

3.2. Composites Characterization

3.2.1. Electrical Characterization of Carbonaceous Filler and Composites

The set-up shown in Figure 1a was used for biochar powders electrical characterization. Around 3 g of carbonaceous powder, which creates a few millimetres distance between copper cylinders was positioned in the chamber. After the closure of the chamber a pressure was applied with the aim of compacting the powder. The pressure range was from 0 to 1500 bar (step of 250 bar). For each step the stabilized value of resistivity was registered such as the distance between the copper cylinders. The same procedure was repeated for composites of few millimetre thickness. Carbonaceous powders and composites decreased their resistance value during compression until they reached a plateau when high pressure was reached. The decreasing of resistance value could be correlated with the decreasing of space between carbon particles as sketched in Figure 7. In the case of powders, the void among particles collapsed with a production of compact carbon agglomerate as shown in Figure 7a. In the case of composites, Figure 7b shows the mechanism where the polymer chains flow let the carbon particles situate.

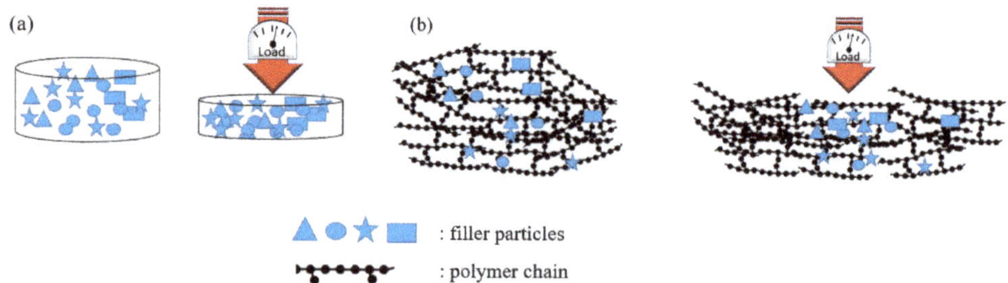

Figure 7. Behaviour of (**a**) powders and (**b**) composites during compression.

The resistance value, R, with the value of surface S and distance l between copper surfaces were used in Ohm law ($\sigma = l/RS$) to evaluate the conductivity σ. The conductivity value of carbon powder and composites were evaluated following this procedure:

(1) A starting value of conductivity was evaluated without any sample in order to measure the value of resistance of the system. This value was subtracted to the resistance value read with samples.

(2) The same quantity of carbon powders (CB and biochar) were positioned between copper cylinders and kept by the Plexiglas hollow cylinder. The measurement was repeated several times in order to have a reliable value.

(3) Composites were positioned between copper cylinders, in this case the Plexiglas hollow cylinder was not necessary and the value of conductivity was measured in different sample portions.

Preliminary results are shown in Figure 8 showing the conductivity of the biochar powders (red line) and percolation curves of related composites.

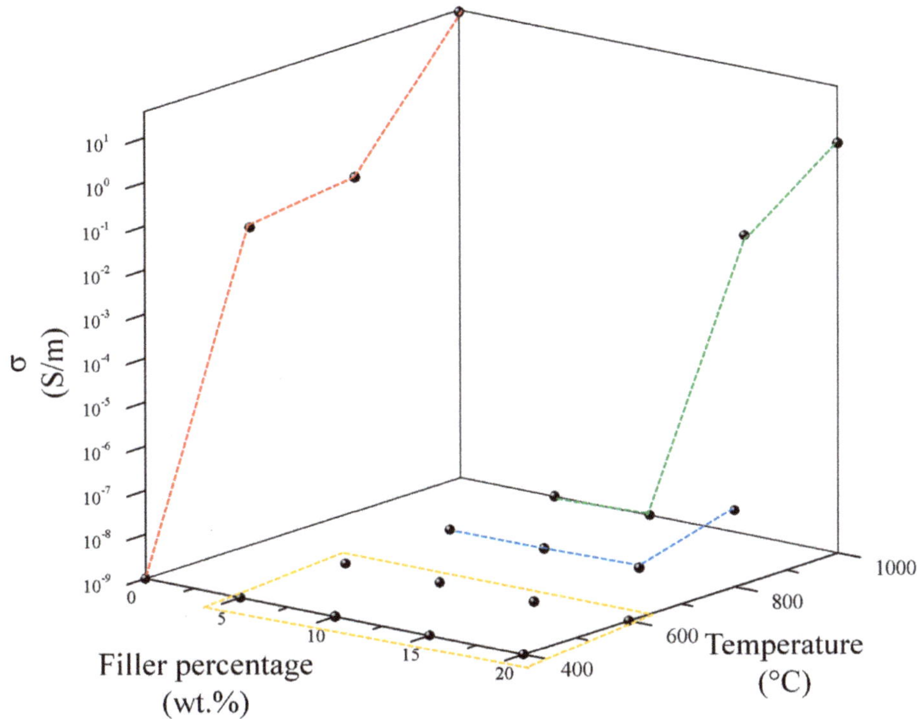

Figure 8. Trends of composites conductivity as function of filler percentage. Red line represents the biochar conductivity trends while yellow, blue and green show respectively the percolation curves of composites containing C400–C600, C800 and C1000.

C400 did not show an appreciable conductivity while an increment of pyrolytic temperature up to 600 °C induced a conductivity of up to 0.02 S/m. Further increments of processing temperature up to 800 and 1000 °C led to a conductivity of up to 0.04 and 35.96 S/m. This remarkable increment of conductivity between 800 and 1000 °C was due the enlargement of aromatic region formed as consequence of high temperature carbonization [64]. This deeply affected the electrical behaviour of related composites. Consequently, C400 and C600 containing composites were not conductive for all the range of filler percentage investigated. CB composites were not conductive until the filler concentration of 15 wt. % reaching a conductivity of 5.4×10^{-8} S/m with a filler loading of 20 wt. %. C1000 composites showed the best performances showing a detectable electrical conductivity with a 15 wt. % of filler and reaching a conductivity of 2.02 S/m with a filler loading of 20 wt. %.

Accordingly, with these data, electrical properties of C1000 and C1000 containing composites were studied under a wide range of static pressures comparing with the related CB and CB composites as shown in Figure 9.

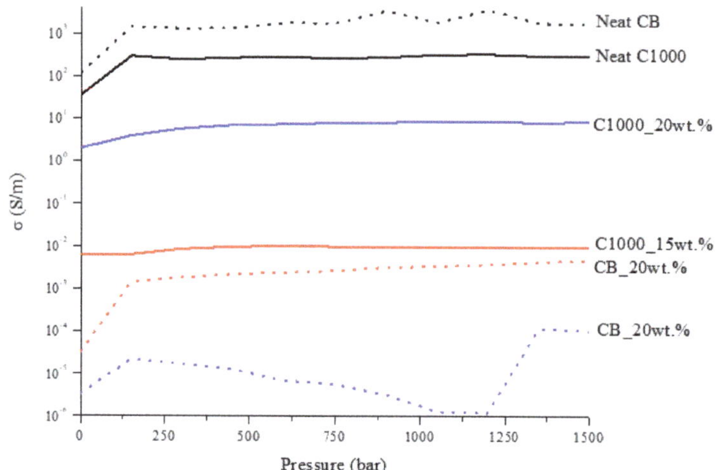

Figure 9. Trends of CB and C1000 powders and related composites conductivity as a function of pressure applied.

CB powder reached a conductivity around 1700 S/m while in the same conditions C1000 reached a conductivity of 300 S/m. Composites containing of CB and C1000 were conductive but the results showed a different trend compared with the relative powders. CB 15 wt. % reached the value of 4×10^{-3} S/m and its conductive value showed an influence of applied pressure in the first compression movement. C1000 15 wt. % reached to 10^{-2} S/m, with an increment around one order of magnitude if compared with CB 15 wt. %. This difference was more relevant for a filler concentration of 20 wt. %. In this case, the conductivity of CB-based composites dropped down to 10^{-5}–10^{-4} S/m in contrast to C1000 which reached ~10 S/m. The high conductive value for coffee biochar could be due to more uniform filler dispersion inside epoxy resin. Dispersion of the filler inside the epoxy matrix was investigated through FE-SEM (Figure 10) after samples were cryo-fractured using liquid nitrogen and compared to composites with a filler loading of 15 wt. % due to the similarity of conductivity. CB containing composites showed a dark and clear area (Figure 10a) with different compositions. The clear ones were rich in CB aggregates (Figure 10b,c) while the darkest were poor (Figure 10d). C1000 containing composites showed smooth surfaces with holes (Figure 10e) due the expulsion of embedded C1000 particles during the fracturing (Figure 10e) as clearly shown in Figure 10g. Particles size analysis (Figure 11) showed clearly that C1000 was composed by two particle populations, one around 100 μm and one around 20 μm. Considering the average size of C1000 particles into the composites was reasonable it was assumed that the bigger ones underwent a disruption during the ultrasonication forming small sized well-dispersed particles. CB particles size showed also that it would be more appropriate speaking of CB aggregates instead of single particles [66]. Aggregates could be justified also from Figure 10c where the CB single particles were less than 100 nm but they created agglomerates that also the particle size analysis (Figure 11) was not able to detect.

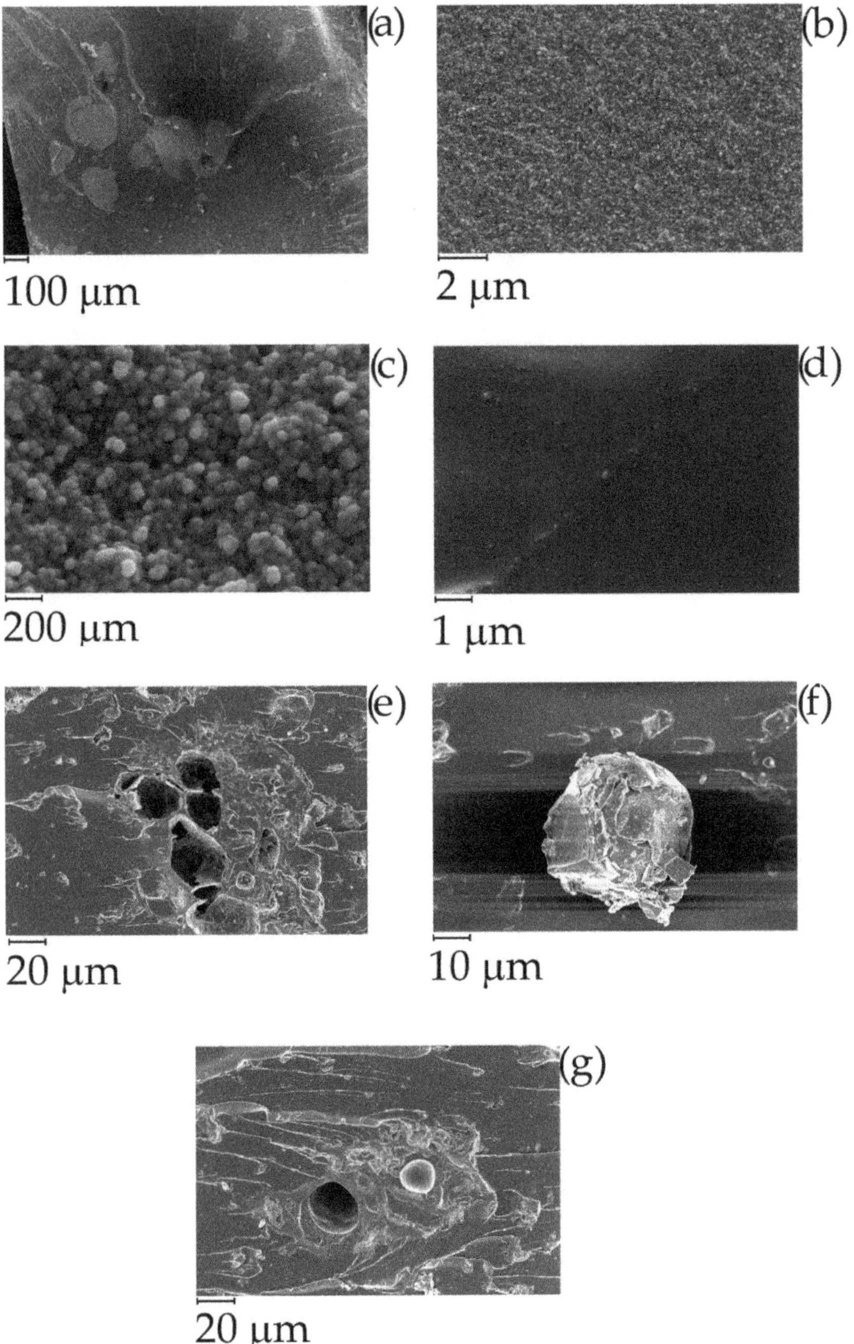

Figure 10. Field emission scanning electrical microscopy (FESEM) of cryo-fractured (**a–d**) CB and (**e–g**) C1000 containing composites with a filler loading of 15 wt. %.

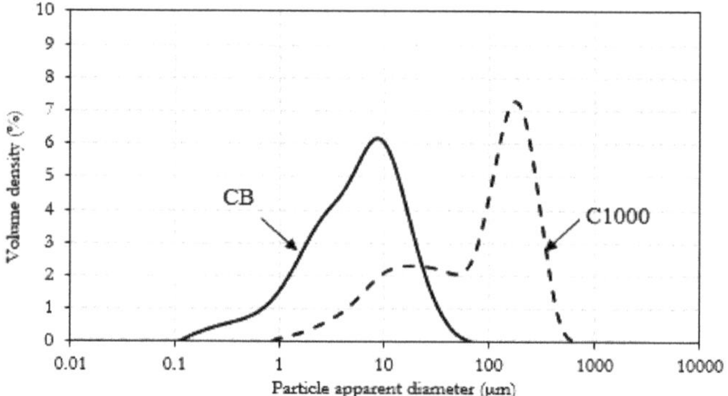

Figure 11. Particle size distribution for CB and C1000.

3.2.2. Composites Mechanical Characterization

With the aim of confirming mechanical consistence of samples, a stress–strain curve was investigated for composites of 15 wt. % for CB and C1000 compared with neat resin. Mechanical tests on dog-bones shaped samples are summarized in the Figure 12.

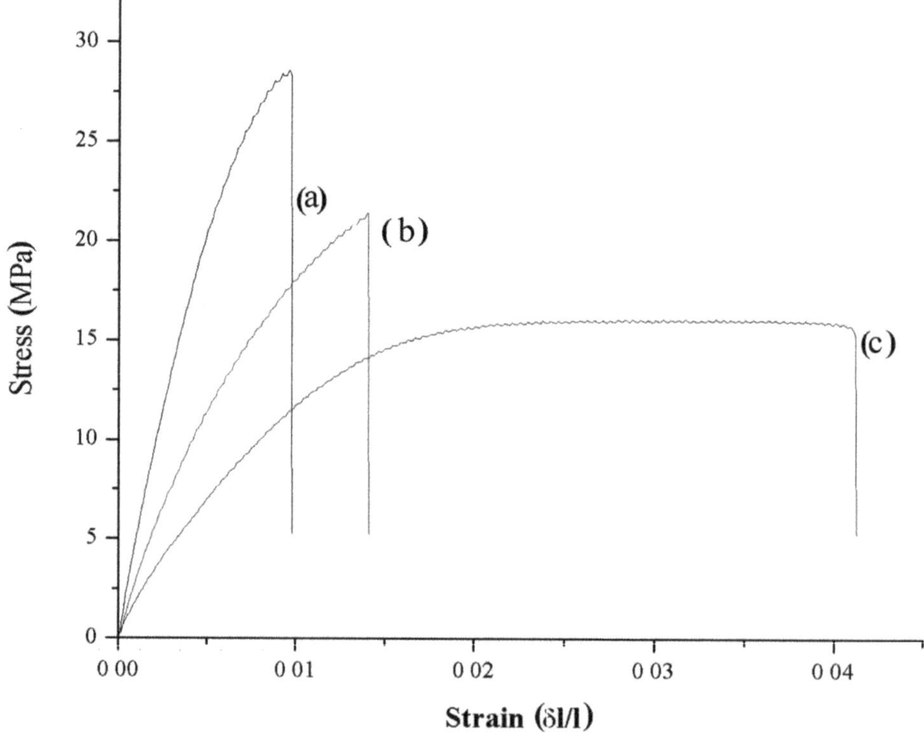

Figure 12. Stress–strain curves of composites containing 15 wt. % of (a) C1000, (b) CB and (c) neat resin.

According to data report in Figure 13, maximum elongation of neat resin (3.50% ± 0.64%) was the highest compared with those of C1000 and CB containing composites (1.16% ± 0.09% and 1.63%

± 0.08% respectively). Neat resin showed also a remarkably higher toughness (0.48 ± 0.03 MJ/m^3) compared with composites that showed values not significantly different from each other close to 0.18 MJ/m^3. Young's modulus (YM) showed a significant difference between C1000-based composites (3258 ± 273 MPa) and CB ones (1940 ± 163 MPa). These last values were quite close to those of neat resin (1510 ± 160 MPa) and a similar trend was observed with ultimate tensile strength with values of CB composites not significantly different from those of neat resin (both close to 19 MPa) and higher values for biochar-based composites (up to 24.9 ± 1.5 MPa).

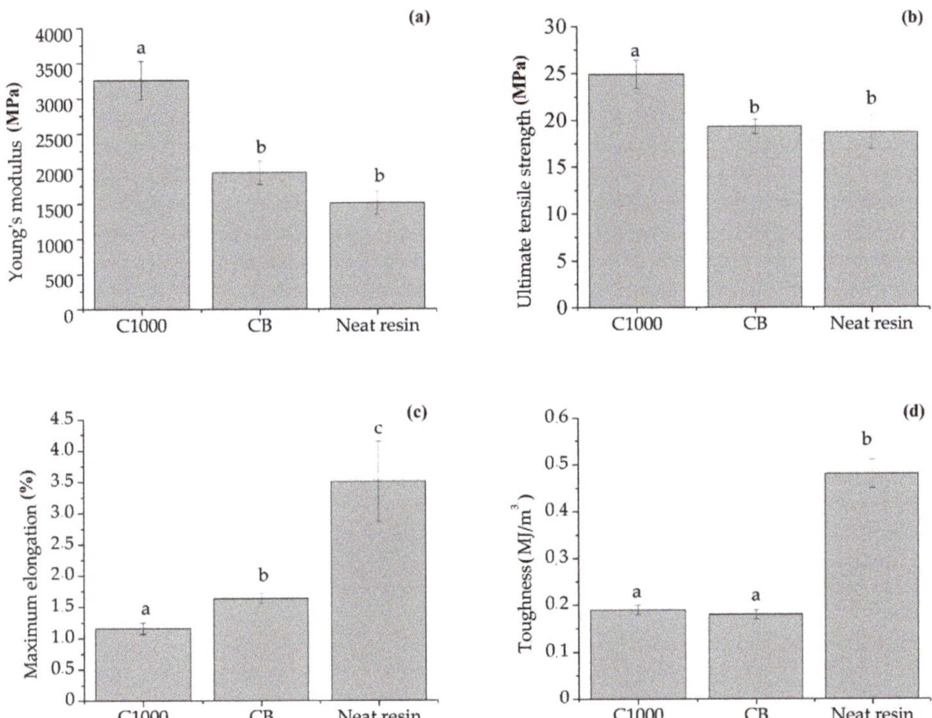

Figure 13. Summary of (**a**) ultimate tensile strength, (**b**) Young's modulus, (**c**) toughness and (**d**) maximum elongation of neat resin, biochar and carbon-based composites. Columns marked with same letters were not significantly different ($p < 0.05$).

Composites behaviour observed during the mechanical tests enlightened the different interaction between different carbonaceous filler with epoxy matrix with magnification of filler–resin interaction, and in the case of biochar-based composites with an increase brittleness and a reduced elongation.

As reported by Chodak et al. [39] about CB containing poly(propylene) composites, the formation of a diffuse particles network is detrimental for the mechanical properties. The same behaviour was observed in the production of CB-based composites which presented a decrement of Ultimate tensile strength compared with CB1000 ones. CB1000 were very close to the percolation threshold (Figure 7) and this induced a very relevant decrement of maximum elongation. Working below the percolation threshold allowed the preservation of some of the appealing properties of a brittle resin (i.e., high Young's modulus and ultimate tensile strength) together with the magnification of electrical conductivity.

4. Conclusions

The coffee waste stream was efficiently used as feedstock for pyrolytic conversion at different temperatures. The effect of process temperature on the properties of biochar was investigated and it was observed that further increments of temperature improved the porous stability and conductivity of the material. This phenomenon was probably due to both the formation of new C–C bonds and to the rearrangement of graphitic and quasi-graphitic domains formed during pyrolysis as shown by Raman characterization.

The most relevant result of this study was that even if neat biochar produced at 1000 °C showed less conductivity with respect to CB when it was dispersed in composite, the electrical properties of a composite containing coffee biochar were some orders of magnitude higher than composites containing CB. In the case of 20 wt. % of C1000, composites showed four orders of magnitudes more that composites containing 20 wt. % of CB. This could be ascribed to the uniform dispersion of coffee biochar, in contrast to CB which creates agglomerations. These agglomerations induced a non-uniform composite structure in the CB containing composites. Mechanical properties of composites with coffee biochar were verified and they were not compromised with respect to composites containing C1000, showing better UTS and YM. Both materials were more brittle than neat resin but C1000 showed some of the properties of high performances resins. Mechanical properties also showed a direct correlation with filler dispersion. Where the filler dispersion was uniform, the mechanical performances were improved.

A new era could be at the door for carbon fillers in polymer composites. Considering the sustainability of coffee biochar production, the results reported show how biomass-derived carbon could be a sound replacement for oil-derived carbon fillers such as CB.

Author Contributions: Conceptualization, M.G.; methodology M.G., M.B.; formal analysis M.G., M.B.; investigation M.G., M.B.; resources, M.G.; data curation, M.G., M.B.; writing—original draft preparation, M.G., M.B.; writing—review and editing, M.G., M.B.; supervision, M.G.

Funding: MODCOMP project (grant 685844, European H2020 program).

Acknowledgments: The authors acknowledge Andrea Marchisio for particle size analysis, Salvatore Guastella for FE-SEM and EDX analysis, Massimo Rovere for Raman spectra, Carlo Rosso for mechanical tests and Alberto Tagliaferro for fruitful discussions.

Conflicts of Interest: The authors declare no conflict of interest.

References

1. Girotto, F.; Alibardi, L.; Cossu, R. Food waste generation and industrial uses: A review. *Waste Manag.* **2015**, *45*, 32–41. [CrossRef]
2. Burkhard, R.; Deletic, A.; Craig, A. Techniques for water and wastewater management: A review of techniques and their integration in planning. *Urban Water* **2000**, *2*, 197–221. [CrossRef]
3. Sänger, C. Value Addition in the Coffee Sector. In Proceedings of the 10th Multi-Year Expert Meeting on Commodities and Development, Geneva, Switzerland, 25–26 April 2018.
4. Battista, F.; Fino, D.; Mancini, G. Optimization of biogas production from coffee production waste. *Bioresour. Technol.* **2016**, *200*, 884–890. [CrossRef] [PubMed]
5. Kusch, S.; Udenigwe, C.C.; Gottardo, M.; Micolucci, F.; Cavinato, C. First-and second-generation valorisation of wastes and residues occuring in the food supply chain. In Proceedings of the 5th International Symposium on Energy from Biomass and Waste, Venice, Italy, 17–20 November 2014.
6. Sena da Fonseca, B.; Vilão, A.; Galhano, C.; Simão, J. Reusing coffee waste in manufacture of ceramics for construction. *Adv. Appl. Ceram.* **2014**, *113*, 159–166. [CrossRef]
7. Lafi, R.; ben Fradj, A.; Hafiane, A.; Hameed, B. Coffee waste as potential adsorbent for the removal of basic dyes from aqueous solution. *Korean J. Chem. Eng.* **2014**, *31*, 2198–2206. [CrossRef]
8. Dominguez, A.; Menéndez, J.; Fernandez, Y.; Pis, J.; Nabais, J.V.; Carrott, P.; Carrott, M.R. Conventional and microwave induced pyrolysis of coffee hulls for the production of a hydrogen rich fuel gas. *J. Anal. Appl. Pyrol.* **2007**, *79*, 128–135. [CrossRef]

9. Tsai, W.-T.; Liu, S.-C.; Hsieh, C.-H. Preparation and fuel properties of biochars from the pyrolysis of exhausted coffee residue. *J. Anal. Appl. Pyrol.* **2012**, *93*, 63–67. [CrossRef]

10. Qian, K.; Kumar, A.; Zhang, H.; Bellmer, D.; Huhnke, R. Recent advances in utilization of biochar. *Renew. Sustain. Energy Rev.* **2015**, *42*, 1055–1064. [CrossRef]

11. Ok, Y.S.; Chang, S.X.; Gao, B.; Chung, H.-J. SMART biochar technology—A shifting paradigm towards advanced materials and healthcare research. *Environ. Technol. Innov.* **2015**, *4*, 206–209. [CrossRef]

12. Das, O.; Kim, N.K.; Hedenqvist, M.S.; Bhattacharyya, D.; Johansson, E.; Xu, Q.; Holder, S. Naturally-occurring bromophenol to develop fire retardant gluten biopolymers. *J. Clean. Prod.* **2019**, 118552. [CrossRef]

13. Das, O.; Kim, N.K.; Sarmah, A.K.; Bhattacharyya, D. Development of waste based biochar/wool hybrid biocomposites: Flammability characteristics and mechanical properties. *J. Clean. Prod.* **2017**, *144*, 79–89. [CrossRef]

14. Ziegler, D.; Palmero, P.; Giorcelli, M.; Tagliaferro, A.; Tulliani, J.-M. Biochars as Innovative Humidity Sensing Materials. *Chemosensors* **2017**, *5*, 35. [CrossRef]

15. Jiang, J.; Zhang, L.; Wang, X.; Holm, N.; Rajagopalan, K.; Chen, F.; Ma, S. Highly ordered macroporous woody biochar with ultra-high carbon content as supercapacitor electrodes. *Electrochim. Acta* **2013**, *113*, 481–489. [CrossRef]

16. Khan, A.; Jagdale, P.; Rovere, M.; Nogués, M.; Rosso, C.; Tagliaferro, A. Carbon from waste source: An eco-friendly way for strengthening polymer composites. *Compos. Part B Eng.* **2018**, *132*, 87–96. [CrossRef]

17. Arrigo, R.; Jagdale, P.; Bartoli, M.; Tagliaferro, A.; Malucelli, G. Structure–Property Relationships in Polyethylene-Based Composites Filled with Biochar Derived from Waste Coffee Grounds. *Polymers* **2019**, *11*, 1336. [CrossRef] [PubMed]

18. Bartoli, M.; Giorcelli, M.; Rosso, C.; Rovere, M.; Jagdale, P.; Tagliaferro, A. Influence of Commercial Biochar Fillers on Brittleness/Ductility of Epoxy Resin Composites. *Appl. Sci.* **2019**, *9*, 3109. [CrossRef]

19. Das, O.; Kim, N.K.; Hedenqvist, M.S.; Lin, R.J.; Sarmah, A.K.; Bhattacharyya, D. An attempt to find a suitable biomass for biochar-based polypropylene biocomposites. *Environ. Manag.* **2018**, *62*, 403–413. [CrossRef]

20. Association, I.C.B. What is Carbon Black? Available online: http://www.carbon-black.org/ (accessed on 19 November 2019).

21. Tchoudakov, R.; Breuer, O.; Narkis, M.; Siegmann, A. Conductive polymer blends with low carbon black loading: Polypropylene/polyamide. *Polym. Eng. Sci.* **1996**, *36*, 1336–1346. [CrossRef]

22. Quosai, P.; Anstey, A.; Mohanty, A.K.; Misra, M. Characterization of biocarbon generated by high-and low-temperature pyrolysis of soy hulls and coffee chaff: For polymer composite applications. *R. Soc. Open Sci.* **2018**, *5*, 171970. [CrossRef]

23. Field, J.L.; Keske, C.M.; Birch, G.L.; DeFoort, M.W.; Cotrufo, M.F. Distributed biochar and bioenergy coproduction: A regionally specific case study of environmental benefits and economic impacts. *Gcb Bioenergy* **2013**, *5*, 177–191. [CrossRef]

24. Dutta, B.; Raghavan, V. A life cycle assessment of environmental and economic balance of biochar systems in Quebec. *Int. J. Energy Environ. Eng.* **2014**, *5*, 106. [CrossRef]

25. Roberts, K.G.; Gloy, B.A.; Joseph, S.; Scott, N.R.; Lehmann, J. Life cycle assessment of biochar systems: Estimating the energetic, economic, and climate change potential. *Environ. Sci. Technol.* **2009**, *44*, 827–833. [CrossRef] [PubMed]

26. Liang-Hsing, L.; Lin, Y.-j.; Lin, Y.-s.; Hwang, K.-Y.; Yi-Sern, W. Reactive Epoxy Compounds and Method for Producing the Same, Core-Shell Type Epoxy Resin Particles, Waterborne Epoxy Resin Composition, and Coating Composition Containing the Reactive Epoxy Compounds. U.S. Patent Application 20190055397, 21 February 2019.

27. Ahmadi, Z. Nanostructured epoxy adhesives: A review. *Prog. Org. Coat.* **2019**, *135*, 449–453. [CrossRef]

28. Liao, J.; Zhang, D.; Wu, X.; Luo, H.; Zhou, K.; Su, B. Preparation of high strength zirconia by epoxy gel-casting using hydantion epoxy resin as a gelling agent. *Mater. Sci. Eng. C* **2019**, *96*, 280–285. [CrossRef] [PubMed]

29. Lall, P.; Dornala, K.; Deep, J.; Lowe, R. Measurement and Prediction of Interface Crack Growth at the PCB-Epoxy Interfaces Under High-G Mechanical Shock. In Proceedings of the 2018 17th IEEE Intersociety Conference on Thermal and Thermomechanical Phenomena in Electronic Systems (ITherm), San Diego, CA, USA, 29 May–1 June 2018; pp. 1097–1105.

30. Rana, S.; Alagirusamy, R.; Joshi, M. A review on carbon epoxy nanocomposites. *J. Reinf. Plast. Compos.* **2009**, *28*, 461–487. [CrossRef]

31. Sela, N.; Ishai, O. Interlaminar fracture toughness and toughening of laminated composite materials: A review. *Composites* **1989**, *20*, 423–435. [CrossRef]

32. Enhua, H.; Kaichang, K.; Lixin, C. Research Progress on Encapsulating Materials of Epoxy Resin. *Chem. Ind. Eng. Progress* **2003**, *22*, 1057–1060.

33. Jin, F.-L.; Li, X.; Park, S.-J. Synthesis and application of epoxy resins: A review. *J. Ind. Eng. Chem.* **2015**, *29*, 1–11. [CrossRef]

34. Rafique, I.; Kausar, A.; Anwar, Z.; Muhammad, B. Exploration of epoxy resins, hardening systems, and epoxy/carbon nanotube composite designed for high performance materials: A review. *Polymer Plast. Technol. Eng.* **2016**, *55*, 312–333. [CrossRef]

35. McIlroy, D.A.; Blaiszik, B.J.; Caruso, M.M.; White, S.R.; Moore, J.S.; Sottos, N.R. Microencapsulation of a Reactive Liquid-Phase Amine for Self-Healing Epoxy Composites. *Macromolecules* **2010**, *43*, 1855–1859. [CrossRef]

36. Yin, T.; Rong, M.Z.; Zhang, M.Q.; Yang, G.C. Self-healing epoxy composites—Preparation and effect of the healant consisting of microencapsulated epoxy and latent curing agent. *Comp. Sci. Technol.* **2007**, *67*, 201–212. [CrossRef]

37. Neisiany, R.E.; Khorasani, S.N.; Lee, J.K.Y.; Ramakrishna, S. Encapsulation of epoxy and amine curing agent in PAN nanofibers by coaxial electrospinning for self-healing purposes. *RSC Advances* **2016**, *6*, 70056–70063. [CrossRef]

38. Sundstrom, D.W.; Chen, S.Y. Thermal conductivity of reinforced plastics. *J. Compos. Mater.* **1970**, *4*, 113–117. [CrossRef]

39. Amarasekera, J. Conductive plastics for electrical and electronic applications. *Reinf. Plast.* **2005**, *49*, 38–41. [CrossRef]

40. Kinloch, A. Toughening epoxy adhesives to meet today's challenges. *MRS Bull.* **2003**, *28*, 445–448. [CrossRef]

41. Chatterjee, S.; Wang, J.; Kuo, W.; Tai, N.; Salzmann, C.; Li, W.-L.; Hollertz, R.; Nüesch, F.; Chu, B. Mechanical reinforcement and thermal conductivity in expanded graphene nanoplatelets reinforced epoxy composites. *Chem. Phys. Lett.* **2012**, *531*, 6–10. [CrossRef]

42. Gojny, F.H.; Wichmann, M.H.; Fiedler, B.; Kinloch, I.A.; Bauhofer, W.; Windle, A.H.; Schulte, K. Evaluation and identification of electrical and thermal conduction mechanisms in carbon nanotube/epoxy composites. *Polymer* **2006**, *47*, 2036–2045. [CrossRef]

43. El-Tantawy, F.; Kamada, K.; Ohnabe, H. In situ network structure, electrical and thermal properties of conductive epoxy resin–carbon black composites for electrical heater applications. *Mater. Lett.* **2002**, *56*, 112–126. [CrossRef]

44. Chekanov, Y.; Ohnogi, R.; Asai, S.; Sumita, M. Electrical properties of epoxy resin filled with carbon fibers. *J. Mater. Sci.* **1999**, *34*, 5589–5592. [CrossRef]

45. Wajid, A.S.; Ahmed, H.T.; Das, S.; Irin, F.; Jankowski, A.F.; Green, M.J. High-Performance Pristine Graphene/Epoxy Composites with Enhanced Mechanical and Electrical Properties. *Macromol. Mater. Eng.* **2013**, *298*, 339–347. [CrossRef]

46. Wang, F.; Drzal, L.T.; Qin, Y.; Huang, Z. Mechanical properties and thermal conductivity of graphene nanoplatelet/epoxy composites. *J. Mater. Sci.* **2015**, *50*, 1082–1093. [CrossRef]

47. Gojny, F.H.; Wichmann, M.H.; Fiedler, B.; Schulte, K. Influence of different carbon nanotubes on the mechanical properties of epoxy matrix composites—A comparative study. *Compos. Sci. Technol.* **2005**, *65*, 2300–2313. [CrossRef]

48. Sandler, J.; Shaffer, M.; Prasse, T.; Bauhofer, W.; Schulte, K.; Windle, A. Development of a dispersion process for carbon nanotubes in an epoxy matrix and the resulting electrical properties. *Polymer* **1999**, *40*, 5967–5971. [CrossRef]

49. Tubes, C. Welcome to Cheap Tubes. Available online: https://www.cheaptubes.com/ (accessed on 9 October 2019).

50. Smail, F.; Boies, A.; Windle, A. Direct spinning of CNT fibres: Past, present and future scale up. *Carbon* **2019**, *152*, 218–232. [CrossRef]

51. Gabhi, R.S.; Kirk, D.W.; Jia, C.Q. Preliminary investigation of electrical conductivity of monolithic biochar. *Carbon* **2017**, *116*, 435–442. [CrossRef]

52. Setter, C.; Silva, F.T.M.; Assis, M.R.; Ataíde, C.H.; Trugilho, P.F.; Oliveira, T.J.P. Slow pyrolysis of coffee husk briquettes: Characterization of the solid and liquid fractions. *Fuel* **2020**, *261*, 116420. [CrossRef]

53. Carrier, M.; Loppinet-Serani, A.; Denux, D.; Lasnier, J.-M.; Ham-Pichavant, F.; Cansell, F.; Aymonier, C. Thermogravimetric analysis as a new method to determine the lignocellulosic composition of biomass. *Biomass Bioenergy* **2011**, *35*, 298–307. [CrossRef]

54. Mishra, R.K.; Mohanty, K. Pyrolysis kinetics and thermal behavior of waste sawdust biomass using thermogravimetric analysis. *Bioresour. Technol.* **2018**, *251*, 63–74. [CrossRef]

55. Bartoli, M.; Rosi, L.; Giovannelli, A.; Frediani, P.; Frediani, M. Pyrolysis of α-cellulose using a multimode microwave oven. *J. Anal. Appl. Pyrol.* **2016**, *120*, 284–296. [CrossRef]

56. Lin, Y.-C.; Cho, J.; Tompsett, G.A.; Westmoreland, P.R.; Huber, G.W. Kinetics and mechanism of cellulose pyrolysis. *J. Phys. Chem. C* **2009**, *113*, 20097–20107. [CrossRef]

57. Collard, F.-X.; Blin, J. A review on pyrolysis of biomass constituents: Mechanisms and composition of the products obtained from the conversion of cellulose, hemicelluloses and lignin. *Renew. Sustain. Energy Rev.* **2014**, *38*, 594–608. [CrossRef]

58. Medalia, A.I.; Rivin, D. Particulate carbon and other components of soot and carbon black. *Carbon* **1982**, *20*, 481–492. [CrossRef]

59. Kambo, H.S.; Dutta, A. A comparative review of biochar and hydrochar in terms of production, physico-chemical properties and applications. *Renew. Sustain. Energy Rev.* **2015**, *45*, 359–378. [CrossRef]

60. Weber, K.; Quicker, P. Properties of biochar. *Fuel* **2018**, *217*, 240–261. [CrossRef]

61. Ferrari, A.C.; Robertson, J. Interpretation of Raman spectra of disordered and amorphous carbon. *Phys. Rev. B* **2000**, *61*, 14095. [CrossRef]

62. Bishui, B. On the Origin of Flurescence in Diamond. *Indian J. Phys.* **1950**, *24*, 441–460.

63. Oberlin, A. Carbonization and graphitization. *Carbon* **1984**, *22*, 521–541. [CrossRef]

64. De Fonton, S.; Oberlin, A.; Inagaki, M. Characterization by electron microscopy of carbon phases (intermediate turbostratic phase and graphite) in hard carbons when heat-treated under pressure. *J. Mater. Sci.* **1980**, *15*, 909–917. [CrossRef]

65. Laskin, A.; Cowin, J.P. Automated single-particle SEM/EDX analysis of submicrometer particles down to 0.1 µm. *Anal. Chem.* **2001**, *73*, 1023–1029. [CrossRef]

66. Schueler, R.; Petermann, J.; Schulte, K.; Wentzel, H.P. Agglomeration and electrical percolation behavior of carbon black dispersed in epoxy resin. *J. Appl. Polym. Sci.* **1997**, *63*, 1741–1746. [CrossRef]

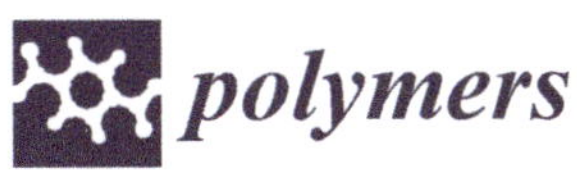 

Article

Durability Analysis of Formaldehyde/Solid Urban Waste Blends

Francesca Ferrari, Raffaella Striani, Paolo Visconti, Carola Esposito Corcione * and Antonio Greco

Department of Engineering for Innovation, University of Salento, 73100 Lecce, Italy;
francesca.ferrari@unisalento.it (F.F.); raffaella.striani@unisalento.it (R.S.); paolo.visconti@unisalento.it (P.V.);
antonio.greco@unisalento.it (A.G.)
* Correspondence: carola.corcione@unisalento.it; Tel.: +39-0832-297326

Received: 28 October 2019; Accepted: 5 November 2019; Published: 8 November 2019

Abstract: Following the innovative research activity carried out in the framework of the POIROT (Italian acronym of dOmotic Platform for Inertization and tRaceability of Organic wasTe) Project, this work aims to optimize the composition of the blends between the organic fraction of municipal solid waste (OFMSW) and formaldehyde-based resins, in order to improve the durability properties. To this aim, in this work, commercial urea-formaldehyde and melamine-formaldehyde powder polymers have been proposed for the inertization of the OFMSW, according to the previous optimized OFMSW-transformation process. A preliminary study about the mechanical properties of the composite panels produced with the different resins was carried out by evaluating compressive, flexural, and tensile performances of the panels. Artificial weathering by cyclic (heating–cooling) and boiling tests were carried out and the mechanical properties were evaluated in order to assess the resistance of the panels to water and humidity. The melamine-formaldehyde based resin had the best performances also when subjected to the weathering tests and despite the higher content of resin in the composites, the panels produced with melamine-formaldehyde have the lowest values of release of formaldehyde minimizing their potential hazard level.

Keywords: solid urban waste; formaldehyde; durability

---

1. Introduction

In the last decade, municipal solid waste management has grown to be one of the main challenges of modern smart cities in the world. Any solution to improve the environmental sustainability represents a welcome candidate that can improve the quality of life for the population. In our previous research [1–3], we reported on the design and realization of an innovative prototype platform able to produce inertized and valorized panels, starting from the organic fraction of the municipal solid waste (OFMSW). The platform is managed by Arduino-based electronic sections for controlling process parameters and integrated with user recognition and a product traceability system based on radio frequency identification (RFID) technology. The POIROT (Italian acronym of dOmotic Platform for Inertization and tRaceability of Organic wasTe) prototype machinery implements a transformation process, constituted by different sequential steps for transforming the organic wastes into a fully inert material. In particular, the main purposes of the transformation process are:

To transform the organic material into bricks, which are inert from a bacteriological point of view and, thus, storable for a long time without any problems in domestic environments;

To modify the composition of the produced inert material by adding specific additives, for giving it appropriate physical and mechanical characteristics, aimed at targeting the specific application of the products, for efficient reuse and recycling;

To label by means of a RFID tag and to properly identify the products for allocating them into storage centers so that they can be recovered and then reutilized in a targeted manner, thanks to the possibility of identification and traceability given by the RFID technology.

Among the other thermos-setting resins, such as epoxy and polyurethane, urea-formaldehyde (UF) systems were previously selected for inertization of OFMSW [1,2], due to their distinct advantages, such as high crosslinking, water solubility, high strength, cost effectiveness, and rapid curing performance [4]. For its own specific characteristics formaldehyde is, in fact, employed for a wide range of applications. The large diffusion of formaldehyde-based materials is due to the high chemical activity and relative cheapness that allows the employment of formaldehyde in several industrial realities. The high bactericidal action is exploited in the medical sector for the conservation of biological material or for sterilizing and disinfecting in pharmacology. Formaldehyde is used in chemical synthesis in the production of detergents, soaps, shampoos, and other cosmetic products as well as for the manufacturing of building materials, such as plywood lacquers, coatings, or glues. Outside of industrial processes, it is also well known that formaldehyde exists naturally in some vegetables and fruits [5]. For instance, pears (38.7–60 mg/kg), grapes (22.4 mg/kg), potatoes (19.5 mg/kg), bananas (16.3 mg/kg), bulb vegetables (11.0 mg/kg), apples (6.3–22.3 mg/kg), carrots (6.7–10.0 mg/kg), and watermelons (9.2 mg/kg) are the aliments that contain the most amount of formaldehyde. Formaldehyde is also employed as a feed hygiene substance in feed for animals [6–8]. The food industry widely uses formaldehyde like food preservatives (identified as E240) especially in smoked food products [9]. The use of formaldehyde as a preservative for food is still an open issue in Europe [7,10], although concentrations equal to 2.5 g/kg are permitted in the United States [9,11]. On one side, people are exposed to small amounts of formaldehyde by eating food, on the other side, the greatest damage occurs by inhalation. Nowadays, automobile and aircraft exhaust emissions, natural gas, fossil fuels, waste incineration, and oil refineries are the major man-made sources of formaldehyde [12]. Formaldehyde is normally present in both indoor [13] and outdoor air [14]. Building materials can be significant emission sources of volatile organic compounds (VOCs), affecting high concentration levels in indoor environments [15,16]. The urea-formaldehyde foam for insulating (UFFI) the buildings, particularly diffused in the 1970s, was substituted, in recent years, by the urea-formaldehyde spray foam (UF) [14], that is dried to remove any volatile compounds, thus less formaldehyde would be expected to be released. A variety of products present in the home can be sources of formaldehydes release, among those more diffused are wood floor finishes, pressed-wood, and wood-based products containing UF resins, such as wallpaper, paints, cigarette smoke, cooking fumes, but also carpets or gypsum board. Due to their porosity, they absorb significant amounts of formaldehyde that could become trapped inside these materials and subsequently released over time in the indoor air. Despite the wide presence of formaldehyde in several materials, it has been classified as cytotoxic, a mutagen, and a human carcinogen by the International Agency for Research on Cancer [17–20]. For such reasons, it is very important to preventively note the risks for human health and the environment by evaluating the hazards in order to limit the exposure. The UF resins have been proven to be efficient, cost and time saving, materials for the inertization of OFMSW. In previous works [1–3], the pre-sterilized OFMSW is mixed with a given amount of UF and water for the production of a pourable slurry. The developed OFMSW-transformation process without any pressure application in the POIROT prototype machinery allows curing of the resin. This approach enables the resin to block the release of odors and percolates from the OFMSW, and at the same time produce low cost bricks or panels, which can be useful in different industrial and building applications.

On the other hand, the aforementioned issues associated with the use of UF-based systems suggest the possibility of replacing them, also in view of their poor water resistance, which can cause degradation of the properties and release of free formaldehyde.

Therefore, this work is aimed at studying the suitableness of different formaldehyde systems (urea/formaldehyde and melamine/formaldehyde) for the inertization of the treated OFMSW. The choice of a proper matrix is mainly based on the durability of the produced panels, with particular emphasis on the resistance towards the effect of water, which can be detrimental for the mechanical properties, and at the same time can promote hydrolysis of the matrix, and consequent release of free formaldehyde.

2. Materials and Methods

Three resins, with decreasing formaldehyde content, were used in this work, in order to allow the intertization of the OFMSW, by producing composite panels with different performances:

1. Urea formaldehyde powder polymer, commercialized by Sadepan as SADECOL P 100N (Sadepan Chimica S.r.l, Viadana, Italy). It is supplied as fine powder/granules with defined grain size and a viscosity from 60 to 130 mPa*s at 20 °C. The resin is characterized by a formaldehyde content lower than 1% [21]. This resin is labelled as HUF (urea-formaldehyde with higher content of free formaldehyde);
2. Urea-formaldehyde powder polymer, commercialized by Sadepan as SADECOL P 410 (Sadepan Chimica S.r.l, Viadana, Italy). The resin is characterized by a formaldehyde content lower than 0.1% [22]. This resin is labeled as LUF (urea-formaldehyde with lower content of free formaldehyde);
3. Melamine-formaldehyde powder polymer, commercialized by Sadepan as SADECOL P 656 (Sadepan Chimica S.r.l, Viadana, Italy), obtained by condensation between melamine and formaldehyde, modified by addition of fillers, additives, and hardeners. The resin is characterized by a formaldehyde content lower than 0.1% [23]. This resin is labelled as MF (melamine–formaldehyde).

An amount of 10 wt% of a proper catalyst was added (Fast sad SD 10, supplied by Sadepan Chimica S.r.l, Viadana, Italy), in order to reduce the time and temperature of the curing process.

The OFMSW used was unsorted food waste collected according to standard Italian regulation. In particular, OFMSW was collected from a local dining room, and apart from food wastes, it contained small amounts (less than 0.5% in weight) of different soft wastes, such as tissues or napkins.

The amount of resin was chosen as the lowest quantity able to reach a complete hardening after the cure (Table 1). Lower amounts involved a partial, or total, desegregation of the sample after the polymerization. Each sample is labelled according to the amount of the organic fraction of municipal solid waste (OFMSW), which is therefore, for three samples, equal to 80% (labelled as OF80), 70% (labelled as OF70) and 50% (labelled as OF50). However, it should be kept in mind that the OFMSW is actually made of about 70% by water, which, however, is removed by evaporation at the end of the curing process. Therefore, in Table 1, the amount of water and the amount of dry OFMSW are also reported. In addition, the composition of cured samples, in which water is completely removed, is reported. The three samples also differ by the type of matrix which was used. Sample OF80_HUF with 80% of wet OFMSW was obtained with the UF SADECOL P 100N matrix (above defined HUF resin) characterized by a higher free formaldehyde content, whereas sample OF70_LUF with 70% of wet OFMSW was obtained with the SADECOL P 410 matrix (above defined LUF resin) characterized by a lower content of free formaldehyde. Finally, the sample OF50_MF with 50% of wet OFMSW was obtained with the melamine-formaldehyde P 656 matrix (above defined MF resin).

Rheological analyses were carried out on a Rheometrics Ares rheometer (TA Instruments, New Castle, DE, USA) on the slurry produced with the different resins. Steady rate tests were carried out at 25 °C, varying the shear rate from 0.05 to 1 s^{-1}, in order to evaluate the possible changes in viscosity due to the use of the different resins. In addition, dynamic temperature ramp tests were performed on all of the mixtures produced, in order to analyze the curing process of the samples during a heating scan from 25 to 120 °C with a heating rate of 5 °C/min, on a parallel geometry plate with a gap of 0.3 mm, constant oscillatory amplitude (1%), and frequency (1 Hz).

Table 1. Composition of the blends: OF80_HUF sample is a blend with 80% of wet OFMSW inertized in 20% of HUF resin, OF70_LUF refers to blend with 70% of wet OFMSW and 30% of LUF resin, finally OF50_MF is a blend with 50% of wet OFMSW and 50% of MF resin.

Formulation	Composition of the Slurry (wt%)			Composition of Cured Samples (wt%)	
	Water	Resin	Dry OFMSW	Resin	Dry OFMSW
OF80_HUF	54.9	21.6	23.5	48	52
OF70_LUF	47.9	31.6	20.5	60	40
OF50_MF	33.6	52	14.4	78	22

Several panels with the different resins of Table 1 (some of them shown in Figure 1) were produced by using the POIROT prototype machinery [1,3]. The latter prototype, besides the possibility to carry out a mechanical and thermal process for the transformation and valorization of the OFMSW, allows us to perform an electronic control aimed to check the correct operation of the different transformation processes. Additionally, a continuous diagnostic is ensured by means of appropriate measurement systems of the physical-chemical parameters related to each transformation phase of the conferred organic waste and of the intermediate and final waste water [3].

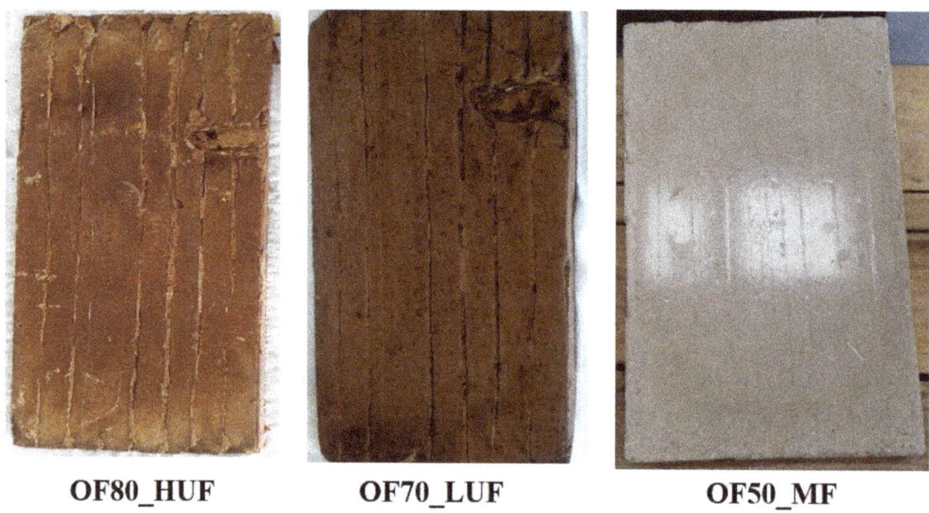

Figure 1. OFMSW-based panel production with different resins.

The dimensions of the produced panels are 400 mm × 200 mm × 30 mm, with an in-plane standard deviation of 3 mm and a through thickness standard deviation of 1 mm (Figure 1).

Compression, tensile, and flexural tests were performed according to, respectively, UNI EN 826 [24], UNI EN 319 [25], and UNI EN 310 [26] Italian standards that implement European directives on the cured panels for building applications. Before mechanical tests, all of the samples were weathered at RH 65%, 20 °C, up to constant weight. Referring to the standards, six samples were extracted from the panels with the following geometry:

UNI EN 826 (compression tests) [24]: sample dimensions equal to $t \times 50$ mm $\times 50$ mm, where t is the thickness of the panel (30 mm). Tests were carried out with a crosshead speed of 0.3 mm/min;

UNI EN 319 (tensile tests) [25]: sample dimensions equal to $L \times 50$ mm $\times t$, where L is 200 mm, corresponding to half of the length of the panel and t is the thickness of the panel (30 mm). The crosshead speed was chosen as 5 mm/min, in order to obtain broken samples in 60 s;

UNI EN 310 (flexural tests) [26]: sample dimensions equal to $t \times 50$ mm $\times 20t$, where t is the thickness of the panel (30 mm). The crosshead speed was chosen as 1.5 mm/min, in order to obtain broken samples in 60 s.

Afterwards, all the samples were subjected to cyclic boiling tests, in order to assess their resistance to water and humidity. In particular, according to UNI EN 321 [27], cyclic tests involved immersion in water for 70 h at 20 °C, followed by freezing at −18 °C for 24 h and heating at 70 °C for 70 h, and finally keeping them at room temperature for 4 h. The whole cycle is repeated three times. At the beginning and at the end of the test, the sample is weathered in a climatic chamber (at 20 °C and 65% RH) until constant weight is reached.

Samples exposed to cyclic tests were then subjected to compression tests, according to UNI EN 826 standard.

Also, boiling tests were performed on samples extracted from the panels, by following the UNI EN 1087-1 standard [28]. Tests consisted of immersion of the samples in neutral water at 20 °C, followed by heating in an oven at 110 °C. After water boiling, the sample is held in the oven for 120 min. After cooling, differently from the cyclic tests, the sample is not weathered.

After boiling, all the samples were subjected to compression, tensile and flexural tests with the procedures described above.

For each test, six repetitions were performed. The six samples were extracted from three different panels, in order to account also for the change of the properties due to different batches of OFMSW.

Finally, the formaldehyde emission for each panel was assessed, according to the UNI EN 717-2 standard [29]. Tests, performed in external laboratories, consisted of placing samples of 400 mm $\times$ 50 mm $\times$ board thickness in a 4-liter cylindrical chamber with controlled temperature (60 $\pm$ 0.5 °C), relative humidity (RH $\leq$ 3%), air flow (60 $\pm$ 3 L/h) and pressure. Air is continuously passed through the chamber at 1 L/min over the test piece, whose edge was sealed with self-adhesive aluminum tape before testing. The determinations were made in duplicate using two different pieces and the actual formaldehyde value is the average of the two pieces after 4 h expressed in mg-HCHO/m^2·h. The formaldehyde amount in the water is then determined photometrically by acetyl acetone spectrophotometric analysis. The determination is based on the Hantzsch reaction, in which aqueous formaldehyde reacts with ammonium ions and acetyl acetone to yield dia-cetyldihydrolutidine (DDL) [30].

3. Results and Discussion

In Figure 2a, the steady state viscosity is reported as a function of shear rate for the different formulations reported in Table 1. In order to attain a low viscosity of the slurry, necessary to allow its pressure-free processing in the POIROT prototype, the matrices used in this work required different amounts of water. In the developed process, the required amount of water was supplied by the OFMSW, which was composed of about 70% water and the remaining fraction (about 30%) is made, above all, of different types of lipids, carbohydrates, and proteins. Reducing the amount of water necessary to attain the same viscosity has the distinct advantage of reducing the porosity of the panels, which in turn involves better mechanical properties [3].

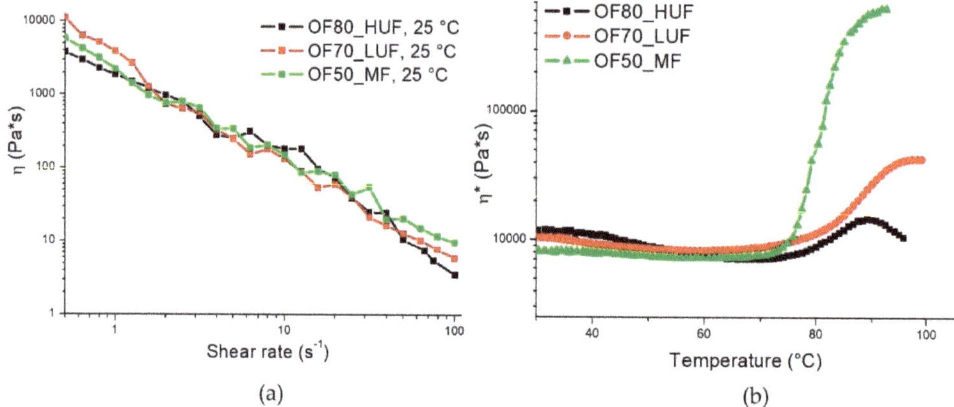

Figure 2. (**a**) Steady rate sweep tests at room temperature and (**b**) dynamic temperature ramp tests on OF80_HUF, OF70_LUF and OF50_MF samples.

The reactivity of the different mixtures was assessed through dynamic rheological analyses. Results, reported in Figure 2b, show that all of the tested formulations are characterized by similar onset temperatures of curing (T_0) (Table 2). On the other hand, as clearly shown in Figure 2b, the melamine/formaldehyde resin shows a much faster curing, as highlighted by the much higher slope of the curve (Table 2). Furthermore, the final viscosity (η_f) that is reached with this sample is much higher than those of the other two blends. This result is related to the lower water content of the mixture, which allows lower reaction times and higher final viscosity values.

Table 2. Rheological data obtained from the curves of Figure 2 (T_0 is the onset temperature of curing, η_f is the samples final viscosity).

Formulation	T_0 (°C)	Slope (Pa*s/°C)	η_f (Pa*s)
OF80_HUF	74.2	0.02688	1.47E4
OF70_LUF	71.5	0.04939	4.41E4
OF50_MF	70.7	0.20397	6.83E6

Typical stress-strain curves from compression, tensile, and flexural tests are reported in Figure 3a–c, respectively. The corresponding values for the strength (σ_R), strain at break (ε_R), and modulus (E) are reported in Table 3. The panels produced by the MF SADECOL P 656 resin are characterized by higher strength and elastic modulus in all of the loading conditions. This is due to the higher value of the density (ρ) of the panels, also reported in Table 3. In turn, the higher density is a direct consequence of the lower amount of water added in the slurry, which reduced the porosity of the panels. All of the samples show much better properties in compression than in tension. This is due to the brittle behavior of all of the panels and to the significant amount of porosity in each sample. As highlighted by the results of Table 3, all of the samples show, in flexural tests, an intermediate behavior between that measured in tension and in compression.

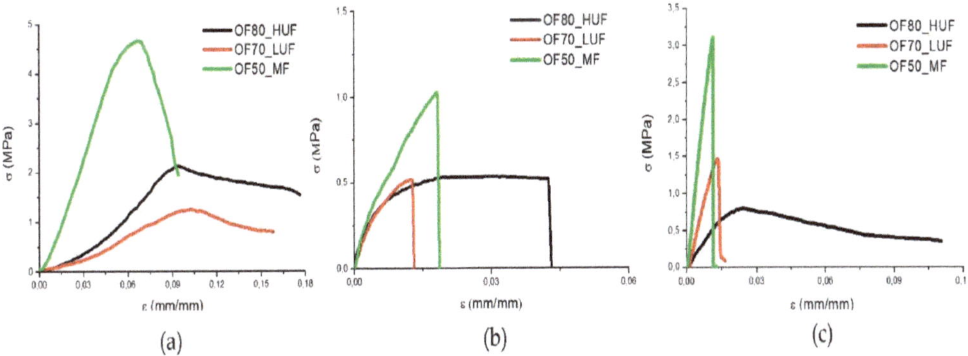

Figure 3. (**a**) Compression, (**b**) tensile, and (**c**) flexural tests on samples realized with different resins.

Table 3. Mechanical tests results (σ_R is the samples strength, ε_R the strain at break, E the modulus and ρ the density).

Formulation	σ_R (MPa)	ε_R (mm/mm)	E (MPa)	ρ (g/cm^3)
Compression Tests				
OF80_HUF	1.75 ± 0.36	0.16 ± 8.5E-02	40.79 ± 3.48	0.76 ± 0.02
OF70_LUF	1.98 ± 0.72	0.27 ± 6E-03	26.61 ± 6.16	0.75 ± 0.04
OF50_MF	5.17 ± 1.04	0.12 ± 1.3E-02	86.48 ± 10.69	0.86 ± 0.05
Tension Tests				
OF80_HUF	0.41 ± 0.13	0.07 ± 1.5E-02	79.14 ± 46.63	
OF70_LUF	0.44 ± 0.10	0.03 ± 1.4E-02	74.07 ± 20.84	
OF50_MF	1.02 ± 0.12	0.02 ± 3E-03	126.69 ± 15.87	
Flexural Tests				
OF80_HUF	0.76 ± 0.24	0.07 ± 0.02	75.87 ± 21.51	
OF70_LUF	1.21 ± 0.17	0.01 ± 2E-03	118.27 ± 17.10	
OF50_MF	2.79 ± 0.41	0.01 ± 2E-03	236.43 ± 41.61	

Compression tests were then carried after subjecting samples to cyclic humidity weathering according to UNI EN 321 standard. Only the OF80_HUF and OF50_MF samples were tested, since OF70_LUF ones were completely broken after the heating–cooling cycles, as detailed in Section 2. Results, reported in Figure 4, indicate, for both systems, a decrease of about 20% in compressive strength. On the other hand, the compressive modulus was strongly affected by the heating-cooling cycles, with a reduction of about 40% for OF50_MF system and a sharp decrease of about 70% for OF80_HUF samples. Therefore, in accordance with the results in Table 3. Mechanical tests resultsthe analysis of the compressive behavior indicates better performances of OF50_MF samples even after cyclic tests.

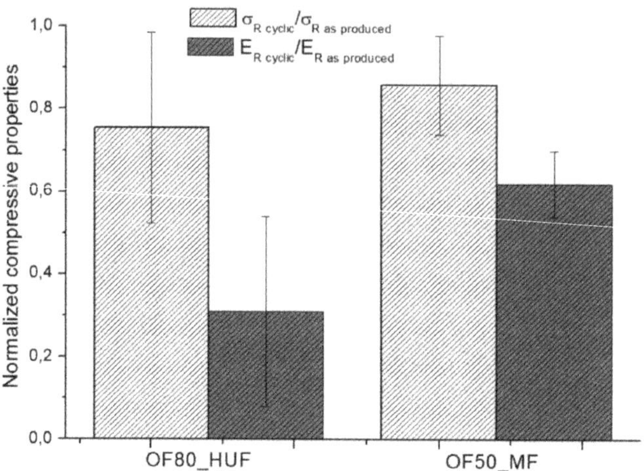

Figure 4. Normalized compressive tests after repeated heating-cooling cycles on OF80_HUF and OF50_MF samples.

Afterwards, all of the produced samples were subjected to boiling tests, according to UNI EN 1087-1. Once extracted, samples were tested in wet conditions with compressive, tensile, and flexural tests.

All of the measured properties showed a significant decrease after the boiling tests. In particular, referring to compressive tests, the results of Figure 5a,b show a reduction of about 30% and 90% respectively for the strength and modulus of the OF80_HUF samples. Referring to compressive properties, the ratio between the strength of the OF50_MF sample after and before the aging are comparable to that of OF70_LUF sample, whereas the ratio of modulus after and before the boiling tests of OF50_MF sample is much higher than that of OF70_LUF sample.

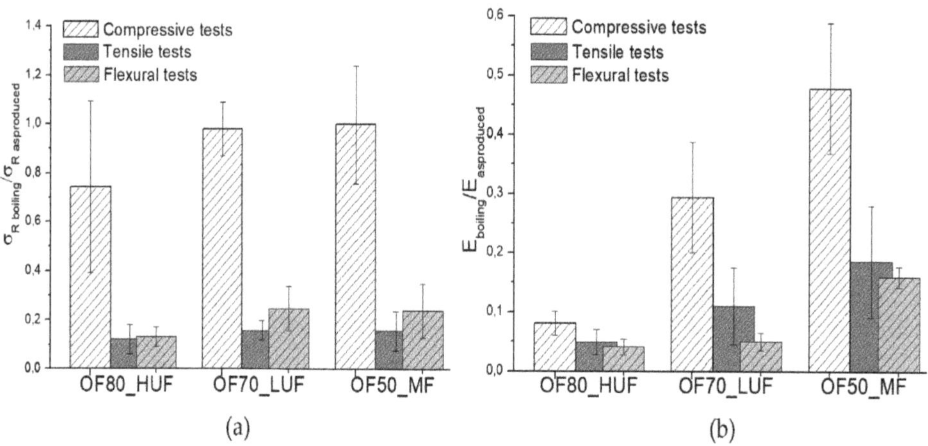

Figure 5. Mechanical response of OF80_HUF, OF70_LUF and OF50_MF samples after the boiling tests; ratio between strength of the samples after and before the boiling tests (**a**), ratio of samples' modulus after and before the boiling tests (**b**).

Tensile and flexural properties are, in general, much more affected by the boiling tests compared to compression properties. In particular, a reduction of about 90% for the strength was found relative to the OF80_HUF sample, both in the flexural and tensile tests. The corresponding modulus decrease is about 95%. Therefore, even for boiling tests, the OF80_HUF sample showed the higher sensitivity.

The strength reduction for the OF70_LUF and OF50_MF samples are substantially equivalent both in tension and flexural tests. In both cases, a reduction of about 75–80% is found. However, the OF50_MF sample showed a better retention of the stiffness compared to the OF70_LUF sample, both in tension and flexural tests. Therefore, even in this case, better performances were reached with the OF50_MF composition.

Formaldehyde release was evaluated according to the UNI EN 717-2 standard, in order to assess whether the formaldehyde emission of the produced panels is lower than the standard limits for manufactured housing at the time of sale, thus allowing a possible commercialization of the realized products. Results in Figure 6 indicate that only OF50_MF samples show a formaldehyde release lower than the threshold limit value (TLV), thus confirming that these samples comply with the UNI EN 717-2 standard. On the other hand, both OF80_HUF and OF70_LUF panels showed a higher formaldehyde release than the TLV, with a higher gap for OF80_HUF samples. This result is attributable to a higher water content of the slurry before the curing process. Higher moisture and water content implies higher formaldehyde emissions, either due to retention of dissolved formaldehyde, less effective cure, or higher hydrolysis rate. Therefore, a lower release could be attained by decreasing the water content of the mixtures before the curing process.

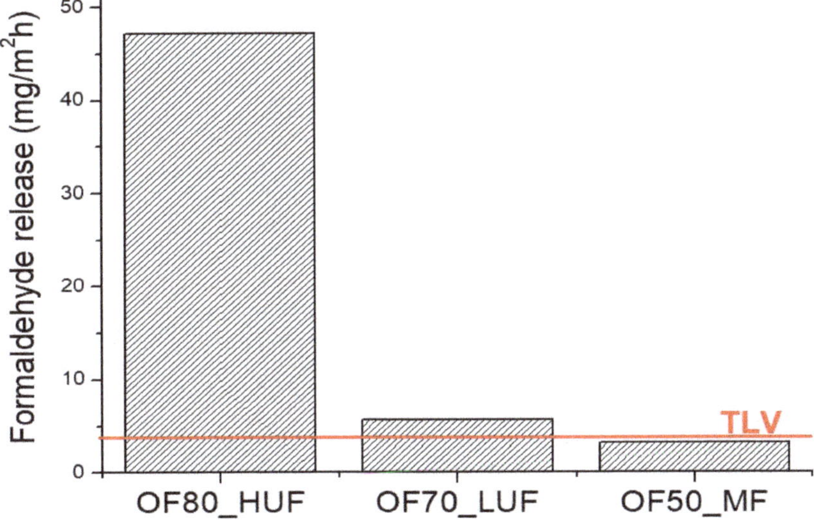

Figure 6. Formaldehyde release according to UNI EN 717-2 standard.

4. Conclusion

An innovative prototype machinery for the stabilization and valorization of the organic fraction of municipal solid waste (OFMSW) has been already assessed in the framework of the POIROT Project research activities, by using a urea–formaldehyde resin containing a formaldehyde content of less than 1 wt% [1,2]. In this paper, the possibility to substitute the previous resin with two different matrices containing a lower amount of free formaldehyde was evaluated. The composition of the blends between OFMSW and formaldehyde resins were, initially, optimized in order to improve the processability of the slurries. Attaining the same viscosity with the three different resins required different amounts of water, and therefore, different amounts of dry OFMWS. In particular, the melamine-formaldehyde SADECOL P 656 resin required the lower amount of water. The panels produced by the POIROT platform showed a density which is strictly correlated to the initial amount of water of the slurry. A lower porosity was found for samples requiring lower amount of water.

As a consequence of this, the melamine–formaldehyde-based blend showed the best mechanical performances on the as-produced samples. However, even after artificial weathering cyclic heating–cooling or boiling tests, the melamine-formaldehyde based blend showed a better retention of the initial mechanical properties. In addition, the panels produced with melamine–formaldehyde showed the lowest values of formaldehyde release, evidencing the lowest hazard level. Starting from the results obtained in this work, the melamine–formaldehyde-based resin was selected as the more suitable matrix for the production of OFMSW-based composites with high mechanical and durability properties and low environmental impact, as achieved with the use of POIROT machinery.

Author Contributions: Investigation, F.F. and R.S.; supervision, C.E.C. and A.G. and P.V.; project administration, P.V.

Funding: This work is funded by the project POIROT with CUP code B89J17000370008 supported by the MISE (Ministry of Economic Development) of Italy within the European Union's Horizon 2020 research and innovation program, in collaboration with Italian companies Medinok SpA (Volla, NA) and Arter Srl (Castello di Cisterna, NA).

Acknowledgments: The authors gratefully acknowledge SADEPAN CHIMICAL S.r.l. for supplying the Urea-Formaldehyde resin and for technical support.

Conflicts of Interest: The authors declare no conflict of interest.

References

1. Esposito Corcione, C.; Ferrari, F.; Striani, R.; Minosi, S.; Pollini, M.; Paladini, F.; Panico, A.; Greco, A. An innovative green process for the stabilization and valorizationof organic fraction of municipal solid waste (OFMSW) I part. *Appl. Sci.* **2019**, *9*, 4516. [CrossRef]

2. Esposito Corcione, C.; Ferrari, F.; Striani, R.; Minosi, S.; Pollini, M.; Paladini, F.; Panico, A.; Greco, A. An innovative green process for the stabilization and valorization of organic fraction of municipal solid waste (OFMSW) II part: Optimization of the curing process. *Appl. Sci.* **2019**, *9*, 3702. [CrossRef]

3. Ferrari, F.; Striani, R.; Minosi, S.; De Fazio, R.; Visconti, P.; Patrono, L.; Catarinucci, L.; Esposito Corcione, C. An innovative IoT-oriented prototype platform for the management and valorization of the organic fraction of municipal solid waste. *Submitt. J. Clean. Prod.*. under review.

4. Conner, A.H. Urea-Formaldehyde Adhesive Resins. *Polym. Mater. Encycl.* **1996**, *128*, 8496–8501.

5. Nowshad, F.; Islam, M.N.; Khan, M.S. Analysis of the Concentration and Formation Behavior of Naturally Occurring Formaldehyde Content in Food. *Int. J. Food Eng.* **2018**, *4*, 71–75. [CrossRef]

6. European Food Safety Authority (EFSA). Scientific Opinion on the safety and efficacy of formaldehyde as a feed hygiene substance in feed for pigs and poultry. *EFSA J.* **2014**, *12*, 3790. [CrossRef]

7. Adiveter, S.L. Scientific Opinion on the safety and efficacy of formaldehyde for all animal species. *EFSA J.* **2014**, *12*, 3562. Available online: https://efsa.onlinelibrary.wiley.com/doi/abs/10.2903/j.efsa.2014.3562 (accessed on 30 October 2019).

8. Regabl, B.V. Scientific opinion on the safety and efficacy of formaldehyde for all animal species based. *EFSA J.* **2014**, *12*, 3561. Available online: https://efsa.onlinelibrary.wiley.com/doi/abs/10.2903/j.efsa.2014.3561 (accessed on 30 October 2019).

9. Wahed, P.; Razzaq, M.A.; Dharmapuri, S.; Corrales, M. Determination of formaldehyde in food and feed by an in-house validated HPLC method. *Food Chem.* **2016**, *202*, 476–483. [CrossRef] [PubMed]

10. Directorate, C. *Update of the Opinion of the Scientific Committee for Animal Nutrition on the Use of Formaldehyde as a Preserving Agent for Animal Feeding Stuffs of 11 June 1999 Scientific Opinions*; European Commission: Brussels, Belgiumc, 1999; pp. 1–19.

11. United States Environmental Protection Agency (USEPA). *Toxicological Review of Formaldehyde-Inhalation Assessment USEPA*; USEPA: Washington, DC, USA, 2010.

12. Liteplo, R.G.; Beauchamp, R.; Chénier, R.; Meek, M.E. *International Programme on Chemical Safety*; World Health Organization: Geneva, Switzerland, 2002.

13. Salthammer, T.; Mentese, S.; Marutzky, R. Formaldehyde in the indoor environment. *Chem. Rev.* **2010**, *110*, 2536–2572. [CrossRef] [PubMed]

14. U.S. Consumer Product Safety Commission. *An Update on Formaldehyde*; U.S. Consumer Product Safety Commission: Bethesda, MD, USA, 2016. Available online: https://www.cpsc.gov/s3fs-public/An-Update-On-Formaldehyde-725_0.pdf (accessed on 30 October 2019).

15. Knudsen, H.N.; Kjaer, U.D.; Nielsen, P.A.; Wolkoff, W. Sensory and chemical characterization of VOC from emissions from building products: Impact of concentration and air velocity. *Atmos. Environ.* **1999**, *33*, 1217–1230. [CrossRef]

16. Calisti, R.; Isolani, L.; Rossano, M. Formaldehyde exposure patterns in a set of Italian indoor workplaces with and without specific emission sources-2011–2018. *Ital. J. Occup. Environ. Hyg.* **2018**, *9*, 165–175.

17. International Agency for Research on Cancer—IARC, Monographs on the Evaluation of Carcinogenic Risks to Humans–Internal report 14/002, France, 2014. Available online: https://monographs.iarc.fr/wp-content/uploads/2018/08/14-002.pdf (accessed on 30 October 2019).

18. Yu, P.H. Deamination of methylamine and angiopathy; toxicity of formaldehyde, oxidative stress and relevance to protein glycoxidation in diabetes. *J. Neural Transm.* **1998**, *52*, 201–216.

19. Feron, V.J.; Til, H.P.; de Vrijer, F.; Woutersen, R.A.; Cassee, F.R.; van Bladeren, P.J. Aldehydes: Occurrence, carcinogenic potential, mechanism of action and risk assessment. *Mutat. Res. Toxicol.* **1991**, *259*, 363–385. [CrossRef]

20. *World Health Organization–International Agency for Research on Cancer–IARC, Overall Evaluation of Carcinogenicity to Humans*; Formaldehyde-Monographs Series; Lyon, France, 2004.

21. Technical Data-Sheet, Sadepan SADECOL P 100N-Product Specifications. 2018. Available online: https://www.sadepan.com/en/products/ (accessed on 30 October 2019).

22. Technical Data-Sheet, Sadepan SADECOL P 410-Product Specifications. 2018. Available online: https://www.sadepan.com/en/products/ (accessed on 30 October 2019).

23. Technical Data-Sheet, Sadepan SADECOL P 656-Product Specifications. 2018. Available online: https://www.sadepan.com/en/products/ (accessed on 30 October 2019).

24. Italian standardization institute, UNI EN 826:2013 Isolanti Termici per Edilizia–Determinazione del Comportamento a Compressione; 2013. Available online: http://store.uni.com/catalogo/uni-en-826-2013, (accessed on 30 October 2019).

25. Italian Standardization Institute. UNI EN 319:1994 Pannelli di Particelle di Legno e Pannelli di Fibra di Legno. Determinazione Della Resistenza a Trazione Perpendicolare al Piano del Pannello; 1994. Available online: http://store.uni.com/catalogo/uni-en-319-1994 (accessed on 30 October 2019).

26. Italian Standardization Institute. UNI EN 310:1994 Pannelli a Base di Legno. Determinazione del Modulo di Elasticità a Flessione e Della Resistenza a Flessione; 1994. Available online: http://store.uni.com/catalogo/index.php/uni-en-310-1994 (accessed on 30 October 2019).

27. Italian Standardization Institute. UNI EN 321:2002 Pannelli a Base di Legno—Determinazione Della Resistenza All Umidità Mediante Prove Cicliche; 2002. Available online: http://store.uni.com/catalogo/index.php/uni-en-321-2002 (accessed on 30 October 2019).

28. Italian Standardization Institute. UNI EN 1087-1:1997 Pannelli di Particelle di Legno. Determinazione Della Resistenza All Umidità. Prova in Acqua Bollente; 1997. Available online: http://store.uni.com/catalogo/uni-en-1087-1-1997 (accessed on 30 October 2019).

29. Italian Standardization Institute. UNI EN 717-2:1996 Pannelli a Base di Legno. Determinazione del Rilascio di Formaldeide. Rilascio di Formaldeide con il Metodo Dell'analisi del Gas; 1996. Available online: http://store.uni.com/catalogo/index.php/uni-en-717-2-1996 (accessed on 30 October 2019).

30. Salem, M.Z.M.; Böhm, M.; Barcík, S.; Beránková, J. Formaldehyde Emission from Wood-Based Panels Bonded with Different Formaldehyde-Based, Resins. *Wood Ind. Drv. Ind.* **2011**, *62*, 177–183. [CrossRef]

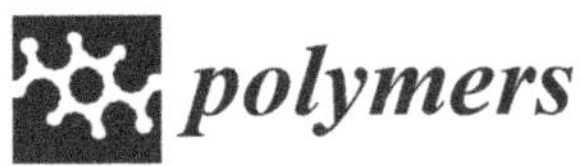 *polymers*

Review

A Review on the Flammability Properties of Carbon-Based Polymeric Composites: State-of-the-Art and Future Trends

Karthik Babu [1], Gabriella Rendén [2], Rhoda Afriyie Mensah [3], Nam Kyeun Kim [4], Lin Jiang [3], Qiang Xu [3], Ágoston Restás [5], Rasoul Esmaeely Neisiany [6], Mikael S. Hedenqvist [2,*], Michael Försth [7], Alexandra Byström [7] and Oisik Das [8,*]

[1] Center for Polymer Composites and Natural Fiber Research, Tamil Nadu 625005, India; karthikbabunitt@gmail.com

[2] Department of Fibre and Polymer Technology, Polymeric Materials Division, School of Engineering Sciences in Chemistry, Biotechnology and Health, KTH Royal Institute of Technology, 100 44 Stockholm, Sweden; grenden@kth.se

[3] School of Mechanical Engineering, Nanjing University of Science and Technology, Nanjing 210094, China; ramensah@ymail.com (R.A.M.); ljiang@njust.edu.cn (L.J.); xuqiang@njust.edu.cn (Q.X.)

[4] Centre for Advanced Composite Materials, Department of Mechanical Engineering, University of Auckland, Auckland 1142, New Zealand; nam.kim@auckland.ac.nz

[5] Department of Fire Protection and Rescue Control, National University of Public Service, H-1011 Budapest, Hungary; Restas.Agoston@uni-nke.hu

[6] Department of Materials and Polymer Engineering, Faculty of Engineering, Hakim Sabzevari University, Sabzevar 9617976487, Iran; r.esmaeely@hsu.ac.ir

[7] Structural and Fire Engineering Division, Department of Civil, Environmental and Natural Resources Engineering, Luleå University of Technology, 97187 Luleå, Sweden; michael.forsth@ltu.se (M.F.); alexandra.bystrom@ltu.se (A.B.)

[8] Department of Engineering Sciences and Mathematics, Luleå University of Technology, 97187 Luleå, Sweden

* Correspondence: mikaelhe@kth.se (M.S.H.); oisik.das@ltu.se (O.D.)

Received: 26 April 2020; Accepted: 1 June 2020; Published: 8 July 2020

Abstract: Carbon based fillers have attracted a great deal of interest in polymer composites because of their ability to beneficially alter properties at low filler concentration, good interfacial bonding with polymer, availability in different forms, etc. The property alteration of polymer composites makes them versatile for applications in various fields, such as constructions, microelectronics, biomedical, and so on. Devastations due to building fire stress the importance of flame-retardant polymer composites, since they are directly related to human life conservation and safety. Thus, in this review. the significance of carbon-based flame-retardants for polymers is introduced. The effects of a wide variety of carbon-based material addition (such as fullerene, CNTs, graphene, graphite, and so on) on reaction-to-fire of the polymer composites are reviewed and the focus is dedicated to biochar-based reinforcements for use in flame retardant polymer composites. Additionally. the most widely used flammability measuring techniques for polymeric composites are presented. Finally. the key factors and different methods that are used for property enhancement are concluded and the scope for future work is discussed.

Keywords: biochar; carbon fillers; nanocomposites; flame retardants; fire

---

1. Introduction

In the forwarded note of World Health Organization (WHO), it is mentioned that burns constitute a major public health problem, especially in low- and middle-income countries, where over 95% of all burn deaths occur. Fire-related burns alone account for over 300,000 deaths per year [1]. The development of safer buildings and appliances is one of the reasons for low death rate in high-income countries. Nowadays, polymers and their composite products are ubiquitous in numerous fields in day-to-day life, such as microelectronics, construction, furniture, automotive, packaging, etc. However, an important limitation is that most polymers are easily flammable [2]. Initially, polymers start to degrade (pyrolyse) when a sufficient amount of heat and oxygen are present. Further. the release of combustible gases, which mixes with atmospheric air, together promote the vigorous burning of substrate and the consequent decomposition of materials. This burn initiation (ignition) depends on flash point and auto-ignition of the material. In brief, polymer decomposition mainly depends on its ignitability, fire spread, and heat release characteristics. The sufficient amount of heat, fuel, and oxygen supply are needed at each and every stage of combustion, and these sources may be ambient or self-induced (especially during material burn. the release of volatile gases and particulates act as a sources for further combustion and create a cyclic process). It is critical to improve the flame retardancy of the polymers and their composites in order to satisfy safety guidelines. Carbon-based materials have demonstrated exceptional thermal, chemical and mechanical properties along with their inherent resistance towards degradation by combustion. Therefore. the enhancement of the flame retardancy of polymer composites by utilizing the carbo-based nano-fillers, such as fullerene, CNTs, graphene, graphene nanosheets (GNSs), Graphene quantum dots (GQDs), graphite, etc., is currently being attempted by numerous researches. Thus, it is worthwhile to gain a holistic view on the effect of carbon-based nano-fillers on flammability characteristics of various multifunctional polymer composites. There are numerous tests that enable the determination of the fire behaviour of polymeric composite materials. For instance. the Limiting Oxygen Index (LOI) test can give information regarding the minimum amount of O_2 that is required by a material to sustain burning. Common polymers like polypropylene (PP) and polyethylene (PE) have LOI ranging between 17 to 19%. This means that the aforementioned materials require 17 to 19% of oxygen concentration for complete material combustion process in 3 min. [3]. In addition, one of the most potent technique to judge the reaction-to-fire properties of materials is cone calorimetry. The fire properties of polymers can be determined at various fire circumstances (Time to ignition/TTI: ignition stage, Heat release rate/HRR: fire developing stage, and Total heat release/THR: fully developed fire stage). The main purpose of the above outline is to emphasise the need for fire testing of polymeric composites and the widely used fire tests, such as LOI test, UL-94 vertical burning test, cone calorimetry, and micro- combustion calorimetry, are discussed in detail in this review. It is envisaged that this review will provide a summative information regarding the flammability properties of carbon-based polymeric composites, thus aiding researchers to gain insight into the efficacy of particular carbon-based additives.

2. Carbon Family Materials

Carbon and its family materials are employed in numerous applications owing to their inherent advantages, such as porosity, high strength and stiffness, conductivity, etc., and the family is comprised of carbon black (low cost), biochar (widely available and eco-friendly) or single and multi-walled carbon nanotubes (sophisticated), and so on. In the past two decades, a significant diversion and development in material science research were noticed when new members entered into the carbon family. This kind of revolutionary development in engineering primarily started with the discovery of fullerene [4]. Subsequently, such developments were propagated by the discovery of carbon nanotubes (CNTs) [5] and graphene [6]. The application of these carbon-based materials to address the issue of flammability in polymeric composites stems from the fact that conventional fire retardants (FRs) are detrimental for the mechanical properties. Additionally, some halogen-based FRs are pernicious towards the environment. Chen et al. used both nanoclays and CNTs FRs in epoxy composites and arrived at the following conclusions. The reduction in the flammability of the polymer composite is primarily due to the formation of network structure layer on the burning surface and, as compared to nanoclays FR, this layer is effectively formed while using CNTs [7]. For instance, Figure 1 shows the importance and role of carbon-based filler addition in polymer. During polymer pyrolysis, a protective layer (Figure 1b) is formed on the polymer composites, which restricts the transfer of combustible gases and heat; thus. the further degradation of materials can be avoided.

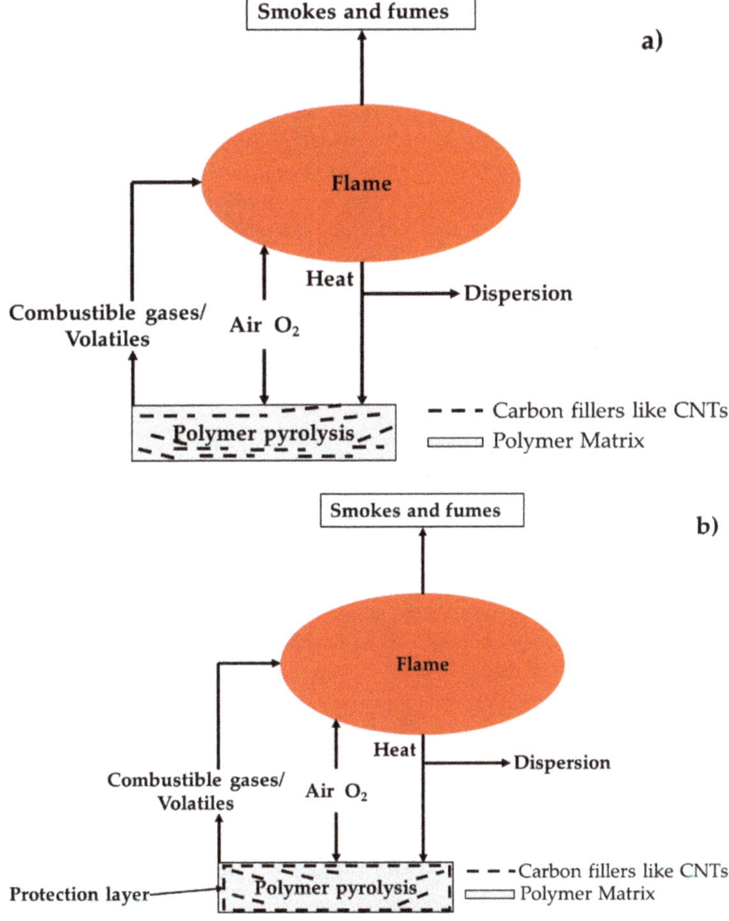

Figure 1. Polymer Combustion (**a**) at initial stage and (**b**) carbon fillers formed protection layer.

2.1. Effect of Various Carbon-Based Materials on Flammability of Polymer Composites

2.1.1. Fullerene

Fullerene, informally called buckyball, is an allotrope of carbon. The fullerene family contain C_{60}, C_{70}, C_{78}, C_{82}, C_{84}, C_{90}, C_{96}, and so on, and, among these. the most important and widely used member in polymer composites is C_{60}, which is spherically shaped carbonaceous nanomaterial having excellent medical benefits and it is also an antioxidant [8]. C60 is called radical sponge i.e., C60 shows high reactivity towards free radicals and can trap more than 34 free radicals during the combustion of polymer [9]. In the past decade. the influence of fullerene reinforcement on the mechanical strength of different polymer matrices has adequately been studied and presented [10–12]. However, there are limited studies available regarding the effect of fullerene on the enhancement of fire resistant properties of polymer matrix composites. For instance, Kausar analysed the effect of polyurethane (PU) coating on various flame retardancy properties of poly (methyl methacrylate) (PMMA), in which the reinforcement used was C_{60}. The continuous reduction in peak heat release rate (pHRR) was recorded while incorporating C_{60} in the PU coated PMMA matrix and compared to neat PU/PMMA, 0.5 wt% of C_{60} added PU/PMMA displayed 61% of reduced pHRR. In addition, C_{60} reinforced composites have showed prolonged time delay to ignition and reduced time to pHRR [13], and the reasons are subsequently explained. Song et al. have demonstrated the flammability behaviour of polypropylene (PP) nanocomposites, in which the reinforcement used was fullerene C_{60}. It was reported that, as compared to neat PP, different amounts of C_{60} reinforced PP composites have showed significantly reduced flammability i.e., notable drop in pHRR and extended time to ignition [14]. Guo et al. prepared the surface functionalised C_{60} using 9, 10-dihydro-9-oxa-10-phosphaphenanthrene-10-oxide in order to further promote the flame retardancy of PP composites. Compared to as received C_{60} nano-fullerene. the reinforcement of 3 wt% surface functionalised filler had exhibited higher TTI and lower combustion duration [15]. The enhancement of flame retardancy of polymers while adding different carbon-based FRs can be perceived from Table 1. Various fire properties are quantitated and their improvement (in %) corresponding to neat polymer value are calculated.

In summary. the combustion mechanism of most of the polymer chains follow a free radical chain reaction via β-scission and the presence of fullerene in polymers might trap these free radicals produced due to thermal degradation of polymers. This process subsequently forms an in-situ crosslinked network and as a result of this network formation the thermal stability and fire-resistant properties are enhanced. Fullerene also shows good synergistic effect with inorganic metal flame retardant (mFR), intumescent flame retardants (iFRs), brominated flame retardants (bFRs), nanoclay, CNTs, graphene oxide (GO), and so on.

2.1.2. Nanotubes

There are generally three classes of carbon nanotubes, namely multi-walled carbon nanotube (MWCNT), double-walled carbon nanotube, and single walled carbon nanotubes (SWCNT). Amongst these. the first discovered MWCNT has two or more tubular shaped graphite fibres and these hollow nanotubes form a concentric cylindrical structure with a space between them that is near to that of the interlayer distance in graphite (0.34 nm) [7]. The MWCNT is a potent FR and its performance is more effective than the organoclays [16]. On the other hand, a single layer tube extending from end to end is called SWCNTs, which has uniform cross section of 0.7–3 nm and their size is close to fullerenes [17]. Both classes of CNTs are widely used as nanofillers in polymer nanocomposites because of their inherent and superior electrical and thermal conductivity and mechanical strength [18]. These varieties of fillers have been used in order to improve the flame retardancy of polymers since the discovery of CNTs in 1991. This is because of the effective formation of continuous thin protective layer on the surface of the polymer, acting as a thermal shield between the oncoming heat/O_2 and underneath virgin polymer. Most importantly, a considerable reduction in HRR might be accomplished with low filler concentrations [19]. P. Patel et al. applied both single and multi-walled CNTs separately in polyether ether ketones (PEEK) matrix and the flammability behaviour of prepared nanocomposites was investigated. The study revealed that the incorporation of small quantity of CNT (0.1 to 1 wt%) showed significant changes of thermal decomposition and flammability of PEEK. Notably. the optimum loading amount of SWCNT (1 wt%) in PEEK is twice than that of MWCNT (0.5 wt%), because the MWCNT showed better dispersion in PEEK than SWCNT [20]. The principal challenge in using nanofillers, like CNTs, is obtaining uniform dispersion and distribution. Since most of the polymers are viscous in nature. the possibility of agglomeration increases with an increasing concentration of nanofillers in the matrix. Kashiwagi et al. showed an impact of CNTs dispersion and concentration on the flammability of PMMA. The better fire resistance performance of the samples during the burning test was found when the dispersion of the SWCNTs is uniform and the filler concentration was between 0.2 to 1 wt%. The PMMA nanocomposites with less than 0.2 wt% of SWCNT showed large number of black discrete islands from which vigorous bubbling during burning occurred. On the other hand.

the samples having more than 0.2 wt% SWCNT exhibited reduced crack formation and the creation of an effective network on the nanocomposite surface during the burning test. The nanocomposites with good network layer and very low discrete islands showed significantly reduced pHRR. The pHRR of the nanocomposite that had good network structured layer is approximately 50% less than those that formed the islands [21]. Therefore. the dispersion of nano-fillers in a matrix can strongly influence the flammability behaviour of the composites and, to improve it further. the functionalization of CNT can also be performed [22]. Mostly, coupling agents were used to functionalize the surface of the CNTs, through which the uniform dispersion was accomplished [23]. For instance, epoxy composites with 9 wt% concentration of vinyltriethoxysilane functionalized CNTs showed 22 to 27% of increase in LOI and V-1 to V-0 rating progress in UL-94. Moreover. the glass transition temperature (T_g) was shifted from 118 to 160 °C and char yield at 750° was increased by 47% for the same level of reinforcement [24]. The uniform dispersion and distribution of CNTs have the main contribution in the formation of continuous barrier layer with the help of high quality char. This high-quality char plays a predominant role in the minimization of pHRR [25,26]. The other commonly used techniques are the use of surfactants [27], controlled sonication of fillers in various solvents [28], and ultra-speed mechanical stirring [29]. Besides. the combined effects of CNTs with other fillers are also demonstrated to enhance the flame retardancy of polymers. For instance. the hybrid filler reinforced composites formed a superior barrier char layer with reduced cracks during cone calorimeter test when compared to individual CNTs or organoclay reinforced composites [30]. In another study, Wen et al. demonstrated the effect of carbon black (CB) on thermal stability and flame retardancy of PP/CNTs ternary nanocomposites. When compared to neat PP and PP/CNTs nanocomposites. the carbon black added nanocomposites (PP/CNTs/CB) displayed improved thermal stability and exhibited lower pHRR and higher LOI. It was concluded that the improvement of flame retardancy was strongly dependent on the concentration of CB [31].

2.1.3. Graphene and Graphene Derivatives

Graphene has the unique structure of one atomically thick and two-dimensional (2D) monolayer composed of sp^2 hybridised carbon atoms. Graphene is the relatively younger member in the carbon family, which was discovered in the year 2004 by the exfoliation of graphite. Graphene has high specific surface area [32], excellent tensile modulus, high strength [33], and superior thermal conductivity [34]. Besides, graphene is an effective FR due to its layered and graphitized structure [35]. Because of these properties of graphene, it is used as a potential reinforcement to enhance the fire retardancy and thermal conductivity of polymer composites. In this section. the recent studies regarding the effect of graphene addition on flammability of various polymer composites are discussed.

Huang et al. prepared poly (vinyl alcohol) (PVA) nanocomposites, in which they have reinforced various amounts of graphene and compared their flammability behaviour with sodium montmorillonite (Na-MMT) and MWCNT reinforced PVA nanocomposites. The PVA filled with 3 wt% of graphene displayed 49% of reduced pHRR when compared to neat PVA. For the same level of filler concentration. the graphene/PVA composites exhibited superior flame retardancy as compared to Na-MMT/PVA and MWCNTs/PVA composites. When compared to neat PVA. the 2 wt% graphene reinforced PVA composites displayed 11 s delay in time to ignition (TTI) and 45% reduction of pHRR, whereas 5 wt% graphene reinforced PVA composites showed 27 s delay in time to ignition (TTI) and 64% reduction of pHRR. When compared to smooth surface of MWCNTs. the presence of oxygen and hydroxyl groups on graphene surface is the main reason for intimate graphene/PVA interactions, which led to the enhancement of flame retardancy of polymer [36]. Attia et al. synthesised graphene while using the ultrasonication process in which maleate diphosphate (MDP) was used as dispersant. Followed by synthesis, acrylonitrile-butadiene-styrene (ABS) composites were fabricated, in which the reinforcements were MDP, graphene-MDP, and graphene-MDP-TiO_2, and their flammability were determined. The TiO_2 nanoparticles decorated graphene reinforced ABS composite exhibited 49% of reduction in both pHRR and total heat release (THR). In addition. the average mass loss rate and emission of CO_2 were significantly reduced by 50% and 37%, respectively. When compared to neat ABS. the nanocomposites exhibited a slow burning rate and the reduction in burning rate was recorded as 71% [37]. In addition. the possibility of using reduced graphene oxide (rGO) as an active synergist for iFR/PP composites was demonstrated by Yuan et al. The reduced heat and smoke release were observed at a lower content (between 0.5 to 1 wt%) of rGO addition in the iFR/PP composites and this was due to the improved char swelling and better insulation by the char. The high thermal conductivity of rGO leads to an increase in thermal conductivity of iFR/PI composites. This caused the enhancement of pHRR from 156 to 262 kW/m^2 while increasing the rGO concentration from 1 to 2 wt%, respectively. This shows that FR synergism is effective at a lower amount of graphene (less than 1 wt%) and higher loading of rGO exhibits antagonistic effect on the iFR. Li et al. prepared epoxy nanocomposites in which the reinforcement used was silane treated graphene oxide nanosheets (GON). The 2-(Diphenylphosphino)ethyltriethoxy silane (DPPES) was grafted onto the surface of the GON while using a condensation reaction, as a result of this synergistic phosphorus/silicon-contained GON FR was obtained. The effect of DPPES-GON addition on the flammability of epoxy was assessed using LOI and UL-94 tests. The 10 wt% of DPPES-GON incorporated epoxy composite displayed significantly enhanced flame retardancy. The LOI of neat and DPPES-GON 10 wt%/epoxy composite is 20% and 36%, respectively. In addition. the UL-94 results changed from no-rating to V-0 rating when 10 wt% of DPPES-GON was added with neat epoxy. The protective layer arrested the flammable gases and acted as a barrier between the heat and unburned epoxy. The synergism of phosphorus and silicon was increased by the effectiveness of the FR system. The presence of phosphorus in compounds formed H_3PO_4 during thermal decomposition and subsequently produced pyrophosphoric acid. The residue from the condensation of phosphoric acid played the major role in curbing combustion. Glass-like phosphorus-containing solid residue is formed due to the condensation of pyrophosphoric acid. This layer would limit the production of volatiles and inhibit the combustion process. Silicon also plays a vital role in this FR system. The low surface energy of silicon caused it to move towards the surface of the protective layer and the protective char layer was evidenced by the char that is obtained from the samples after the LOI test. As the sample was burned. the silicon oxidized into inorganic silicon dioxide due to high heat generation and it formed a thermally stable protective char layer [38]. Thus. the incorporation of low concentration of graphene into iFR/PP composites leads to the formation of closed chamber in the char residue and, consequently, an enhanced char swelling was accomplished that imparted flame retardancy [39].

GQD is zero-dimensional graphene nanofragments, which consists of one or a few layers of graphene and their lateral dimension is less than 100 nm [40]. GQDs exhibit different properties as compared to bulk graphene, like easy functionalization, good physical and chemical stability, high surface to mass ratio, and offer many benefits for energy storage applications [41]. More recently, GQDs have been used as a FR material in polymer composites. Mostly. the hydrothermal method was adopted to synthesise GQDs from nitrogen, nitrogen phosphorous, GO, etc., since this method is economical, sustainable, and the resultant FR will be effective [42–44]. Rahimi-Aghdam, et al. prepared two types of GQDs, in which the first one is nitrogen doped GQDs (NGQDs) and the other one is nitrogen and phosphorous co-doped GQDs (NPGQDs). They have used the hydrothermal method to perform the synthesis process. Subsequently. the polyacrylonitrile (PAN) nanocomposites were prepared with the two types of GQDs as additives and their flammability behaviours were recorded. PAN/NGQDs and PAN/NPGQDs nanocomposites both achieved V-0 rating in UL-94 test [42]. The same group of authors have synthesised ZnAl layered double hydroxide and mixed with NPGQDs. The resultant hybrid fillers were used as reinforcement in PAN nanocomposites and their flammability performance was assessed through cone calorimeter (Table 1) [43]. Khose et al. effectively synthesised functionalized FR GQDs while using GO and phosphorous source via a hydrothermal treatment and recommended for textile applications. It was reported that the transparency of prepared carbon-based GQDs FR retains the colour of the cloth. The flame test results showed that the FR GQDs coated cloth initially emitted very low smoke and did not ignite for more than 300 s while retaining its shape. On the contrary. the normal cloth ignited in just in 5 s and it was completely burnt within 15 s [44].

In summary, graphene and newly found graphene derivatives, like GO, rGO, and GQDs, are potent FRs. These can be employed individually (graphene alone) and in hybrid form (graphene with conventional FRs or inorganic nanofillers) in order to enhance the flame retardancy of the polymer composites.

2.1.4. Graphite

Graphite, which is also known as plumbago or black lead, is an important allotrope of elemental carbon. It is a layered mineral set (can be natural or synthetic) that is made up of stacked GNSs, in which the carbon atoms in the layer forms hexagonal rings through covalent bonds and the successive carbon layers are connected together by weak Van der Waals forces. The usage of as received graphite as FRs in polymers is limited, since the infiltration of viscous polymer resins is very difficult in natural graphite. Therefore. the chemically treated natural graphite, known as expandable graphite (EG), has been extensively used as FR for a variety of polymers. Chemicals, such as sulfuric acid (H_2SO_4) or nitric acid (HNO_3), may be inserted between the graphite layers [45]. Hence, EG acting as an intumescent additive is also a graphite intercalation compound. When EG is exposed to a heat source. the decomposition of H_2SO_4 occurs, which is followed by a redox reaction process (Equation (1)) between H_2SO_4 and the graphite will produce the blowing gaseous products, such as CO_2, SO_2, and H_2O [46].

$$C + 2H_2SO_4 = CO_2 \uparrow + 2H_2O \uparrow + 2SO_2 \uparrow \tag{1}$$

As stated above. the EG contains treated flake graphite with intercalation reagents, such as H_2SO_4. When EG material is exposed to high heat. the H_2SO_4 starts to decompose and release gaseous products. This process leads to an increase in inter-graphene layer pressure and generates sufficient strong push, which keeps graphite layers apart. As a result of high heat and successive pressure development. the material starts to expand and the volume of the EG increases about 10 to 100 times the initial volume, known as the blowing effect. The expansion suffocates the flame, acts as a good smoke suppressant, and restricts mass transfer from the polymers, which prevents further degradation of the underneath virgin materials [47,48]. Therefore, when a material is exposed to high temperature, EG expands and produces a voluminous protective layer, thus providing FR performance to various polymeric matrices [49]. Lee et al. demonstrated the enhancement of flame retardancy and self-extinguishing properties of polyketone (PK) nanocomposites. The authors have reinforced hybrid fillers in PK matrix, which has EG and MWCNTs, in order to achieve superior flame retardancy. The addition of small quantity of MWCNTs (1 wt%) with EG led to better protection network formation in PK and, as a result. the thermal stability and LOI were significantly enhanced. This network that formed during combustion acted as a barrier and restricted the polymer degradation. From the experimental results. the LOI and pHRR for neat PK is 25% and 464.4 kW/m^2, respectively. The reinforcement of 30 wt% of EG in PK displayed 35% of LOI and 182.7 kW/m^2 of pHRR. Further addition of 1 wt% of MWCNTs with 40 wt% of EG hybridisation showed the LOI of 45% and pHRR of 118.4 kW/m^2. This tremendous enhancement of flame retardancy was due to the formation of bridging network by EG and MWCNTs. During combustion. the exfoliation of EG is restricted while adding 1 wt% of MWCNTs and degradation of underneath materials are also prevented [47]. Zhu et al. analysed the synergistic effect of adding EG and ammonium polyphosphate (APP) on flame retardancy of poly lactic acid (PLA)-based composites. The prepared PLA composites contained 15 wt% of APP/EG (1:3 ratio) that exhibited 36.5% of LOI and rated V-0 in UL-94 test. The PLA containing same combination of filler showed 38.3% reduced pHRR than neat PLA. The synergism between APP and EG was advantageous, since they together formed a stable and more dense char protective layer. This layer formation avoided the further combustion of underlying substrate [50].

In summary, EG is widely used as an effective FR in various polymeric materials. Compared to individual EG as FR. the synergistic effect of two fillers produced significant enhancement of flame retardancy in polymeric composites. Addition of small amount of CNTs with EG performed well during combustion.

2.1.5. Biochar (BC)

The limitations of inorganic carbon family based (fullerene, CNT, etc.) FRs are their high cost, since they need advanced synthesis techniques. There is a huge demand for green, sustainable, eco-friendly, and renewable alternative materials for composite applications due to the increased environmental awareness. Hence, carbon rich filler materials derived from the renewable source is an appropriate substitute for this issue and can be used as reinforcement in polymer composites preparation to enhance various physical, mechanical, and FR properties.

BC, or biocarbon, is a carbonaceous material that is made by heating virtually any biomass in a neutral environment. This carbon rich material has been recently used as the reinforcement in polymer composites and led the way for the production of eco-friendly composites with enhanced mechanical and FR properties [51,52]. Instead of using the organic wastes directly in the manufacturing of biocomposites, BC derived from various biomasses, such as Rice Husk [53], bamboo [54], paunch grass, pine wood saw dust, date palm [55], poultry litter, and sewage and dewatered sludge [56], are utilized as a reinforcement. Thermo-chemical conversion technology of slow pyrolysis is the main process that is used for the generation of high yield BC. The physical and chemical properties of the BC are highly dependent on the selected biomass and thermal processing conditions, such as pyrolysis temperature, residence time, heating rate, sweep gas flow rate, etc. [57,58]. These properties include density, surface area, microscopic changes, like pore growth (size and volume), hardness/modulus, and pulverisability [59]. The density of the selected biomass as a feedstock has strong influence on the density of BC. For instance. the high-density BC could be produced using high density biomass [60]. The BC has better thermal stability than the natural fibres [61]. The macro, meso, and micro-pores on its surface provide better physical bonding with matrix (Figure 2) [59]. Chemically treated and untreated BCs reveal different functional groups on their surfaces, which consist of carboxyl ($-COOH$) and hydroxyl ($-OH$) groups. These functionalities are sensitive to carbonisation temperature and, with the increase in temperature, they start to dwindle [62]. Better bonding and good compatibility between matrix and BC could be obtained while the surface area and pore volume are high. Typically. the BC surface has a porous honeycomb structure consisting of a high concentration of carbon. These porous honeycomb structures of BC filler allow for the infiltration of the molten polymer during processing and they create a physical bonding, which could result in an improvement of mechanical properties of the composites [63]. Liu et al. studied the combustion characteristics of bamboo-BCs at the heat flux of 35 kW/m^2, which were produced at three different pyrolysis temperatures (200, 250, and 300 °C) and at three different residence times (1, 1.5, and 2 h). For all temperatures. the TTI is shortened with an increase in residence time, while all of the bamboo-BCs (produced at different temperatures and residence time) displayed a shorter TTI when compared to bamboo materials. It was observed that with an increase in test time, bamboo-BCs exhibited random cracks on its surface due to the differential thermal stability of their compositions. With the help of these cracks, some volatile material is released and caused faster ignition. The pHRR of bamboo-BCs was lower when compared to bamboo materials, which indicated a lower content of moisture content and volatile matter [64]. Zhao et al. analysed the flammability of BC, which were produced from different feedstocks, namely, corn, wood, dairy manure with rice husk, and bull manure with sawdust at different pyrolysis temperatures and as a function of time post production. In general, BCs made at higher pyrolysis temperatures had lower flammability. All four BCs used in the study also displayed the highest surface area at high temperature. The study also confirmed that none of the tested biochar samples qualified as flammable substances, which extend its application in manufacturing FR polymer composites [65]. Das et al. fabricated BC/PP biocomposites and their mechanical, thermal stability, and flammability behaviour were evaluated. The reinforced biocomposites showed increased thermal stability, reduced pHRR, and lower smoke release when compared to neat PP. The strong covalent bonding of carbon atoms makes them difficult to be separated during combustion, thereby increasing fire resistance. The BC had high thermal stability, as observed from TGA by the authors and, thus, their addition in composites also increased the thermal stability of the polymer due to the additive effect. It was concluded that both tensile and flexural modulus of the biocomposites were increased with an increase in concentration of BC. Additionally. the major reasons for this enhancement of mechanical properties are better compatibility and good physical bonding between the BC and the matrix, owing to the porous structure of the BC [66]. Elnour et al. used lignocellulosic biowaste from date palm, which were pyrolysed at different temperatures and their effect on physical structure and surface morphology was studied. The authors manufactured BC/PP composites and concluded that the reinforced PP showed better thermal stability and enhanced stiffness. In particular, as compared to neat PP. the BC added PP displayed reduced thermal decomposition and lower maximum degradation temperature. Moreover. the authors suggested surface functionalization of filler for the further enhancement of the mechanical, thermal, and flammability properties [67]. Ikram et al. studied the mechanical and flammability characteristics of wood/pine wood BC/PP biocomposites and demonstrated the properties with respect to neat PP/and maleated anhydride polypropylene. It was concluded that the addition of MAPP coupling agent and wood particles have significantly enhanced the tensile and flexural properties, but the pHRR remained unaffected [68]. Elsewhere. the hybridisation technique was followed by Das et al., where the authors used a mixture of BC and wool. Biocomposites with hybrid fillers significantly minimised the pHRR and smoke release when compared to neat PP. The char layer limited the heat and fuel transfer between the ambient air and underneath polymer. The LOI value was enhanced because of the hybridisation with wool [69]. When BC is made at high temperature, all of the volatiles escape from its surface, leaving behind a carbon skeleton. The absence of these flammable volatiles does not provide the fuel for combustion to occur [67,70]. Two batches of wood dust (WD)/BC/PP composites were fabricated, in which two types of conventional FRs such as APP and magnesium hydroxide $Mg(OH)_2$, were individually added and their reaction-to-fire properties were assessed through cone calorimeter test by Das et al. The TTI and pHRR of neat PP was found as 29 s and 1054 kW/m^2, respectively. Whereas, adding 20 wt% of APP and $Mg(OH)_2$ with WD (10 wt%)/BC (24 wt%)/PP composites significantly reduced the pHRR to 376.2 kW/m^2 and 333.3 kW/m^2, respectively. In both cases. the PP composite, which has higher BC concentration and less WD, actively reduced the pHRR. The carbonaceous layer that formed by the thermally stable BC with other constituents have restricted the transport of fuel and O_2, which led to improved fire properties, such as LOI, pHRR, and THR. In addition. the tensile strength was unaffected, whereas other mechanical properties, such as tensile modulus, flexural strengths, and flexural modulus, were considerably improved when compared to neat PP. However, some of the FR particles got trapped inside the biochar pores, thereby somewhat reducing their efficiencies. APP was more affected, because it relies on condensed phase reaction requiring contact with the polymer [71]. In summary, enhanced flame retardancy of polymer composites using BC reinforcement is one of the best techniques, since both environmental sustainability and low-cost are considered. The BC that is derived from wood, grasses and agricultural wastes using any suitable thermo-chemical conversion technique can be effectively used in biocomposites fabrication, which significantly reduces the landfilling of agro wastes and also provides a new platform for the development of new materials [72,73].

Table 1. Cone calorimetry and Limiting Oxygen Index (LOI) test data for various neat polymers and its nanocomposites.

Type of Composites	TTI (s)	% of Ignition Time Delay from Neat Polymer Increased (↑) or Decreased (↓)	pHRR (kW/m^2)	% of pHRR Increased (↑) or Decreased (↓) w.r. to Neat Polymer	Time to pHRR (s)	THR (MJ/m^2)	LOI (%)	Ref.
PU/PMMA	70	—	343	—	189	—	—	[13]
PU/PMMA/Full-C$_{60}$ 0.5	101	44% (↑)	131	61.8% (↓)	150	—	—	
Neat PP	29	—	1054	—	—	97	—	[63]
TCP 900 *	20	31% (↓)	473.68	55% (↓)	—	86.95	—	
Neat PP	29 ± 2	—	1054 ± 120	—	120 ± 18	97 ± 14	18 ± 0.1	[69]
BC + PP + APP	14 ± 0	51.7% (↓)	277.82 ± 2.4	73.6% (↓)	120 ± 29.6	88.75 ± 0.7	22.08 ± 0.1	
Neat PP	30	—	1261	—	335	208	18	[31]
PP + 3wt% CB	20	33.3% (↓)	584	53.6% (↓)	355	192	22.6	
PP + 3wt% CNT + 5wt% CB	25	16.6% (↓)	314	75.1% (↓)	70	180	27.6	
Neat PVA	18 ± 2	—	373 ± 6	—	—	58 ± 0.6	—	[36]
PVA + 3wt% Na-MMT	20 ± 2	11.1% (↑)	263 ± 7	29.4% (↓)	—	58 ± 0.4	—	
PVA + 3wt% MWCNT	24 ± 2	33.3% (↑)	241 ± 8	35.3% (↓)	—	52 ± 0.4	—	
PVA + 3wt% GNS	33 ± 2	83.3% (↑)	190 ± 6	49% (↓)	—	45 ± 0.3	—	
PVA + 5wt% GNS	45 ± 3	150% (↑)	133 ± 5	64.3% (↓)	—	38 ± 0.5	—	
Neat ABS	43 ± 1.5	—	1385 ± 92	—	—	145 ± 11	—	[37]
ABS-MDP	32 ± 1.5	25.5% (↓)	821 ± 55	40.7% (↓)	—	97 ± 9	—	
ABS-GRP-MDP	18 ± 1	58.1% (↓)	812 ± 54	41.3% (↓)	—	91 ± 6	—	
ABS-GRP-MDP- TiO$_2$NP-5	35 ± 1.2	18.6% (↓)	720 ± 48	48% (↓)	—	75 ± 6	—	
Neat PAN	10	—	609	—	25	9.1	—	[42]
PAN/NGQDs	15	50% (↑)	565	7.2% (↓)	30	8.5	—	
PAN/NPGQDs	20	100% (↑)	515	15.4% (↓)	35	7.7	—	
Neat PAN	10	—	609	—	25	9.1	—	[43]
PAN/ZnAl LDH	20	100% (↑)	462	24.1% (↓)	40	7.9	—	
PAN/ZnAl LDH-NPGQD	25	150% (↑)	435	28.5% (↓)	45	7.4	—	

* pine wood-based BC were produced at pyrolysis temperatures of 900 °C. Note: The expansion for all used abbreviations is available in the running text.

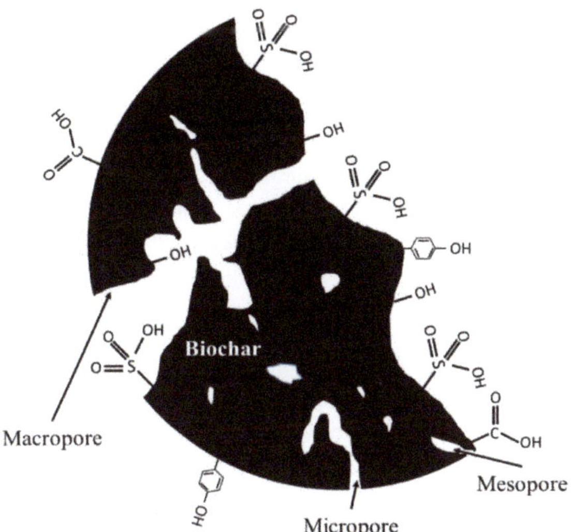

Figure 2. Schematic of biochar (BC) with different functional groups and pores. Adapted with permission from Ref. [59].

2.1.6. Other Carbon-Based Materials

Apart from all of the aforementioned fillers, nanosized carbon black (CB) is a low cost, abundantly available, electrically conductive, and low-density reinforcement that has been widely used to enhance properties of polymer composites. A few studies reported its effect on the flammability of various polymeric matrices (Table 1) Studies revealed that the CB filled composites not only exhibited good flame retardancy, but also the thermal stability was improved [74–76]. Yang et al. studied the effect of CB incorporation on flame retardancy and thermal decomposition of PP/carbon fibre (CF) composites. The authors confirmed the uniform dispersion of CB fillers in PP/carbon fibre composites while using morphological analysis. LOI of neat PP was 18.2% and the individual effect of 3 wt% of CF and 5 wt% of CB reinforced PP composites on LOI was 19.9% and 24.6%, respectively. The hybrid form of CF and CB fillers showed significant beneficial effect on the flammability of PP composites when compared to individual CF and CB reinforcement. LOI of 3 wt% CF and 5 wt% CB reinforced PP composites was recorded as 25.7%. Importantly. the pHRR assessed from cone calorimeter for neat PP and hybrid fillers reinforced composites is 1212 kW/m^2 and 361 kW/m^2, respectively. This synergistic effect of hybrid form of CB and CF have shown better flame retardancy in the PP matrix as compared to the individual performance of CB and CF. The one-dimensional (1-D) CF and zero-dimensional (0-D) CB together formed a strong three-dimensional (3-D) network in PP matrix. The developed network had significant role in the formation of compact carbonaceous protection layer during pyrolysis. As a result, as compared to neat PP, PP/CF, and PP/CB composites, a significant enhancement of flame retardancy of PP/CF/CB composites was obtained [75].

3. Flammability Measuring Techniques

Experiments for measuring material flammability and fire behaviour are classified into small, bench, and large scales, depending on the sample size required [77]. However, these tests involve a great deal of expertise in their operation. Hence, standard protocols, like ASTM and ISO, have been provided for easy application. The following sections briefly reports some commonly used small-scale and bench-scale experiments for flammability analysis.

3.1. LOI Test

The LOI test (Figure 3a) is a laboratory scale test process that provides a measure of the lowest amount of oxygen needed to ignite a vertically positioned sample of size $80 \times 10 \times 4$ mm^3 in an oxygen and nitrogen mixed environment [78,79]. The test procedure and calibrations can be found in ASTM D2863, ISO 4589-2, and NES 714. According to the standards. the gas stream flows in an upward direction to the vertically oriented sample in a chimney, whilst a propane gas flame ignites the upper part of the material. Thus. the sample's burning length and time are determined for flammability analysis. LOI can be calculated by following mathematical expression:

$$\text{LOI} = \left(\frac{[O_2]}{([O_2] + [N_2])} \right) \times 100$$

where, $[O_2]$ and $[N_2]$ are the flow rate of oxygen and nitrogen in L/min. respectively.

Therefore, a material, which demands more oxygen, will display higher LOI. In addition, a higher index indicates that the material is more flame resistant. Given that atmospheric air has 21% of O_2. the risk of burning of polymer materials is high, whose LOI value is less than 21; however, materials with an LOI above 21 are categorized as self-extinguishing because their combustion cannot be retained at standard atmosphere without the support of an external source [80].

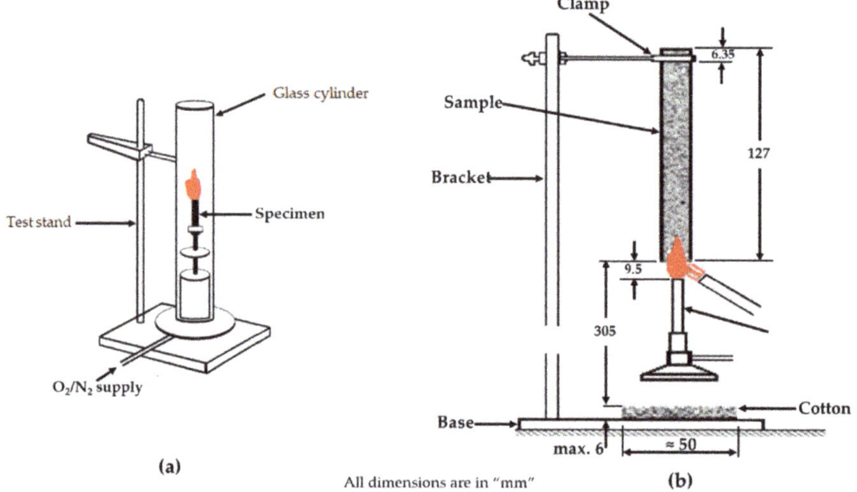

Figure 3. Experimental set-up: (**a**) LOI and (**b**) UL-94 vertical burn tests. Adapted with permission from Ref. [81].

In summary, LOI is a common characterisation method and it has been used in numerous studies. The main objective of most of the studies is fabricating polymer composites with increased flame retardancy, so that it will demand higher oxygen percentage in order to combust.

3.2. Vertical Burn Test (UL 94)

This test has been developed by Underwriters Laboratory Inc. for testing the flammability of plastics. In practice. the UL 94 vertical burning test is common technique (Figure 3b), which provides the rating for the test specimens based on its ignition and flame spread of materials exposed to a small flame [82]. The test protocol is designated in ASTM D3801. The procedure involves the preparation and exposure of the said sample of size $127 \times 13 \times 3$ mm^3 to a carefully controlled flame for 10 s. Any burning action after the removal of the flame is monitored and recorded. If the specimen self-extinguishes. the flame is then reapplied for another 10 s, and then removed. For improved accuracy and reliability, at least five samples are tested for each material combination. The burning time for flame exposures and afterglow are recorded. The qualitative ranks for evaluating the test results of the experiment are no-rating, V-0, V-1, and V-2, as shown in the test protocol. Materials are classified into these three categories, depending on satisfaction of the conditions that are mentioned in Table 2. If the sample continue to burn upon initial flame application, it is given no rating.

Table 2. Conditions for UL-94 classifications [83].

Specific Flaming Characteristics	Rating		
	V-0	V-1	V-2
Total flaming combustion time (in seconds)			
• for each specimen	≤10	≤30	≤30
• for all five specimens of any set	≤50	≤250	≤250
Flaming and glowing combustion for each specimen after second burner flame application	≤30	≤60	≤60
Cotton ignited by flaming drips from any specimen	No	No	Yes
Glowing or flaming combustion of any specimen to holding clamp	No	No	No

3.3. Micro-Scale Combustion Calorimetry (MCC)

MCC is used to characterize the fire behaviour of materials. It uniquely measures the heat release capacity (HRC), which is a combination of thermal stability and combustion properties of materials that can be used to categorise the flammability [77,84]. The HRC is also described as a rate-independent flammability parameter and, from thermodynamics point of view, it is an intensive property and can be measured from chemical structure of a material [84]. The HRC (in $J\ g^{-1}\ K^{-1}$) is defined as the maximum HRR per unit mass in the test (in $W\ g^{-1}$) divided by the average heating rate over the measurement range ($K\ s^{-1}$), and this is the single best measure of the fire hazard of a material [85].

The test apparatus has two separate stages. the pyrolysis and combustion phases. In ASTM D7309-19. the experimental procedure consists of two selectable pyrolysis modes, namely Method-A and Method-B, which are used for controlled thermal and thermal oxidative decomposition, respectively. Samples of mass 0.5–50 mg are pyrolysed in inert gas and the volatile effluent is mixed with excess oxygen prior to combustion in Method-A, while pyrolysis in Method-B occurs in a mixture of oxygen and inert gas [86]. The heat release rate from the test is obtained from oxygen consumption calorimetry. Other derived parameters are the total heat release rate, time, and temperature at pHRR. MCC curves are represented by plots of heat release rate against temperature or time.

3.4. Cone Calorimetry

Cone calorimetry is the most widely applied bench-scale fire experiment. The test method measures the heat release rate, ignition time, mass loss rate, combustion or extinction time, smoke production, soot yield, and quantities of CO and CO_2. The sample size for the test is within $0.1 \times 0.1 \times 0.001$–$0.05\ m^3$ and the applicable heat flux ranges from 10 to 100 kW/m^2. The prepared samples are wrapped in aluminium foils and positioned horizontally or vertically under the cone-shaped heater, according to the designated standards (ASTM E1354, ISO 5660). A load cell measures the weight change and the pyrolysate is ignited by an electric pilot spark igniter. The smoke is collected in the hood of the equipment for further analysis. The cone calorimeter operates on the principles of oxygen consumption calorimetry. Extrinsic factors, such as geometry and orientation of the sample, sample thickness, ignition source, ventilation, and temperature, affect the measurement accuracy [77,87].

1. TTI or tig in s: describes the ease of ignition of the polymeric material by measuring how fast the flaming combustion occurs when the polymeric material is exposed to incident heat flux (in kW/m^2) and in oxygen-controlled ambient environment. Hence, polymeric material with a high TTI indicates material that is difficult to be ignited. In the case of flame-retarding polymer composites, sometimes the addition of FRs lead to advance decomposition and, thereby. the reduction of TTI (Table 1. Thus. the shorter TTI is not an indication of worsening flame retardancy of a material.

2. HRR in kW/m^2: is known as the heat release per unit time and unit surface area during the cone calorimetry test. Mainly. the amount of peak HRR (pHRR) and time taken to reach the pHRR are used to measure the fire performance of polymeric materials.

3. Total heat release (THR in kJ/m^2): is the total quantity of calorific value released per unit area after the combustion of materials, and this can be determined according to the integration of the HRR vs. time.

4. Fire growth rate (FGR in KW/(s.m^2)): FGR is mathematically calculated as FGR = pHRR/(pHRR)$_{ti}$ [81], (pHRR)$_{ti}$ is the time taken to reach the pHRR. The faster FGR indicates the shorter the time that is taken to notice the fire [88].

5. Mass loss rate (MLR in g/s): is the amount of mass loss of polymeric material per unit time during combustion.

4. Conclusions and Scope for Future Research

In summary, carbon and its family materials are potential FR reinforcement in polymer composites and these reinforcements are attractive alternatives for conventional FRs. The carbon-based fillers actively reduce the flammability of the polymer composites by (1) the formation of protective char layer and (2) absorbing free radicals. In the first case. the contact of atmosphere and flame with underlying materials is reduced, since the char layer acts as a shield between them. The second is the internal process, which minimise the reaction rate and, as a result. the combustion is disrupted. In addition, carbon-based filler reinforcements are able to improve thermal stability, mechanical properties, and thermal conductivity of the polymers. Chemicals, like silane, can be grafted onto the surface of the carbon-based fillers and analysing their performance on flame retardancy in polymers without compromising the mechanical strength still has scope for further research.

The high purity nanofillers are cost intensive; therefore, achieving the flame retardancy at lowest filler concentration is desired. This could be achieved by using carbon based nanofillers (like CNTs, Fullerene, graphene sheets, etc.). Either one of following methods was followed in order to obtain further enhancement of flame retardancy in polymer composites: surface functionalization, coupling agents, and hybridisation of fillers. In most of the cases. the best FR synergism with iFRs can be achieved at less than 1 wt% of CNTs or graphene concentration and further loading of carbon-based fillers exhibits adverse effects. Recently, there is a huge scope for the enhancement of fire safety of polymers using GQDs based FRs and limited studies are available in this field.

Most importantly, eco-composites that were produced with BC reinforcement exhibited acceptable FR effect and this BC could be derived from various biomasses (feedstock is agro and forestry wastes) by the pyrolysis process, which also reduces the landfilling of agro wastes. However, scientific studies are required to understand the synergistic effect of BC with other fillers (it might be a carbon-based filler or inorganic particle in hybrid form) on FR properties of polymer composites. BC is merely a fire-resistant additive that indicates further research is needed to make them fire retarding. Furthermore, BC could be used in conjunction with other FRs. In addition. the incorporation of FRs reduces the mechanical properties, such as tensile and flexural strength of composites, and the use of BC with conventional FRs could conserve the strength and enhance fire resistance. However. the application of BC in the field of FR polymer composites is mainly at the stage of laboratory experiments or at infant stages in some industrial application. However, in the future, such a situation might be entirely changed, since BC could be produced in large quantity at a low cost.

Author Contributions: Conceptualization, O.D.; data curation, K.B., G.R., R.A.M., N.K.K., R.E.N.; writing—original draft preparation, K.B., G.R., R.A.M., R.E.N., O.D., N.K.K.; writing—review and editing, O.D., M.F., M.S.H., Á.R., L.J., Q.X., A.B.; supervision, O.D., M.S.H., M.F.; All authors have read and agreed to the published version of the manuscript.

Funding: The funders had no role in the design of the study; in the collection, analyses, or interpretation of data; in the writing of the manuscript, or in the decision to publish the results.

Acknowledgments: The authors express their gratitude towards STINT/NFSC grant (CH2018-7733).

Conflicts of Interest: The authors declare no conflict of interest.

References

1. World Health Organization. *A WHO Plan for Burn Prevention and Care*; World Health Organization: Geneva, Switzerland, 2008.

2. Younis, A. Flammability properties of polypropylene containing montmorillonite and some of silicon compounds. *Egypt. J. Pet.* **2017**, *26*, 1–7. [CrossRef]

3. Fang, L.; Lu, X.; Zeng, J.; Chen, Y.; Tang, Q. Investigation of the Flame-Retardant and Mechanical Properties of Bamboo Fiber-Reinforced Polypropylene Composites with Melamine Pyrophosphate and Aluminum Hypophosphite Addition. *Materials* **2020**, *13*, 479. [CrossRef] [PubMed]

4. Kroto, H.W.; Heath, J.R.; O'Brien, S.C.; Curl, R.F.; Smalley, R.E. C60: Buckminsterfullerene. *Nature* **1985**, *318*, 162–163. [CrossRef]

5. Ribeiro, B.; Botelho, E.C.; Costa, M.L.; Bandeira, C.F. Carbon nanotube buckypaper reinforced polymer composites: A review. *Polímeros* **2017**, *27*, 247–255. [CrossRef]

6. Novoselov, K.; Geim, A.K.; Morozov, S.; Jiang, D.; Zhang, Y.; Dubonos, S.V.; Grigorieva, I.V.; Firsov, A.A. Electric Field Effect in Atomically Thin Carbon Films. *Science* **2004**, *306*, 666–669. [CrossRef]

7. Gürünlü, B.; Yücedağ, Ç.T.; Bayramoğlu, M.R. Green Synthesis of Graphene from Graphite in Molten Salt Medium. *J. Nanomater.* **2020**, *2020*, 1–12. [CrossRef]

8. Pan, Y.; Guo, Z.; Ran, S.; Fang, Z. Influence of fullerenes on the thermal and flame-retardant properties of polymeric materials. *J. Appl. Polym. Sci.* **2019**, *137*, 47538. [CrossRef]

9. Krusic, P.J.; Wasserman, E.; Keizer, P.N.; Morton, J.R. Radical reactions of C60. *Science* **1991**, *254*, 1183–1185. [CrossRef]

10. Zuev, V.V. Polymer Nanocomposites Containing Fullerene C60 Nanofillers. *Macromol. Symp.* **2011**, *301*, 157–161. [CrossRef]

11. Zuev, V.V.; Kostromin, S.; Shlykov, A.V. The effect of fullerene fillers on the mechanical properties of polymer nanocomposites. *Mech. Compos. Mater.* **2010**, *46*, 147–154. [CrossRef]

12. Ogasawara, T.; Ishida, Y.; Kasai, T. Mechanical properties of carbon fiber/fullerene-dispersed epoxy composites. *Compos. Sci. Technol.* **2009**, *69*, 2002–2007. [CrossRef]

13. Kausar, A. Adhesion, morphology, and heat resistance properties of polyurethane coated poly (methyl methacrylate)/fullerene-C60 composite films. *Compos. Interface* **2017**, *24*, 649–662. [CrossRef]

14. Song, P.; Zhu, Y.; Tong, L.; Fang, Z. C60reduces the flammability of polypropylene nanocomposites byin situforming a gelled-ball network. *Nanotechnology* **2008**, *19*, 225707. [CrossRef] [PubMed]

15. Guo, Z.; Wang, Z.; Fang, Z. Fabrication of 9,10-dihydro-9-oxa-10-phosphaphenanthrene-10-oxide-decorated fullerene to improve the anti-oxidative and flame-retardant properties of polypropylene. *Compos. Part B Eng.* **2020**, *183*, 107672. [CrossRef]

16. Beyer, G. Short communication: Carbon nanotubes as flame retardants for polymers. *Fire Mater.* **2002**, *26*, 291–293. [CrossRef]

17. Ajayan, P.M. Nanotubes from Carbon. *Chem. Rev.* **1999**, *99*, 1787–1800. [CrossRef]

18. Al Sheheri, S.Z.; Al-Amshany, Z.M.; Al Sulami, Q.A.; Tashkandi, N.Y.; Hussein, M.A.; El-Shishtawy, R.M. The preparation of carbon nanofillers and their role on the performance of variable polymer nanocomposites. *Des. Monomers Polym.* **2019**, *22*, 8–53. [CrossRef]

19. Kausar, A.; Rafique, I.; Anwar, Z.; Muhammad, B. Recent Developments in Different Types of Flame Retardant and Effect on Fire Retardancy of Epoxy Composite. *Polym. Technol. Eng.* **2016**, *55*, 1512–1535. [CrossRef]

20. Patel, P.; A Stec, A.; Hull, R.; Naffakh, M.; Díez-Pascual, A.M.; Ellis, G.; Safronava, N.; Lyon, R.E. Flammability properties of PEEK and carbon nanotube composites. *Polym. Degrad. Stab.* **2012**, *97*, 2492–2502. [CrossRef]

21. Kashiwagi, T.; Du, F.; Winey, K.I.; Groth, K.; Shields, J.R.; Bellayer, S.P.; Kim, H.; Douglas, J.F. Flammability properties of polymer nanocomposites with single-walled carbon nanotubes: Effects of nanotube dispersion and concentration. *Polymer* **2005**, *46*, 471–481. [CrossRef]

22. Wang, X.; Kalali, E.N.; Wan, J.-T.; Wang, D.-Y. Carbon-family materials for flame retardant polymeric materials. *Prog. Polym. Sci.* **2017**, *69*, 22–46. [CrossRef]

23. Guadagno, L.; De Vivo, B.; Di Bartolomeo, A.; Lamberti, P.; Sorrentino, A.; Tucci, V.; Vertuccio, L.; Vittoria, V. Effect of functionalization on the thermo-mechanical and electrical behavior of multi-wall carbon nanotube/epoxy composites. *Carbon* **2011**, *49*, 1919–1930. [CrossRef]

24. Kuan, C.-F.; Chen, W.J.; Li, Y.-L.; Chen, C.-H.; Kuan, H.; Chiang, C.-L. Flame retardance and thermal stability of carbon nanotube epoxy composite prepared from sol–gel method. *J. Phys. Chem. Solids* **2010**, *71*, 539–543. [CrossRef]

25. Gao, F.; Beyer, G.; Yuan, Q. A mechanistic study of fire retardancy of carbon nanotube/ethylene vinyl acetate copolymers and their clay composites. *Polym. Degrad. Stab.* **2005**, *89*, 559–564. [CrossRef]

26. Ma, H.; Tong, L.; Xu, Z.; Fang, Z. Synergistic effect of carbon nanotube and clay for improving the flame retardancy of ABS resin. *Nanotechnology* **2007**, *18*, 375602. [CrossRef]

27. Vaisman, L.; Marom, G.; Wagner, H.D. Dispersions of Surface-Modified Carbon Nanotubes in Water-Soluble and Water-Insoluble Polymers. *Adv. Funct. Mater.* **2006**, *16*, 357–363. [CrossRef]

28. Schaefer, D.W.; Zhao, J.; Brown, J.M.; Anderson, D.P.; Tomlin, D.W. Morphology of dispersed carbon single-walled nanotubes. *Chem. Phys. Lett.* **2003**, *375*, 369–375. [CrossRef]

29. Martin, C.; Sandler, J.; Shaffer, M.S.P.; Schwarz, M.-K.; Bauhofer, W.; Schulte, K.; Windle, A. Formation of percolating networks in multi-wall carbon-nanotube–epoxy composites. *Compos. Sci. Technol.* **2004**, *64*, 2309–2316. [CrossRef]

30. Beyer, G. Carbon Nanotubes—A New Class of Flame Retardants for Polymers. *Int. Polym. Sci. Technol.* **2003**, *30*, 1–6. [CrossRef]

31. Wen, X.; Tian, N.; Gong, J.; Chen, Q.; Qi, Y.; Liu, Z.; Liu, J.; Jiang, Z.; Chen, X.; Tang, T. Effect of nanosized carbon black on thermal stability and flame retardancy of polypropylene/carbon nanotubes nanocomposites. *Polym. Adv. Technol.* **2013**, *24*, 971–977. [CrossRef]

32. Dai, J.F.; Wang, G.J.; Ma, L.; Wu, C.K. Surface properties of graphene: Relationship to graphene-polymer composites. *Rev. Adv. Mater. Sci.* **2015**, *40*, 60–71.

33. Lee, J.-U.; Yoon, D.; Cheong, H. Estimation of Young's Modulus of Graphene by Raman Spectroscopy. *Nano Lett.* **2012**, *12*, 4444–4448. [CrossRef] [PubMed]

34. Ghosh, S.; Calizo, I.; Teweldebrhan, D.; Pokatilov, E.P.; Nika, D.L.; Balandin, A.A.; Bao, W.; Miao, F.; Lau, C.N. Extremely high thermal conductivity of graphene: Prospects for thermal management applications in nanoelectronic circuits. *Appl. Phys. Lett.* **2008**, *92*, 151911. [CrossRef]

35. Szeluga, U.; Pusz, S.; Kumanek, B.; Olszowska, K.; Kobyliukh, A.; Trzebicka, B. Effect of graphene filler structure on electrical, thermal, mechanical, and fire retardant properties of epoxy-graphene nanocomposites—A review. *Crit. Rev. Solid State Mater. Sci.* **2020**, 1–36. [CrossRef]

36. Huang, G.; Gao, J.; Wang, X.; Liang, H.; Ge, C. How can graphene reduce the flammability of polymer nanocomposites? *Mater. Lett.* **2012**, *66*, 187–189. [CrossRef]

37. Attia, N.F.; El-Aal, N.A.; Hassan, M. Facile synthesis of graphene sheets decorated nanoparticles and flammability of their polymer nanocomposites. *Polym. Degrad. Stab.* **2016**, *126*, 65–74. [CrossRef]

38. Li, K.-Y.; Kuan, C.-F.; Kuan, H.; Chen, C.-H.; Shen, M.-Y.; Yang, J.-M.; Chiang, C.-L. Preparation and properties of novel epoxy/graphene oxide nanosheets (GON) composites functionalized with flame retardant containing phosphorus and silicon. *Mater. Chem. Phys.* **2014**, *146*, 354–362. [CrossRef]

39. Yuan, B.; Fan, A.; Yang, M.; Chen, X.; Hu, Y.; Bao, C.; Jiang, S.; Niu, Y.; Zhang, Y.; He, S.; et al. The effects of graphene on the flammability and fire behavior of intumescent flame retardant polypropylene composites at different flame scenarios. *Polym. Degrad. Stab.* **2017**, *143*, 42–56. [CrossRef]

40. Smith, A.T.; Lachance, A.M.; Zeng, S.; Liu, B.; Sun, L. Synthesis, properties, and applications of graphene oxide/reduced graphene oxide and their nanocomposites. *Nano Mater. Sci.* **2019**, *1*, 31–47. [CrossRef]

41. Tian, P.; Tang, L.; Teng, K.S.; Lau, S.P. Graphene quantum dots from chemistry to applications. *Mater. Today Chem.* **2018**, *10*, 221–258. [CrossRef]

42. Rahimi-Aghdam, T.; Shariatinia, Z.; Hakkarainen, M.; Haddadi-Asl, V. Nitrogen and phosphorous doped graphene quantum dots: Excellent flame retardants and smoke suppressants for polyacrylonitrile nanocomposites. *J. Hazard. Mater.* **2019**, *381*, 121013. [CrossRef] [PubMed]

43. Rahimi-Aghdam, T.; Shariatinia, Z.; Hakkarainen, M.; Haddadi-Asl, V. Polyacrylonitrile/N,P co-doped graphene quantum dots-layered double hydroxide nanocomposite: Flame retardant property, thermal stability and fire hazard. *Eur. Polym. J.* **2019**, *120*, 109256. [CrossRef]

44. Khose, R.V.; Pethsangave, D.A.; Wadekar, P.H.; Ray, A.K.; Some, S. Novel approach towards the synthesis of carbon-based transparent highly effective flame retardant. *Carbon* **2018**, *139*, 205–209. [CrossRef]

45. Wang, G.; Yang, J. Influences of expandable graphite modified by polyethylene glycol on fire protection of waterborne intumescent fire resistive coating. *Surf. Coat. Technol.* **2010**, *204*, 3599–3605. [CrossRef]

46. Ye, L.; Meng, X.-Y.; Ji, X.; Li, Z.-M.; Tang, J.-H. Synthesis and characterization of expandable graphite–poly(methyl methacrylate) composite particles and their application to flame retardation of rigid polyurethane foams. *Polym. Degrad. Stab.* **2009**, *94*, 971–979. [CrossRef]

47. Lee, S.; Kim, H.M.; Seong, D.G.; Lee, D. Synergistic improvement of flame retardant properties of expandable graphite and multi-walled carbon nanotube reinforced intumescent polyketone nanocomposites. *Carbon* **2019**, *143*, 650–659. [CrossRef]

48. Hong, L.; Hu, X. Mechanical and Flame Retardant Properties and Microstructure of Expandable Graphite/Silicone Rubber Composites. *J. Macromol. Sci. Part B* **2016**, *55*, 175–187. [CrossRef]

49. Guo, C.; Zhou, L.; Lv, J. Effects of Expandable Graphite and Modified Ammonium Polyphosphate on the Flame-Retardant and Mechanical Properties of Wood Flour-Polypropylene Composites. *Polym. Polym. Compos.* **2013**, *21*, 449–456. [CrossRef]

50. Zhu, H.; Zhu, Q.; Li, J.; Tao, K.; Xue, L.; Yan, Q. Synergistic effect between expandable graphite and ammonium polyphosphate on flame retarded polylactide. *Polym. Degrad. Stab.* **2011**, *96*, 183–189. [CrossRef]

51. Savi, P.; Jose, S.P.; Khan, A.A.; Giorcelli, M.; Tagliaferro, A. Biochar and carbon nanotubes as fillers in polymers: A comparison. In Proceedings of the 2017 IEEE MTT-S International Microwave Workshop Series on Advanced Materials and Processes for RF and THz Applications (IMWS-AMP), Pavia, Italy, 20–22 September 2017; pp. 1–3. [CrossRef]

52. Das, O.; Sarmah, A.K.; Bhattacharyya, D. A novel approach in organic waste utilization through biochar addition in wood/polypropylene composites. *Waste Manag.* **2015**, *38*, 132–140. [CrossRef]

53. Zhang, Q.; Yi, W.; Li, Z.; Wang, L.; Cai, H. Mechanical Properties of Rice Husk Biochar Reinforced High Density Polyethylene Composites. *Polymer* **2018**, *10*, 286. [CrossRef] [PubMed]

54. You, Z.; Li, D. Highly filled bamboo charcoal powder reinforced ultra-high molecular weight polyethylene. *Mater. Lett.* **2014**, *122*, 121–124. [CrossRef]

55. Poulose, A.M.; Elnour, A.Y.; Anis, A.; Shaikh, H.; Al-Zahrani, S.; George, J.; Al-Wabel, M.I.; Usman, A.R.; Ok, Y.S.; Tsang, D.C.; et al. Date palm biochar-polymer composites: An investigation of electrical, mechanical, thermal and rheological characteristics. *Sci. Total. Environ.* **2017**, *619*, 311–318. [CrossRef]

56. Srinivasan, P.; Sarmah, A.K.; Smernik, R.; Das, O.; Farid, M.; Gao, W.; Smernik, R. A feasibility study of agricultural and sewage biomass as biochar, bioenergy and biocomposite feedstock: Production, characterization and potential applications. *Sci. Total. Environ.* **2015**, *512*, 495–505. [CrossRef] [PubMed]

57. Das, O.; Sarmah, A.K. Mechanism of waste biomass pyrolysis: Effect of physical and chemical pre-treatments. *Sci. Total. Environ.* **2015**, *537*, 323–334. [CrossRef]

58. Das, O.; Sarmah, A.K. Value added liquid products from waste biomass pyrolysis using pretreatments. *Sci. Total Environ.* **2015**, *538*, 145–151. [CrossRef]

59. Lee, J.; Kim, K.-H.; Kwon, E.E. Biochar as a Catalyst. *Renew. Sustain. Energy Rev.* **2017**, *77*, 70–79. [CrossRef]

60. Byrne, C.; Nagle, D. Carbonization of wood for advanced materials applications. *Carbon* **1997**, *35*, 259–266. [CrossRef]

61. Das, O.; Capezza, J.A.; Mårtensson, J.; Dong, Y.; Neisiany, E.R.; Pelcastre, L.; Jiang, L.; Xu, Q.; Olsson, T.R.; Hedenqvist, S.M. The Effect of Carbon Black on the Properties of Plasticised Wheat Gluten Biopolymer. *Molecules* **2020**, *25*, 2279. [CrossRef]

62. Netravali, A.N.; Mittal, K.L. *Interface/Interphase in Polymer Nanocomposites*; John Wiley & Sons: Hoboken, NJ, USA, 2016.

63. Das, O.; Sarmah, A.K.; Bhattacharyya, D. Biocomposites from waste derived biochars: Mechanical, thermal, chemical, and morphological properties. *Waste Manag.* **2016**, *49*, 560–570. [CrossRef] [PubMed]

64. Liu, Z.; Fei, B.; Jiang, Z.; Yang, X. Combustion characteristics of bamboo-biochars. *Bioresour. Technol.* **2014**, *167*, 94–99. [CrossRef]
65. Zhao, M.Y.; Enders, A.; Lehmann, J. Short- and long-term flammability of biochars. *Biomass Bioenergy* **2014**, *69*, 183–191. [CrossRef]
66. Das, O.; Bhattacharyya, D.; Hui, D.; Lau, K.T. Mechanical and flammability characterisations of biochar/polypropylene biocomposites. *Compos. Part B Eng.* **2016**, *106*, 120–128. [CrossRef]
67. Elnour, A.Y.; Alghyamah, A.A.; Shaikh, H.M.; Poulose, A.M.; Al-Zahrani, S.M.; Anis, A.; Al-Wabel, M.I. Effect of Pyrolysis Temperature on Biochar Microstructural Evolution, Physicochemical Characteristics, and Its Influence on Biochar/Polypropylene Composites. *Appl. Sci.* **2019**, *9*, 1149. [CrossRef]
68. Ikram, S.; Das, O.; Bhattacharyya, D. A parametric study of mechanical and flammability properties of biochar reinforced polypropylene composites. *Compos. Part A Appl. Sci. Manuf.* **2016**, *91*, 177–188. [CrossRef]
69. Das, O.; Kim, N.K.; Sarmah, A.K.; Bhattacharyya, D. Development of waste based biochar/wool hybrid biocomposites: Flammability characteristics and mechanical properties. *J. Clean. Prod.* **2017**, *144*, 79–89. [CrossRef]
70. Zhang, Q.; Khan, M.U.; Lin, X.; Cai, H.; Lei, H. Temperature varied biochar as a reinforcing filler for high-density polyethylene composites. *Compos. Part B Eng.* **2019**, *175*, 107151. [CrossRef]
71. Das, O.; Kim, N.K.; Kalamkarov, A.L.; Sarmah, A.K.; Bhattacharyya, D. Biochar to the rescue: Balancing the fire performance and mechanical properties of polypropylene composites. *Polym. Degrad. Stab.* **2017**, *144*, 485–496. [CrossRef]
72. Behazin, E.; Misra, M.; Mohanty, A.K. Sustainable biocarbon from pyrolyzed perennial grasses and their effects on impact modified polypropylene biocomposites. *Compos. Part B Eng.* **2017**, *118*, 116–124. [CrossRef]
73. Zhang, Q.; Zhang, D.; Xu, H.; Lu, W.; Ren, X.; Cai, H.; Lei, H.; Huo, E.; Zhao, Y.; Qian, M.; et al. Biochar filled high-density polyethylene composites with excellent properties: Towards maximizing the utilization of agricultural wastes. *Ind. Crop. Prod.* **2020**, *146*, 112185. [CrossRef]
74. Wen, X.; Wang, Y.; Gong, J.; Liu, J.; Tian, N.; Wang, Y.; Jiang, Z.; Qiu, J.; Tang, T. Thermal and flammability properties of polypropylene/carbon black nanocomposites. *Polym. Degrad. Stab.* **2012**, *97*, 793–801. [CrossRef]
75. Yang, H.; Gong, J.; Wen, X.; Xue, J.; Chen, Q.; Jiang, Z.; Tian, N.; Tang, T. Effect of carbon black on improving thermal stability, flame retardancy and electrical conductivity of polypropylene/carbon fiber composites. *Compos. Sci. Technol.* **2015**, *113*, 31–37. [CrossRef]
76. Liu, Z.; Li, Z.; Yang, Y.-X.; Zhang, Y.-L.; Wen, X.; Li, N.; Fu, C.; Jian, R.-K.; Li, L.; Wang, D.-Y. A Geometry Effect of Carbon Nanomaterials on Flame Retardancy and Mechanical Properties of Ethylene-Vinyl Acetate/Magnesium Hydroxide Composites. *Polymer* **2018**, *10*, 1028. [CrossRef]
77. Mensah, R.A.; Xu, Q.; Asante-Okyere, S.; Jin, C.; Bentum-Micah, G. Correlation analysis of cone calorimetry and microscale combustion calorimetry experiments. *J. Therm. Anal. Calorim.* **2018**, *136*, 589–599. [CrossRef]
78. Van Krevelen, D.W.; Te Nijenhuis, K. *Properties of Polymers: Their Correlation with Chemical Structure; Their Numerical Estimation and Prediction from Additive Group Contributions*; Elsevier: Amsterdam, The Netherlands, 2009.
79. Shrivastava, A. *Introduction to Plastics Engineering*; William Andrew: Cambridge, MA, USA, 2018; pp. 1–16.
80. Papaspyrides, C.D.; Kiliaris, P. *Polymer Green Flame Retardants*; Newnes: Waltham, MA, USA, 2014.
81. Laoutid, F.; Bonnaud, L.; Alexandre, M.; Lopez-Cuesta, J.-M.; Dubois, P. New prospects in flame retardant polymer materials: From fundamentals to nanocomposites. *Mater. Sci. Eng. R Rep.* **2009**, *63*, 100–125. [CrossRef]
82. Hsinjin, E.Y. *Quantitative Microscale Assessment of Polymer Flammability*; Plastics Research: Gansu, China, 2015.
83. Wan, L.; Deng, C.; Zhao, Z.-Y.; Chen, H.; Wang, Y.-Z. Flame Retardation of Natural Rubber: Strategy and Recent Progress. *Polymer* **2020**, *12*, 429. [CrossRef]
84. Xu, Q.; Jin, C.; Majlingova, A.; Restas, A. Discuss the heat release capacity of polymer derived from microscale combustion calorimeter. *J. Therm. Anal. Calorim.* **2017**, *133*, 649–657. [CrossRef]
85. Xu, Q.; Jin, C.; Majlingova, A.; Zachar, M.; Restas, A. Evaluate the flammability of a PU foam with double-scale analysis. *J. Therm. Anal. Calorim.* **2019**, *135*, 3329–3337. [CrossRef]
86. *Standard Test Method for Determining Flammability Characteristics of Plastics and Other Solid Materials Using Microscale Combustion Calorimetry*; ASTM D7309; American Society for Testing and Materials: West Conshohocken, PA, USA, 2019.

87. Babrauskas, V. Development of the cone calorimeter? A bench-scale heat release rate apparatus based on oxygen consumption. *Fire Mater.* **1984**, *8*, 81–95. [CrossRef]
88. Holborn, P.; Nolan, P.; Golt, J. An analysis of fire sizes, fire growth rates and times between events using data from fire investigations. *Fire Saf. J.* **2004**, *39*, 481–524. [CrossRef]

Review

A Review on Barrier Properties of Poly(Lactic Acid)/Clay Nanocomposites

Shuvra Singha * and Mikael S. Hedenqvist *

KTH Royal Institute of Technology, School of Engineering Sciences in Chemistry, Biotechnology and Health, Department of Fibre and Polymer Technology, SE-100 44 Stockholm, Sweden
* Correspondence: shuvras@kth.se (S.S.); mikaelhe@kth.se (M.S.H.)

Received: 30 March 2020; Accepted: 1 May 2020; Published: 11 May 2020

Abstract: Poly(lactic acid) (PLA) is considered to be among the best biopolymer substitutes for the existing petroleum-based polymers in the field of food packaging owing to its renewability, biodegradability, non-toxicity and mechanical properties. However, PLA displays only moderate barrier properties to gases, vapors and organic compounds, which can limit its application as a packaging material. Hence, it becomes essential to understand the mass transport properties of PLA and address the transport challenges. Significant improvements in the barrier properties can be achieved by incorporating two-dimensional clay nanofillers, the planes of which create tortuosity to the diffusing molecules, thereby increasing the effective length of the diffusion path. This article reviews the literature on barrier properties of PLA/clay nanocomposites. The important PLA/clay nanocomposite preparation techniques, such as solution intercalation, melt processing and in situ polymerization, are outlined followed by an extensive account of barrier performance of nanocomposites drawn from the literature. Fundamentals of mass transport phenomena and the factors affecting mass transport are also presented. Furthermore, mathematical models that have been proposed/used to predict the permeability in polymer/clay nanocomposites are reviewed and the extent to which the models are validated in PLA/clay composites is discussed.

Keywords: barrier properties; poly(lactic acid); clay; nanocomposite; permeability

1. Introduction

Amid the growing environmental concern about the decreasing fossil resources and the increasing plastic footprint, biopolymers obtained from renewable resources such as agricultural products represent a promising alternative to the non-degradable petroleum-based polymers for short-life range applications, for example food packaging [1–5]. Poly (lactic acid) (PLA) has emerged as the frontrunner among the many biopolymers in this regard owing to its many eco-friendly attributes such as low energy consumption during production, availability and low cost of the raw material, biodegradability in soil and water and being non-toxic to the environment [6–12]. Although the most successful application of PLA is in the containers and food packaging industry, other applications include biodegradable scaffolds for tissues, bioresorbable implants, surgical equipment, intravenous administration of antivirals, cardiovascular stents and controlled drug delivery. PLA is also used for making fibers in the textile industry and mulching materials for agriculture. The important properties which make PLA a promising candidate for food packaging is that it possesses sufficient thermal stability, i.e., the onset degradation temperature lies in the range of 330–350 °C [13,14] and good mechanical properties: tensile strength of ca. 50–70 MPa, Young's modulus of ca. 3 GPa, elongation at break of ca. 4% and impact strength of around 2.5 kJ/m^2, making it a useable substitute for single-use plastics such as PE, PP and PET [10,15].

PLA is derived from renewable agro-resources such as corn, cassava, potato, cane molasses and sugar beet, and hence is considered as an eco-friendly thermoplastic. The polymer is produced from the monomer of lactic acid (LA), the simplest hydroxy acid, which is obtained either biologically by the fermentation of carbohydrates by lactic bacteria belonging to the Lactobacillus genus or by chemical synthesis [16–18]. PLA is produced through two important routes—(a) direct polycondensation (DP) of LA and (b) ring opening polymerization (ROP) of the cyclic dimer of LA, i.e., lactide (Figure 1B). The DP route is an equilibrium reaction which demands high temperature, long reaction times and continuous removal of water from the reaction vessel, often leading to low molecular weight PLA [19]. Hence, ring opening polymerization (ROP) of lactide (cyclic dimer of LA) in the presence of a robust catalyst–initiator, tin(II)bis(2-ethylhexanoate) ($Sn(Oct)_2$) and alcohol, is the preferred synthesis route in industry, which can result in high molecular weight PLA [7,20,21].

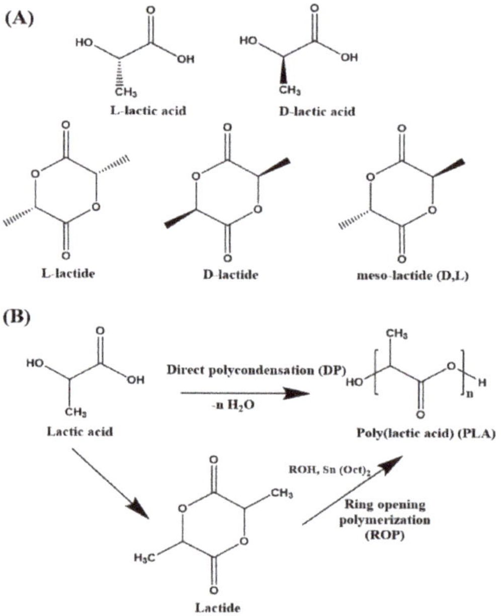

Figure 1. (**A**) Chemical structures of L-lactic acid, D-lactic acid, L-lactide, D-lactide and mesolactide, (**B**) Schematic representation of direct polycondensation (DP) of lactic acid and ring opening polymerization (ROP) of lactide.

LA is a chiral molecule and exists in two stereo-isomeric forms (optical isomers) L-lactic acid (L-LA) and D-lactic acid (D-LA). Two optically inactive forms are also available which are the meso-LA and racemic mixture (50:50) of L-LA and D-LA (Figure 1A) [16]. PLAs formed from the isotactic sequence of L-LA and D-LA are referred to as PLLA and PDLA, respectively. PLA prepared from a racemic mixture of both the enantiomers and from meso-LA is referred to as PDLLA [22]. The final properties of PLA largely depend on the ratio and distribution of the LA enantiomers in the polymer chains. A high L-isomer in the chains results in a crystalline matrix, whereas a high D-isomer (>15%) results in an amorphous matrix. The meso-form (atactic PDLLA) is also amorphous. The polymer chain orientation and packing affect the crystallinity, crystal thickness, spherulite size and morphology [12]. These are important factors which influence two important physical properties, i.e., mechanical and barrier performance [23–25]. Although PLA is among the best biodegradable and nontoxic polymers with high thermal stability and good mechanical stability (although with low extensibility without a plasticizer), its limiting property is its permeability to low molecular weight gases, vapors and organic molecules [26]. Permeation of oxygen and water vapor through polymer films can drastically decrease the service performance of a packaging material, thereby making it difficult to maintain food quality throughout its shelf life [27,28]. Research on mass transfer in polymers is, therefore, of high importance. Inclusion of two-dimensional (2D) platelets or disk-shaped nanoparticles in the polymer matrix has proven to be a good strategy to significantly decrease gas/liquid permeation in polymers. The 2D inclusions act as physical obstacles in the diffusion path of the permeant molecule creating a tortuosity effect. This helps to enhance the barrier performance of the polymer and increases the food shelf-life [29–32]. Nanoclays such as mica, saponite, montmorillonite and kaolinite are widely used 2D nanoparticles for improving the barrier properties in many polymers [33]. Thousands of publications can be found on PLA/Clay nanocomposites which have largely focused on improving the thermal [34–36], mechanical [37–39] and optical properties [40,41] and biodegradability [42,43]. However, only a meagre number of publications have been devoted to the study of mass transfer in PLA, and just a handful on PLA/clay nanocomposites. This encourages further review so as to update the trends, accomplishments and recurring challenges in this field. This review intends to highlight the usefulness of clay platelets for improving the barrier properties of PLA. First, the current methods of fabrication of PLA/clay nanocomposites are summarized, and then the barrier performance of the nanocomposites is reviewed. A brief introduction to the theory, mechanism and factors affecting mass transport in polymers is presented followed by a description of some of the important mathematical models that have been proposed to predict permeability in polymer/clay nanocomposites. The validation of some of the models in PLA/clay nanocomposites is reviewed.

2. PLA/Layered Silicate Nanocomposites

The 2D layered inorganic nanofillers like clays and silicates, owing to their abundance, low cost, high aspect ratio, rich intercalation chemistry, high strength and stiffness and thermal stability, provide favorable synergetic effects that help to significantly improve many polymer properties, especially mechanical and barrier properties [44–47]. However, the extent of dispersion of the clay layers and the morphology thus achieved in the polymer matrix (intercalation, exfoliation, mixed intercalation and exfoliation, aggregation, etc.) greatly affect the gas barrier properties. To achieve a high level of exfoliation and desired orientation of the platelets has remained a challenging task [48–51]. Good dispersion can be realized by increasing the affinity between the clay layers and the polymer through organic modification of the interlayer galleries with organic ammonium, sulfonium or phosphonium cations. A detailed and extensive list of common organic modifiers has been reported by Nordqvist and Hedenqvist [33]. The common routes to achieve dispersion of the organo-modified layered silicates in a PLA matrix are solution intercalation, melt processing and in situ polymerization (Figure 2) [52].

2.1. Solution Intercalation

Solution intercalation is one of the easiest techniques on a laboratory scale to prepare nanocomposites. In this technique, clay platelets are first exfoliated in a solvent in which the polymer is also soluble. The polymer solution is then mixed with the clay suspension, where the polymer chains intercalate/are adsorbed on the surface of the platelets and form a clay–polymer complex. The solvent is later removed by evaporation. This method is considered environmentally unfriendly because of the use of organic solvents [51]. Maharana et al. [12] demonstrated the preparation of PLA/clay nanocomposites using a solution intercalation method and showed improved mechanical and barrier properties of the nanocomposites. The effect of the structure of different organic modifiers of clay nanoparticles, Cloisite-15A, -25A and -30B modified with dimethyl dihydrogenated tallow quaternary ammonium, dimethyl hydrogenated tallow-2-ethylhexyl ammonium and methyl tallow-bis-2-hydroxyethyl quaternary ammonium, respectively, was studied by Pochan and Krikorian [53] to determine the extent of exfoliation of the nanoclay in a PLA matrix by solvent intercalation. Cloisite is a montmorillonite (MMT) clay: the term 'Cloisite' followed by an alphanumeric sequence refers to the commercially available clay, whereas the organically modified MMT (OMMT) refers to the tailor-made clay prepared by individual research groups. Cloisite 30B containing an organic diol in the inter-galleries established favorable interactions with the carbonyl functionality of PLA, leading to significant intercalation of

PLA chains into the clay spacing. Hence, PLA/Cloisite 30B formed the best nanocomposites in terms of maximum intercalation.

2.2. Melt Intercalation

This is a widely used technique to fabricate PLA/clay nanocomposites. The method involves mixing organo-modified nanoclay and the polymer and heating the mixture above the melting temperature of the polymer, either under shear or no shear. Due to the high temperatures and mechanical forces used, polymer chains are forced to diffuse into the clay galleries, giving rise to either intercalated or exfoliated nanostructures depending on the amount of polymer chains diffused into the silicate layers [32]. The main advantage of the technique is the specificity for the polymer intercalation into the clays as there is no solvent in the system that can give rise to competing clay–solvent or polymer–solvent interactions [54]. Most of the PLA/clay systems prepared by melt processing have resulted in intercalated structures. To achieve further exfoliation, Sabet and Katbab [55] investigated the role of oligo(ε-caprolactone) as compatibilizer. Although the effort did not result in complete exfoliation, it did result in flocculation of the clay layers due to hydroxylated edge-edge interactions and, therefore, better parallel stacking of the layers. However, fully exfoliated nanostructures were achieved by Chen et al. [56] who performed a second time functionalization of Cloisite 25A using an epoxy containing organic modifier—(glycidoxypropyl)trimethoxysilane. Melt processing of the nanoclay with PLA yielded fully exfoliated nanostructures when the epoxy content in the clay was high (about 0.36 mmol/g). PLA nanocomposites with epoxy containing Cloisite 25A showed better mechanical properties than those of the unmodified Cloisite composites. Melt blending of PLA in the presence of nanoclay with other polymers have been reported by several authors [57–59].

2.3. In Situ Polymerization

This technique is the most effective to obtain well exfoliated clay platelets in the polymer matrix. First, the clay is swollen in a suitable monomer melt or monomer solution. Then, polymerization is carried out induced by heat or radiation or by pre-intercalated initiators or catalyst. During the polymerization reaction, polymeric chains are formed inside the clay galleries which force delamination of the platelets in the matrix. Melt intercalation of LA monomer in the clay galleries followed by in situ ROP of PLA was found to be an efficient route to prepare high molecular weight PLA composites. Here, the silicate inter-galleries are considered as "nano-reactors" yielding high molecular weight PLA. Cloisite-Na$^+$, Cloisite 20A, Cloisite 30B and organo-modified montmorillonite (abbr. as OMMT) (modified with hexadecyltrimethyl ammonium bromide and dioctadecyldimethyl ammonium bromide) nanoclays were used to first form the LA monomer–clay intercalated mixture. The mixture was subjected to ROP using Sn(Oct)$_2$ as the catalyst for 2 h at 120–180 °C. High molecular weight PLA composite, ca. 126,000 g/mol, was obtained [60]. Katiyar and Nanavati [61] demonstrated a novel solid-state polymerization route to prepare high molecular weight PLA using a two-step in situ ROP process. The PLA prepolymer was first synthesized via ROP inside the clay layers followed by solid-state polymerization at 150–160 °C. Another innovative approach using in situ coordination insertion polymerization was reported by Paul et al. [62], referred to as the "grafting-from" approach. In this method, aluminum oxide reactive species was first formed in situ by reacting triethylaluminum with hydroxyl groups of the ammonium cation (organic modifier) of Cloisite 30B. ROP of the intercalated monomer was then carried out at the site of active species in the presence of initiator and catalyst.

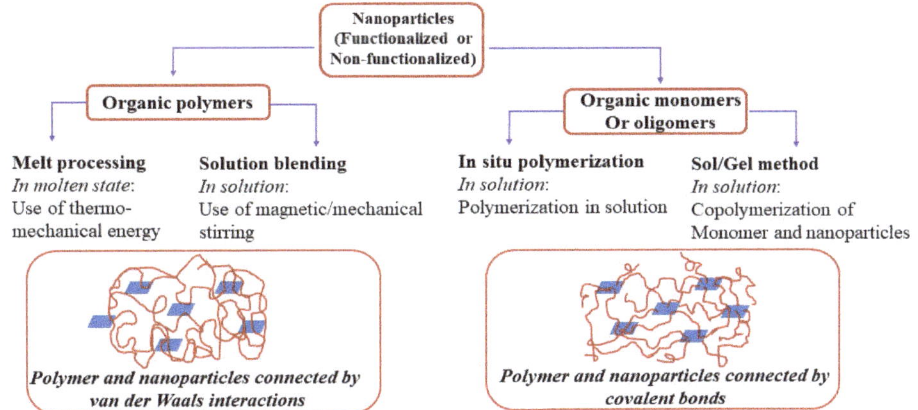

Figure 2. Outline of polymer nanocomposite preparation techniques.

3. Barrier Performance

The effects of the type of organo-modifier used, clay volume fraction, aspect ratio and dispersion on the barrier properties of PLA/organoclay nanocomposite films have been investigated. The nanocomposite films demonstrated improved barrier properties compared to the neat PLA film. Gorrasi et al. [63] prepared PLA nanocomposites with Cloisite 30B by the melt blending method where PLA, clay and polyethylene glycol (PEG) as plasticizer and stabilizer were mixed together in a counter rotating mixer. Films were then produced by compression molding the blend. These authors also prepared nanocomposite by in situ polymerization of PLA from the monomer–swollen clay. Fully exfoliated nanostructures were obtained from the in situ process that had significantly lower water solubility and diffusivity than the melt processed composite. It was also observed that the water vapor zero concentration diffusivity measured at 30 °C was decreased in the case of in situ blends by about two orders of magnitude compared to that of the melt-processed blend and neat PLA films.

The oxygen permeability of PLA nanocomposites prepared with different clay modifications (Cloisite 25A, OMMT modified with dodecyltrimethyl ammonium cation and OMMT modified with hexadecylamine) was investigated by Chang et al. [64]. The permeability of all the composites was found to be less than that of pure PLA films and, at a clay content of 10 wt%, the permeability decreased to less than half the value of the neat PLA film. The barrier performance was determined from the barrier improvement factor (BIF) which is the ratio of transmission through the neat film to the transmission through the composite film. The BIF values are tabulated in Table 1. Maiti et al. [65] studied the effect of chain length of organic modifier in different types of clay, smectite, mica and OMMT. The clays were modified with a phosphonium ion containing three butyl branches and an alkyl chain the length of which was varied from 1–16 carbon atoms. The composite containing the C16 modifier was only evaluated for oxygen permeability. Smectite clays showed better dispersion than other clays and, therefore, showed better barrier properties than the mica and MMT loaded films (BIF for the highest clay loading of 4 wt% is shown in Table 1). The mica system with stacked clay layers exhibited poor barrier performance and low modulus.

Ray et al. [66–70] measured the oxygen permeability of PLA/organo-modified clays in a series of papers where the nanocomposites were prepared by melt extrusion in a twin-screw extruder followed by compression molding the granulated material. In Ref [66], three different clays with three different organic modifiers were used to assess the PLA nanocomposite properties. MMT was modified with octadecyl ammonium and octadecyltrimethyl ammonium cations, saponite was modified with hexadecyltributyl phosphonium cation, and synthetic fluorine mica (SFM) was modified with dipolyoxyethylene alkyl(coco) methyl ammonium cation. Although the saponite system showed the best dispersion, higher barrier properties were obtained with the mica system. Synthetic fluorine mica modified with hydroxy functional ammonium cations was dispersed in the PLA matrix, resulting in intercalated stacks and a fairly large number of exfoliated layers being revealed in TEM micrographs. The highest barrier performance in terms of oxygen permeability was seen at 10 wt% mica content [67]. OMMT modified with octadecyltrimethyl ammonium cation [68] and also with the linear analog (octadecyl ammonium cation) [69], was blended with PLA. Interestingly, the linear surfactant-containing composite displayed higher barrier properties at 7 wt% clay than with the trimethyl functional surfactant.

Investigation of the effect of different processing parameters and techniques such as compounding and blown-film processing using a co-rotating twin screw extruder, were carried out by Thellen et al. [71]. The nanocomposite films showed 48% improvement in oxygen barrier and 50% improvement in water vapor barrier properties compared to the neat PLA film. Their results indicated that barrier property enhancements can be achieved by conventional processing techniques.

Lagaron et al. [72] used food contact approved nanoclays and amorphous polylactic acid (aPLA) to fabricate aPLA/organoclay nanocomposites. The organoclays used were Nanoter C1 (kaolinite) and AE21 (MMT) from nano-biomatters S.L. (Spain) (organo-modifiers not disclosed). TEM analyses revealed a mixed morphology of exfoliation and agglomeration of the clay platelets in the OM-kaolinite/aPLA matrix. The oxygen permeability BIF was higher (1.8) for the kaolinite system at 40% RH and 21 °C, whereas for the MMT system BIF was 1.1 (Table 1). The nanocomposites displayed very little swelling in water compared to the neat unfilled aPLA.

The solvent casting method was used to prepare nanocomposite films of PLA with Cloisite Na^+, Cloisite 30B and Cloisite 20A nanoclays by Rhim et al. [73]. Among the clays used, Cloisite 20A showed the highest water vapor barrier performance with a BIF of 1.5 measured at 99% RH, but the tensile properties were sacrificed. The BIF values for the Cloisite Na^+ and Cloisite 30B systems were 0.8 and 1.05, respectively. Zenkiewicz and Richert [74] used Cloisite 30B and Nanofil 2 nanoclays, poly(methyl methacrylate) (PMMA) and ethylene–vinyl alcohol copolymer as modifiers and polycaprolactone and poly(ethylene glycol) (PEG) as compatibilizers to fabricate a series of 27 PLA nanocomposite samples, and studied the effect on oxygen, water vapor and carbon dioxide permeability. Cloisite 30B samples improved the barrier properties much more than Nanofil clay—water vapor, oxygen and CO_2 permeability decreased by 60%, 55% and 90%, respectively, at 5 wt% clay content. All the modifiers and compatibilizers decreased the CO_2 transmission rate, whereas the oxygen and water vapor transmission rates were reduced only with the modifiers and not with the compatibilizers. The same group also studied the effect of blow molding ratio of PLA/MMT nanocomposite films containing (i) MMT, (ii) MMT with PMMA as modifier and iii) MMT and PEG as plasticizer on water vapor, oxygen and CO_2 transmission. Among them, the MMT system showed the highest barrier performance with reduction in the transmission rates of water vapor, oxygen and CO_2 by 40%, 40% and 80%, respectively. A further decrease by 10% to 27% was achieved by extrusion blow molding. The least permeable films were obtained at a blow molding ratio of 4 [75].

Koh et al. [76] prepared PLA nanocomposites with Cloisite 15A, Cloisite 20A and Cloisite 30B clays using the solution intercalation method. The Cloisite 30B system revealed exfoliated morphology in the TEM analysis and, consequently, outstanding gas (O_2, CO_2 and N_2) barrier properties were obtained compared to the other clay systems and neat PLA. It was also found that the gas permeability decreased with decreasing kinetic diameter of the molecule: CO_2 (3.3 Å) > O_2 (3.46 Å) > N_2 (3.64 Å), i.e., the films showed highest permeability for CO_2 owing to its comparatively small size. The BIF values are shown in Table 1. Nanocomposites of aPLA and aPLA/polycaprolactone blends with organo-modified kaolinite (OM-kaolinite) were reported by Cabedo et al. [77]. Addition of OM-kaolinite drastically decreased the O_2 permeability of aPLA (BIF 1.8, Table 1). The BIF of aPLA/PCL was 0.44 and the addition of OM-kaolinite was not as effective as in a neat aPLA matrix because of the effect of the interfaces in blend systems which provide a path for permeation. Trifol et al. [78] explored the synergistic effect of Cloisite 30B and cellulose nanofibers (CNF) on PLA barrier properties. A combination of 5 wt% Cloisite 30B and 5 wt% CNF showed a reduction of 90% in oxygen transmission rate and 76% in water vapor transmission rate compared to the neat PLA film. Even at a low filler content, 1 wt% of both materials, significant reductions in oxygen transmission rate, OTR (74%) and water vapor transmission rate, WVTR (57%), were achieved. However, in another study [79] 1 wt% Cloisite 30B in PLA showed only a 26% decrease in OTR and 43% in WVTR. Many groups have used Cloisite 30B as filler material in PLA matrix and demonstrated improved barrier properties, and the BIF values are shown in Table 1 [80–84]. Darie et al. [85] used nanoclays (Cloisite 93A modified with methyl dihydrogenated tallow quaternary ammonium and Dellite HPS, a hydrophilic smectite clay) varying in their degree of hydrophilicity to prepare PLA nanocomposites by melt processing. The O_2 and CO_2 transmission rates reduced by half in the Cloisite 93A matrix, and an even more drastic reduction occurred in the hydrophilic Dellite HPS matrix. Rhim [86] performed lamination of PLA films by using agar/κ-carrageenan modified Cloisite-Na$^+$. The double layer and multilayer nanocomposite films showed a large decrease in OTR. Jalalvandi et al. [87] prepared nanocomposites of PLA/starch blend using unmodified MMT. The barrier properties of the nanocomposites were studied in terms of water uptake of the films. Neat starch films showed an uptake of 38%, whereas the nanocomposite with the highest clay loading of 7 wt% showed only 2% uptake. Othman et al. [88] varied the MMT clay content from 1 wt% to 9 wt% in the PLA matrix and obtained the best barrier performance at 3 wt% (BIF 1.5). In a similar study, Mohsen and Ali [89] varied the clay content and achieved best barrier properties at 4 wt%–6 wt% of nanoclay in a PLA matrix. A novel silver based organo-modified MMT (Bactiblock$^®$ from Nanobiomatters, Spain) was used by Busolo et al. [90] to prepare antimicrobial PLA nanocomposite coating for food packaging. However, a reduction in the permeability of water vapor by only 20% was achieved at a clay loading as high as 10 wt%. Sengül et al. [91] used MMT clay modified with different organic modifiers (Table 1) and investigated the effect of clay modification and ratio on the barrier properties of PLA nanocomposites. The oxygen permeability decreased by 22% to 49% and the water vapor permeation by 46% to 80%. Chowdhury [92] investigated the effect of clay aspect ratio and the degree of dispersion on the barrier properties of PLA nanocomposites. Three different clays modified with the same organic modifier were used in the study. The trend in permeability depended on the aspect ratio, dispersion and the degree of disorder of the clays in the matrix. Jorda-Beneyto et al. [93] prepared PLA/MMT nanocomposite bottles by injection stretch blow molding and obtained decreased oxygen and water vapor permeability compared to the neat PLA bottle.

The best improvement in O_2 permeability of a PLA/clay system to date was achieved with layer-by-layer (Lbl) technique. Svagan et al. [94] prepared transparent films of PLA/MMT nanocomposites using Lbl techniques that showed tunable O_2 barrier properties (Figure 3). Very thin laminar multilayer structures of chitosan and MMT were constructed by an Lbl process (driven by electrostatic interactions) on extruded PLA films. Light transmittance analysis revealed high optical clarity for the coated PLA films, and TEM images showed well-ordered laminar structures of the bilayers. When 70 bilayers were used, the oxygen permeability coefficient of the coated PLA reduced by 99% and 96% at 20% and 50% RH, respectively. The data correspond to better oxygen barrier properties than PET at these humidity levels. Federico et al. [95] developed quadlayers (QL) and hexalayers (HL) of alternating branched poly(ethylene imine), Nafion and MMT on PLA thin film by Lbl technique. The oxygen permeability reduced by 98% and 97% in dry and humid conditions, respectively for 10 HL and QL layers, whereas the water vapor transmission reduced by 78%. HL films displayed efficient barrier properties than the QL films.

Table 1. Barrier improvement factors (BIFs) for PLA/clay nanocomposites.

Matrix	Nanoclay	Name and Formula of Organic Modifier	Penetrant	Clay Content	BIF	Ref
PLA	MMT	Dodecyltrimethyl ammonium, $(Me)_3(C_{12}H_{25})N^+$	O_2	10 wt%	2.3	[64]
	MMT	Hexadecyl ammonium, $(C_{16}H_{33})NH_3^+$	O_2	10 wt%	2.4	
	Cloisite 25A	Dimethyloctyl tallow amine $(Me)_2(C_8H_{17})TN^+$	O_2	10 wt%	2.3	
PLA	Smectite	Hexadecyltributyl phosphonium $(C_4H_9)_3(C_{16}H_{33})P^+$	O_2	4 wt%	1.7	[65]
PLA	MMT	Octadecyl ammonium $C_{18}H_{37}NH_3^+$	O_2	4 wt%	1.2	[66]
	MMT	Octadecyltrimethyl ammonium $(Me)_3(C_{18}H_{37})N^+$	O_2	4 wt%	1.1	
	Saponite	Hexadecyltributyl phosphonium $(C_4H_9)_3(C_{16}H_{33})P^+$	O_2	4 wt%	1.7	
	Synthetic fluorine mica (SFM)	Dipolyoxyethylene alkyl (coco) methyl ammonium $(CH_2CH_2O)_xH(CH_2CH_2O)_yH(Me)R(coco)N^+$	O_2	4 wt%	2.8	
PLA	SFM	N-(cocoalkyl)-N,N-[bis(2-hydroxyethyl)]-N-methyl ammonium $(Me)(EtOH)_2R(cocoalkyl)N^+$	O_2	10 wt%	5.5	[67]
PLA	MMT	Octadecyltrimethyl ammonium $(Me)_3(C_{18}H_{37})N^+$	O_2	7 wt%	1.2	[68]
PLA	MMT	Octadecyl ammonium $C_{18}H_{37}NH_3^+$	O_2	7 wt%	1.5	[69]
PLA	SFM	N-(cocoalkyl)-N,N-[bis(2-hydroxyethyl)]-N-methyl ammonium $(Me)(EtOH)_2R(cocoalkyl)N^+$	O_2	4 wt%	2.8	[70]
PLA	Cloisite 25A	Dimethyl hydrogenated tallow-2-ethylhexyl ammonium $(Me)_2(C_8H_{17})(HT)N^+$	O_2	5 wt%	1.7	[71]
			H_2O	5 wt%	2.7	
aPLA	Kaolinite	Not disclosed	O_2	4 wt%	1.8	[72]
	MMT	Not disclosed	O_2	4 wt%	1.1	
PLA	Cloisite 20A	Dimethyl dihydrogenated tallow quaternary ammonium $(Me)_2(HT)_2N^+$	H_2O	5 pph	1.5	[73]
	Cloisite 30B	Methyltallow-bis-2-hydroxyethyl quaternary ammonium $(Me)(CH_2CH_2OH)_2(T)N^+$	H_2O	5 pph	1.0	
	Cloisite Na^+	Unmodified	H_2O	5 pph	0.8	
PLA	Cloisite 15A	Dimethyl dihydrogenated tallow quaternary ammonium $(Me)_2(HT)_2N^+$	CO_2	0.8 wt%	2.0	[76]
			O_2	0.8 wt%	1.4	
			N_2	0.8 wt%	1.5	
	Cloisite 20A	Dimethyl dihydrogenated tallow quaternary ammonium $(Me)_2(HT)_2N^+$	CO_2	0.8 wt%	1.4	
			O_2	0.8 wt%	1.1	
			N_2	0.8 wt%	1.5	
	Cloisite 30B	Methyl tallow-bis-2-hydroxyethyl quaternary ammonium $(Me)(CH_2CH_2OH)_2(T)N^+$	CO_2	0.8 wt%	2.0	
			O_2	0.8 wt%	1.3	
			N_2	0.8 wt%	2.0	
aPLA	Kaolinite	Not disclosed	O_2	4 wt%	1.8	[77]

Table 1. *Cont.*

Matrix	Nanoclay	Name and Formula of Organic Modifier	Penetrant	Clay Content	BIF	Ref
PLA	Cloisite 30B	Methyl tallow-bis-2-hydroxyethyl quaternary ammonium $(Me)(CH_2CH_2OH)_2(T)N^+$	O_2	5 wt%	1.6	[78]
			H_2O	5 wt%	2.1	
PLA	Cloisite 30B	Methyl tallow-bis-2-hydroxyethyl quaternary ammonium $(Me)(CH_2CH_2OH)_2(T)N^+$	O_2	3 phr	1.5	[80]
PLA	Cloisite 30B	Methyl tallow-bis-2-hydroxyethyl quaternary ammonium $(Me)(CH_2CH_2OH)_2(T)N^+$	O_2	1 wt%	187.0	[81]
			H_2O	1 wt%	1.25	
PLA	Cloisite 30B	Methyl tallow-bis-2-hydroxyethyl quaternary ammonium $(Me)(CH_2CH_2OH)_2(T)N^+$	O_2	2 wt%	1.6	[82]
			H_2O	1 wt%	1.2	
PLA	Cloisite 30B	Methyl tallow-bis-2-hydroxyethyl quaternary ammonium $(Me)(CH_2CH_2OH)_2(T)N^+$	H_2O	5 wt%	2.8	[83]
PLA	Cloisite 30B	Methyl tallow-bis-2-hydroxyethyl quaternary ammonium $(Me)(CH_2CH_2OH)_2(T)N^+$	O_2	3 wt%	1.3	[84]
PLA	Cloisite 93A	Methyl dihydrogenated tallow quaternary ammonium $(Me)(HT)_2NH^+$	O_2	3 wt%	2.0	[85]
			CO_2	3 wt%	3.45	
	Dellite HPS	Not disclosed	O_2	3 wt%	18.4	
			CO_2	3 wt%	30.2	
PLA	Cloisite-Na$^+$	Agar/κ-carrageenan	O_2	5 wt%	516.0	[86]
PLA	MMT	unmodified	H_2O	7 wt%	19.0	[87]
PLA	MMT	Not disclosed	O_2	3 wt%	1.5	[88]
PLA	Clay name not mentioned	Not disclosed	O_2	4 wt%	2.6	[89]
			H_2O	6 wt%	3.1	
PLA	Ag-based MMT	Not disclosed	H_2O	10 wt%	1.2	[90]
PLA	MMT	Dimethyldialkyl ammonium $(Me)_2(R)_2N^+$	O_2	10 wt%	2.0	[91]
			H_2O	10 wt%	4.8	
		Aminopropyltriethoxysilane $(CH_3CH_2O)_3Si(C_3H_6)NH_2$	O_2	10 wt%	1.5	
			H_2O	10 wt%	2.7	
		Distearyldimethyl ammonium $(C_{18}H_{37})_2(Me)_2N^+$	O_2	10 wt%	1.9	
			H_2O	10 wt%	5.0	
		Hydrogenated tallow quaternary ammonium $(HT)_4N^+$	O_2	10 wt%	1.7	
			H_2O	10 wt%	2.3	
PLA	MMT (Southern clay)	Octadecyl ammonium $C_{18}H_{37}NH_3^+$	O_2	5 wt%	1.8	[92]
	MMT (Nanocor)	Octadecyl ammonium $C_{18}H_{37}NH_3^+$	O_2	5 wt%	1.3	
	SFM	Octadecyl ammonium $C_{18}H_{37}NH_3^+$	O_2	5 wt%	2.1	
PLA	MMT	Hexadecyltrimethyl ammonium $(Me)_3(C_{16}H_{33})N^+$	H_2O	4 wt%	1.6	[93]
			O_2	4 wt%	1.7	

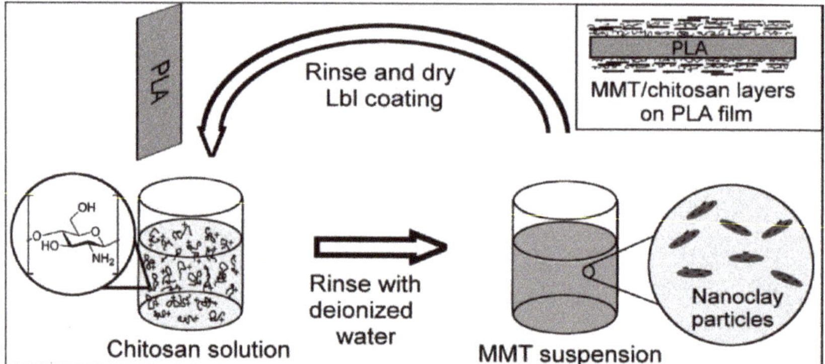

Figure 3. Schematic representation of layer-by-layer (Lbl) deposition of chitosan and montmorillonite (MMT) on extruded poly(lactic acid) (PLA) film. Reprinted with permission from Ref [94]. Copyright (2012) American Chemical Society.

4. Mass Transfer in Polymers

Small molecules, such as O_2, CO_2, H_2O, N_2, permeate through a polymer membrane due to a gas chemical potential gradient through the membrane. The chemical potential difference acts as the driving force for the molecules to permeate from the high chemical potential side to the side of low chemical potential. The phenomenon of permeant transport in polymers is described using the solution–diffusion model. According to this model, the permeation in polymers consists of three steps, as depicted in Figure 4: (a) sorption of the permeant from the high concentration side onto the membrane/film surface, (b) diffusion of the permeant along the concentration gradient through the membrane and (c) desorption through evaporation from the low concentration surface of the membrane. Deviations from a gradient with a straight line can be observed when the permeating molecule interacts with the polymer and is categorized as non-Fickian diffusion, which is described by the diffusion–relaxation model [96,97].

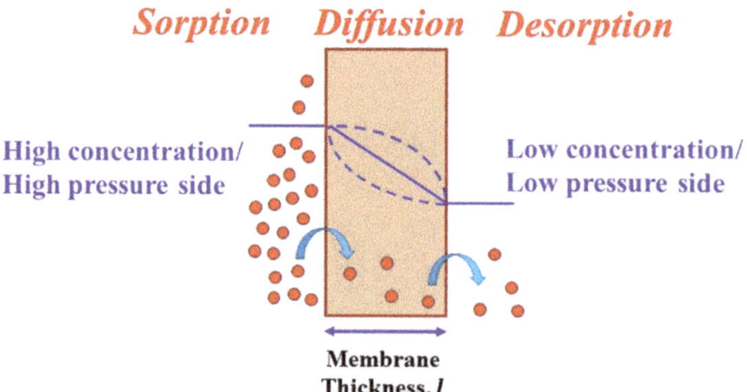

Figure 4. Schematic illustration of the solution diffusion model.

Based on the assumption that the diffusion takes place in the x-direction of a flat membrane/film, the process is described by Fick's first law of diffusion that gives the relationship between flux (F) and the concentration gradient (dc/dx) [96]:

$$F = -D\frac{dc}{dx} \tag{1}$$

where D is the diffusion coefficient, c is the concentration and x is the direction of the moving permeant. This equation is used in steady-state conditions, i.e., when the permeant concentration does not change with time. The flux at steady state is defined as the amount of the permeant (no. of moles or weight) passing through a surface of unit area (perpendicular to the flow direction) per unit time:

$$F = \frac{q}{At} \tag{2}$$

where q is the amount of the permeant, A is the membrane area and t is the time. At steady state, the permeant concentration is constant on both sides (just inside the material), $c1$ and $c2$ for high and low concentration, respectively, of the film. Therefore, Equation (1) can be integrated across the total thickness (L) of the membrane which gives:

$$F = D\frac{(c_1 - c_2)}{L} \tag{3}$$

Equations (2) and (3) can be equated to get q:

$$q = D\frac{(c_1 - c_2)At}{L} \tag{4}$$

For gases, it is convenient to measure the partial pressure (p) of the gas that is in equilibrium with the polymer rather than the concentration. Henry's law is applied [96] at sufficiently low concentration and, when the interaction between the permeating molecule and the polymer is small:

$$c = Sp \tag{5}$$

where S is the solubility coefficient of the permeant in the polymer. Hence, assuming there is no interaction between the permeant and the polymer, Equation (4) can be expressed as:

$$q = DS\frac{(p_1 - p_2)At}{L} \tag{6}$$

which, can be rearranged as:

$$DS = \frac{qL}{(p_1 - p_2)At} \tag{7}$$

Equation (7) is nothing, but the permeability, P, of the permeant at steady state:

$$P = \frac{qL}{At\Delta p} \tag{8}$$

Therefore, from Equations (7) and (8) permeability can be expressed as the product of the diffusion coefficient D and the solubility coefficient S:

$$P = DS \tag{9}$$

D is the kinetic term (from Equation (1)) describing the mass flux of permeant through the film in response to a concentration gradient and S is the thermodynamic factor arising due to the interactions between the polymer and permeant molecules, which, is the ratio of equilibrium permeant concentration at the high concentration side of the film to the permeant partial pressure [98]:

$$S = \frac{c}{p} \tag{10}$$

Equations (7) and (8) are very simplistic and can be applied to penetrants in rubbery polymers which typically exhibit Fickian behavior at low concentrations. For glassy polymers, deviation from

Fickian behavior can be observed due to the restricted chain mobility, leading to slow polymer chain reorganization in comparison to permeant-induced swelling [99]. As a consequence, a dual-mode sorption model is used to describe gas sorption in polymers at temperature below the glass transition temperature (T_g) [100].

The unsteady state portion of the mass transfer permeation process is described by Fick's second law, given by [101]:

$$\frac{dc}{dt} = \frac{d}{dx}\left(D\frac{dc}{dx}\right) \tag{11}$$

When D is position-, concentration- and time-independent, Equation (11) is expressed as:

$$\frac{dc}{dt} = D\frac{d^2c}{dx^2} \tag{12}$$

When there is a strong interaction between the polymer and the permeant, D becomes dependent on time, position and concentration and Equation (11) is solved using numerical methods [102].

4.1. Measurement of Mass Transport Properties

Two basic methods are used to determine the permeability of gases or vapors in polymer films: (a) isostatic and (b) quasi-isostatic methods [103]. In the isostatic method (continuous flow method), one side of the film is exposed to a constant concentration of the permeant and zero concentration is maintained on the other side. On the zero-concentration side, inert gas is purged to carry the permeant to the detector for quantification. In the quasi-isostatic method (lag-time method) constant permeant concentration is maintained on one side and the permeant is allowed to accumulate on the other side to a very low concentration of <5 wt% of the concentration on the feed side. The permeant from the accumulated side is removed at regular time intervals and quantified to generate a plot of permeant quantity vs time. By applying specified initial conditions (concentration throughout the film to be equal to zero) and specified boundary conditions with constant permeant concentration on the feed side and zero permeant concentration on the permeate side, a mathematical expression can be derived to describe the situation [104]:

$$q = \frac{Dc_1}{L}\left(t - \frac{L^2}{6D}\right) - \frac{2Lc_1}{\pi^2}\sum_{n=1}^{\infty}\frac{(-1)^n}{n^2}\exp\left(\frac{-Dn^2\pi^2t}{L^2}\right) \tag{13}$$

When steady state is reached, t becomes sufficiently large, i.e., $t\rightarrow\infty$, then the exponential term in Equation (13) becomes negligibly small and hence the Equation reduces to:

$$q = \frac{Dc_1}{L}\left(t - \frac{L^2}{6D}\right) \tag{14}$$

A plot of q vs t gives a straight line with an intercept on the time-axis. The slope of the straight-line curve is the steady state flux ($F = DC/L$, from Equation (1)) and the intercept is the time-lag (t_{lag}) (the intercept is an extrapolation from the straight-line curve to the time-axis, thus it is a shorter time to reach the steady state):

$$t_{lag} = \frac{L^2}{6D} \tag{15}$$

The diffusion coefficient D can then be calculated from the above Equation as:

$$D = \frac{L^2}{6t_{lag}} \tag{16}$$

This is a simple time-lag analysis and may result in errors when measuring diffusion coefficient in concentration-dependent cases where t_{lag} may vary with pressure differences across the membrane. In this case, a concentration-averaged diffusion coefficient can be estimated from the plot of normalized permeant flux, i.e., ratio of the flux at time t to the flux at equilibrium (steady state) as a function of time. The diffusion coefficient can then be estimated using the relationship [105]:

$$D = \frac{L^2}{7.199 t_{1/2}} \tag{17}$$

where $t_{1/2}$ is the time required to reach half of the steady state value. The permeability coefficient can be calculated using Equations (7) and (8). There is also another method to determine D, whereby the equation:

$$\frac{Q}{Q_\infty} = \frac{4}{\sqrt{\pi}} \sqrt{\frac{l^2}{4Dt}} \exp\left(\frac{-l^2}{4Dt}\right) \tag{18}$$

is fitted to Q/Q_∞ vs t curve using a simplex search algorithm [106]. Q is the flow rate at time t and Q_∞ is the steady-state flow rate. Equation (18) can be obtained from dynamic flow rate permeation experiments [107,108]. Assuming Henry's law is valid, the solubility, S, can be calculated using:

$$S = \frac{Q_\infty l}{Dp} \tag{19}$$

It is also possible to obtain D and S and then P from a gravimetric method, but it is not considered here [109].

4.2. Factors Affecting Mass Transport

One important factor affecting the mass transfer in polymers is the free volume of the polymer. Free volume holes are created due to Brownian motion and thermal perturbations of the polymer chains. During the sorption process, the permeant molecule occupies a free volume hole and then diffuses by short "jumps" into neighboring holes. It can also occur through gradual motion into a new hole that develops next to the first hole due to Brownian motion. The latter process is not really thermally activated since there is no barrier in energy to get across. Thus, the transport depends on the static free volume (number and size of the holes) and dynamic free volume (frequency of jumps). The static free volume is independent of the thermal motions of the polymer chains and is related to the permeant solubility, S, whereas the dynamic free volume is due to the segmental motions of the chains and is related to permeant diffusivity, D. The solubility coefficient S is related to specific free volume by [110]:

$$v_{sp} = v - v_0 = \frac{S}{\rho_{gas}} \tag{20}$$

where v_{sp} is the specific free volume, v is the specific volume, v_0 is the occupied specific volume and ρ_{gas} is the density of the gas. The fractional free volume, v_f, is given by:

$$v_f = \frac{v_{sp}}{v} \tag{21}$$

Assuming that the holes are identical spheres arranged in a cubic lattice with lattice constant 'a', the average radius of the holes, R can be calculated by:

$$R = a \sqrt[3]{\frac{3v_f}{4\pi}} \tag{22}$$

The gas diffusivity depends on the dynamic free volume of the matrix, size of the gas molecules (molecular diameter, d') and the velocity of the gas molecules (u) by [111]:

$$D = g\,d'\,u\,\exp\!\left(-\frac{\gamma v_0}{v_{sp}}\right) \tag{23}$$

where, g is a geometric factor and γ is the overlap free volume factor, i.e., the degree to which more than one molecule can access the same free volume site. Therefore, after regrouping, the constants in Equation (23) become:

$$D = A\,\exp\!\left(-\frac{B}{v_f}\right) \tag{24}$$

The higher the fractional free volume, the larger will be the diffusivity. The dependence of solubility on v_f is weaker than the diffusivity. Thus, permeability often follows a similar dependence on free volume as the gas diffusivity.

The effect of temperature on permeability, diffusivity and solubility is modeled using the Arrhenius equation [104]:

$$P = P_0\,\exp\!\left(\frac{-E_p}{RT}\right) \tag{25}$$

$$D = D_0\,\exp\!\left(\frac{-E_D}{RT}\right) \tag{26}$$

$$S = S_0\,\exp\!\left(\frac{-\Delta H_s}{RT}\right) \tag{27}$$

where P_0, D_0 and S_0 are the pre-exponential factors, E_P and E_D are the activation energies for permeation and diffusion, respectively and ΔH_S is the heat of dissolution of the permeant molecule in the polymer. Based on Equation (9), E_P can be given as:

$$E_p = E_D + \Delta H_s \tag{28}$$

E_D is always positive, ΔH_S can be positive for light gases like H_2, O_2 and N_2 and negative for condensable vapors like water, C_3H_8 and C_4H_{10}.

Other factors which affect the transport phenomenon include polymer chain structure (flexibility, polarity), crystallinity, chain orientation and packing, permeant solubility and humidity [28,104]

5. Modeling of Permeability of Polymer/Clay Nanocomposites

The mass transport mechanism in polymers containing platelet fillers (like nanoclays, graphene, etc.) is similar to that in semi-crystalline polymers. In semi-crystalline polymers, the content, shape and size of the crystals and the superstructure they form (spherulites, axialites) affect the transport properties. Thus, the crystals are considered as the gas-impermeable phase in an otherwise permeable amorphous matrix. There is, however, an important difference between the effects of crystals and impermeable platelets. It is only in special cases that the crystals are randomly dispersed in the amorphous matrix, e.g., in ultra-high molar mass polyethylene. Normally, the spherulitic structure gives rise to "dead-ends" at points where the crystals splay, and all amorphous parts are not necessarily reachable by the permeant [112,113]. The gas sorption in amorphous polymers at low to moderate uptake is given by Equation (10) (Henry's law) and for semi-crystalline polymers it is given, assuming that all the amorphous parts are accessible by the permeant, by [114]:

$$S = S_0(1 - \phi_c) \tag{29}$$

where S_0 is the solubility coefficient of the amorphous phase and ϕ_c is the volume fraction of the crystalline phase. For a "theoretically" 100% crystalline polymer, $S = 0$. In nanocomposites, the clay platelets are the non-permeable phase dispersed in the permeable polymer phase. The three main factors that influence the transport properties in clay/polymer nanocomposites are (a) the volume fraction of the nanoparticles (ϕ), (b) aspect ratio (l/w) of the platelets and (c) platelet orientation with respect to the direction of diffusion [45,51]. Incorporation of nano-platelets results in a decrease in the permeability of the polymer due to the permeant having to circumvent the platelets (leading to a tortuous diffusion path, or, in other words, a labyrinth effect) and this reduced permeability, represented as the ratio of composite permeability to the neat matrix permeability (P/P_0) or the 'relative permeability,' is plotted as a function of the filler volume fraction (ϕ) to describe the transport properties in several models. A typical plot displays the nonlinear decay in (P/P_0) with increasing filler volume fraction [33]. The volume fraction, which is the main input parameter in all mathematical models, can be calculated with [49]:

$$\phi = \frac{\dfrac{w_{np}}{\rho_{np}}}{\dfrac{w_{np}}{\rho_{np}} + \dfrac{1 - w_{polymer}}{\rho_{polymer}}} \tag{30}$$

where w_{np} and ρ_{np} are, respectively, the weight fraction and density of the nanoparticles and $w_{polymer}$ and $\rho_{polymer}$ are the weight fraction and density of the polymer matrix. The main assumptions in most of the models are that the platelets have a regular geometry (thin rectangular or circular shaped platelets) and form an ordered array in space arranged either parallel to each other or display a distribution of orientation [45]. The average orientation is assumed to be at a particular angle to the direction of diffusion of the permeant molecules. Some of the important and common models can be grouped into three categories of spatial arrangement (i) parallel arrangement, (ii) random positioning and (iii) arrangements at an angle $\theta \neq 90°$ and these are discussed below.

5.1. Periodic Arrangement of Parallel Nanoplatelets

A simple permeability model was proposed by Nielsen [115]. In this model the platelets are considered to have a rectangular shape with a finite length (l) and thickness (w) and are dispersed evenly in the polymer matrix with orientation perpendicular to the diffusion direction. The basic theory of the model is that the presence of impermeable platelets forces the permeant molecules to follow a longer diffusion path by traversing around the platelets. Therefore, this is also called the 'tortuous path' model, as shown in Figure 5.

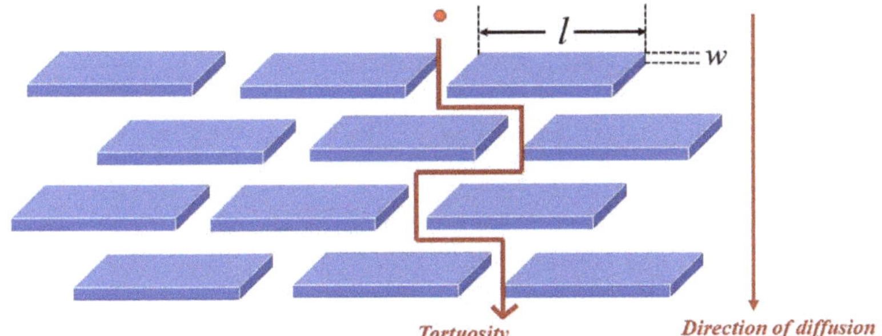

Figure 5. Schematic illustration of the tortuous path model.

The solubility coefficient, S, of this clay/polymer composite can be arrived at, from Equation (29) as:

$$S = S_0(1 - \phi) \tag{31}$$

where S_0 is the solubility coefficient of the neat polymer and ϕ is the volume fraction of the clay nano-filler. The diffusion coefficient, being influenced by the tortuous path, is given by:

$$D = \frac{D_0}{\tau} \tag{32}$$

where D_0 is the diffusion coefficient of the neat polymer and τ is the tortuosity factor that depends on the platelet shape, aspect ratio and its orientation in the matrix. It is defined as:

$$\tau = \frac{d'}{L} \tag{33}$$

where, d' is the distance that the permeant molecules must travel through the film in the presence of platelets and L is the actual distance the molecule would have traveled in the absence of platelets, i.e., thickness of the membrane. From Equations (9) and (31), we have;

$$\frac{P}{P_0} = \frac{1-\phi}{\tau} \tag{34}$$

If $\langle N \rangle$ is the average number of platelets that the permeant molecule encounters during diffusion and if each platelet enhances the diffusion length by $l/2$ on average, then the tortuous path length (prolonged diffusion length) is given by:

$$d' = L + \langle N \rangle \frac{l}{2} \tag{35}$$

Since, $\langle N \rangle = \frac{L\phi}{w}$, the tortuosity factor, τ becomes:

$$\tau = 1 + \frac{l}{2w}\phi \tag{36}$$

Combining Equations (34) and (36) gives:

$$\frac{P}{P_0} = \frac{1-\phi}{1 + \frac{\alpha}{2}\phi} \tag{37}$$

where $\alpha = l/w$ is the aspect ratio of the clay platelets. This is Nielsen's equation which shows that the relative permeability decreases with increase in α and ϕ in the nanocomposite membrane [115]. However, it can be used as a rough estimate only up to a threshold limit in filler content, $\phi \leq 10\%$, beyond which the particles may aggregate leading to increased permeation. The Nielsen equation was remarkably successful in validating the permeability reduction in many polymer systems. Figure 6 shows the predicted permeability decay curves for Nielsen's model at different aspect ratios. However, it should be highlighted that incomplete exfoliation or orientation of the platelets and the occurrence of voids will result in systems deviating from the Nielsen model [115].

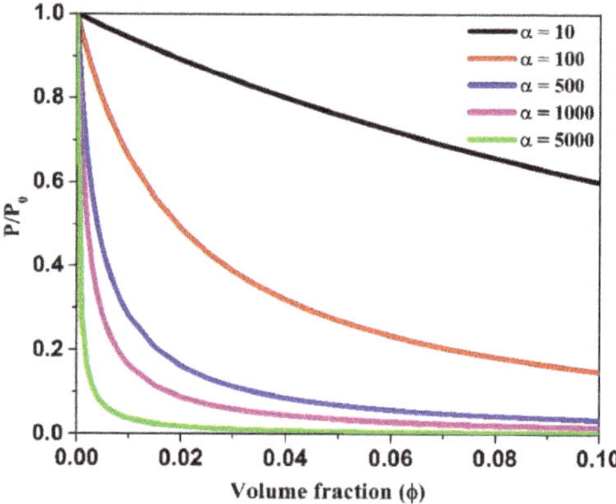

Figure 6. Prediction plot of Nielsen's model at different aspect ratios.

A second model where the resistance to diffusion arising from the tendency of the permeant molecule to get constricted in the slits (distance between two adjacent platelets) along with the contribution from the platelet length is given by Cussler et al. [116]. In this model, platelets are considered to be arranged parallel in multiple layers with a narrow-slit separation (s) between the platelets in each layer. In this case the following equation was derived:

$$\frac{P}{P_0} = \left(1 + \frac{da}{s(a+b)} + \frac{d^2}{b(a+b)} + \frac{2b}{L}ln\left(\frac{d}{2s}\right)\right)^{-1}$$

(38)

where L is the film thickness and other parameters are as defined in Figure 7.

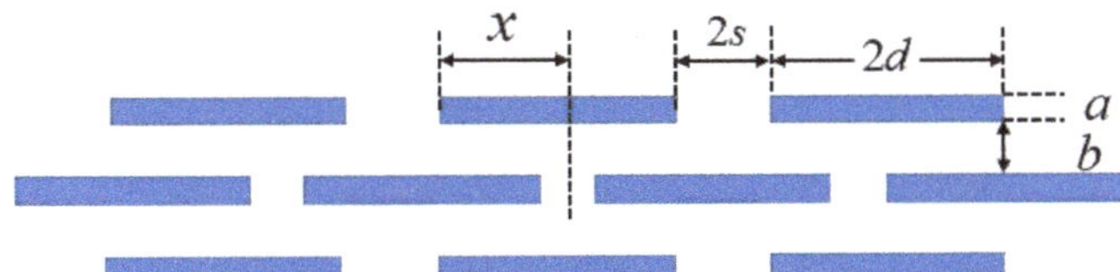

Figure 7. Ribbon arrangement of platelets.

Here, the volume fraction and aspect ratio are given by:

$$\phi = \frac{da}{(d+s)(a+b)}, \quad \alpha = \frac{d}{a}$$

(39)

In this model, d is half the platelet length, and hence the aspect ratio is half that of the Nielsen model. Since the slit is considered to be very narrow, the second term was neglected, and the simplified expression of the relative permeability is given as:

$$\frac{P}{P_0} = \left(1 + \frac{\alpha^2\phi^2}{1-\phi}\right)^{-1}$$

(40)

This model predicts a rapid reduction in relative permeability at low volume fraction, as opposed to Nielsen's model which requires high volume fraction or aspect ratio to achieve the same reduction in permeability.

5.2. Random Arrangement of Parallel Nanoplatelets

Brydges et al. [117] described the relative permeability considering random positioning of the parallel platelets in each layer and used a stacking parameter $\gamma' = x/2d$ to account for the deviation from periodicity, i.e., it defines the horizontal offset of each ribbon layer with respect to the platelet layer beneath it. A case of $\gamma' = 1/2$ is when the platelets in one ribbon layer are positioned at the center of the slit gaps of the layer underneath, and thus gives the lowest permeability. For very high aspect ratio, $\alpha > 100$, this model gives:

$$\frac{P}{P_0} = \left(1 + \frac{\alpha^2\phi^2}{1-\phi}\gamma'(1-\gamma')\right)^{-1}$$

(41)

In another case, Lape et al. [118] also considered platelets of the same aspect ratio arranged in a random fashion in the parallel ribbons. The reduced permeability is given by the product of reduced area and increased diffusion path length:

$$\frac{P}{P_0} = \left(\frac{A}{A_0}\right)\left(\frac{d'}{L}\right)$$

(42)

The distance that the permeant has to diffuse through the nanocomposite films is given by:

$$d' = L + \langle N \rangle \langle n \rangle \tag{43}$$

This equation is similar to Equation (35) except that $l/2$ is replaced by $\langle n \rangle$, which is the average distance the permeant travels to reach the platelet edge. Using statistical considerations, d' is estimated to be:

$$d' = \left(1 + \frac{1}{3}\alpha\phi\right)L \tag{44}$$

The area available for diffusion is calculated by dividing the volume available for diffusion by the distance traversed to cross the membrane:

$$\frac{A}{A_0} = \frac{\left(V_{tot} - V_{np}\right)/d'}{V_{tot}/L} \tag{45}$$

where V_{tot} is the total volume of the membrane and V_{np} is the volume of the nanoplatelets. Using Equation (44), the relative permeability is then given by:

$$\frac{P}{P_0} = \frac{1-\phi}{\left(1 + \frac{1}{3}\alpha\phi\right)^2} \tag{46}$$

Fredrickson and Bicerano [119] modeled the case of circular shaped nanoplatelets with length $2R$ and thickness w having an aspect ratio $\alpha = R/w$. Two situations were considered, as shown in Figure 8, (a) when the average distance between the platelets exceeds R due to low volume fraction and aspect ratio ($\alpha\phi \ll 1$), i.e., in the dilute regime, the relative diffusivity is given by:

$$\frac{D}{D_0} = \frac{1}{1 + \kappa\alpha\phi} \tag{47}$$

where $\kappa = \pi/ln\,\alpha$ and (b) in the semi-dilute regime when the circular disks overlap due to higher aspect ratios ($\alpha\phi \gg 1$), the relation is given by:

$$\frac{D}{D_0} = \frac{1}{1 + \mu\alpha^2\phi^2} \tag{48}$$

where μ is a geometric factor.

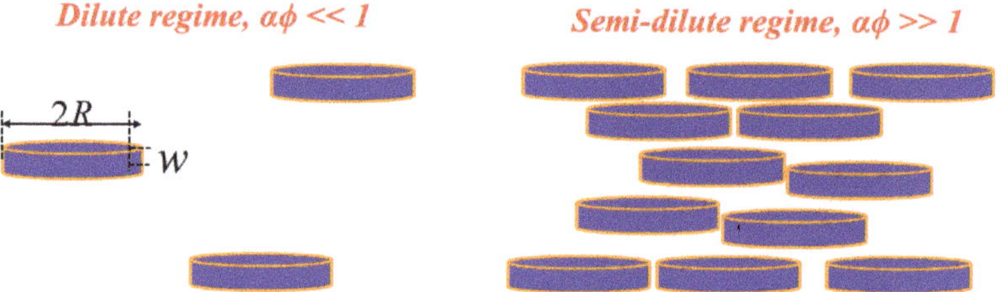

Figure 8. Schematic illustration of the dilute and the semi-dilute regimes of the oriented disk-shaped platelets.

Gusev and Lusti [120] developed periodic three-dimensional computer models containing a random dispersion of disk platelets in an isotropic matrix and solved the Laplace's equation for the local chemical potential μ, $(\nabla P(\mathrm{r}).\nabla u = 0)$. The expression developed for the relative permeability is given as:

$$\frac{P}{P_0} = \exp\left[-\left(\frac{\alpha\phi}{x_0}\right)^{\beta}\right]$$ (49)

The values of β and x_0 are 0.71 and 3.47, respectively [98].

In Figure 9, prediction curves for Nielsen, Cussler, Fredrickson and Bicerano and Gusev and Lusti models are compared for three different aspect ratios, $\alpha = 10$, 100 and 1000.

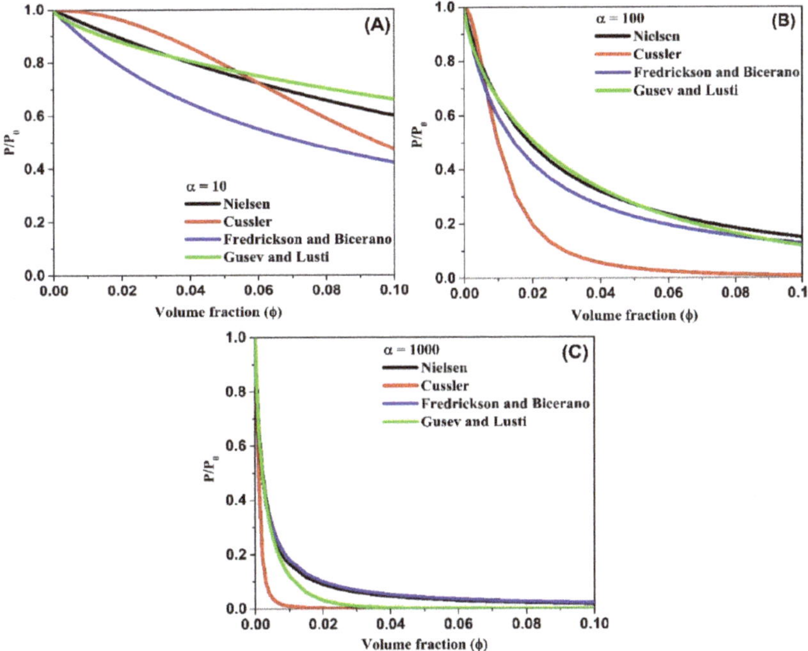

Figure 9. Prediction plots of relative permeability for different models at a fixed aspect ratio, (**A**) α = 10, (**B**) α = 100 and (**C**) α = 1000.

It is observed that the predictions for the decrease in relative permeability is different for different models, particularly in the region of low ϕ values. To avoid this anomaly, P/P_0 vs $\alpha\phi$ can be plotted [46]. Cussler and Gusev's models predict dramatic decrease in P/P_0 to almost zero permeability for $\phi \geq 0.02$ at very high aspect ratios. The plots also show that, at low aspect ratios, the models predict the need for a large volume fraction to achieve a significant decrease in permeability.

5.3. Platelet Arrangement at an Angle $\theta \neq 90°$ to the Diffusion Direction

The main assumption in all the models discussed above is that the platelets are aligned perpendicular to the diffusion direction and hence the tortuosity is the highest. However, Bharadwaj [121] described the case where the platelets can be oriented at different angles ($\neq 90°$) with respect to the direction of diffusion. For describing this nonuniformity in alignment, Nielsen's model was modified accordingly by introducing an order parameter which gives the degree of orientation of the platelets to the diffusion direction:

$$S' = \frac{1}{2}\left(3\cos^2\theta - 1\right)$$ (50)

where θ is the angle between direction of diffusion and the unit vector normal to the nanoplatelets' large surface. When the platelets are oriented perpendicular to the direction of diffusion (i.e., $\theta = 0$), then $S' = 1$, whereas when platelets are oriented parallel to the diffusion direction (i.e., $\theta = \pi/2$, then $S' = -1/2$. For a random degree of orientation, $S' = 0$. The modified Nielsen's equation is then given by:

$$\frac{P}{P_0} = \frac{1 - \phi}{1 + \frac{\alpha\phi}{2}\frac{2}{3}\left(S' + \frac{1}{2}\right)} \tag{51}$$

The case of $S' = 1$ presents maximum tortuosity, and hence, the greatest reduction in relative permeability can be observed. The values of the order parameter for the different orientations are shown in Figure 10.

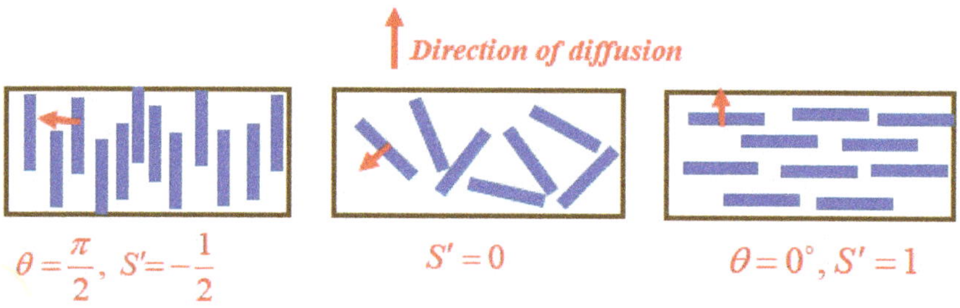

Figure 10. Different orientations of the platelets with the corresponding order parameter.

6. Model Validation for PLA/Clay Nanocomposites

Although a large body of literature is available describing the effects of two-dimensional clay sheets on reducing the water vapor permeability and gas permeability in PLA/clay nanocomposites, only a handful is available where the mathematical models have been successfully validated to account for the experimental results. Ray et al. [67] prepared PLA nanocomposites with organically modified (N-(coco alkyl)-N,N-[bis(2-hydroxyethyl)]-N-methyl ammonium cation) synthetic fluorine mica by melt extrusion using a twin-screw extruder. Films were prepared by compression molding at 190 °C. The WAXD and TEM analysis revealed intercalation of the clay platelets. However, the reduction in oxygen permeability with increasing clay concentration could not be explained by an intercalated nano-structure. Nevertheless, HRTEM revealed the co-existence of mixed intercalated and exfoliated structures that were found to be responsible for the improved oxygen barrier. Nielsen's tortuosity model was found to match the experimental results well, which confirmed the presence of exfoliated mica sheets in large amounts in the matrix, with negligible role of the intercalated structures in the observed gas barrier properties.

Guo et al. [122] used two different modifications of Cloisite-Na$^+$; (i) Cloisite 30B modified with bis(2-hydroxyethyl) methyl hydrogenated tallow quaternary ammonium cation and (ii) Cloisite- RDP modified with resorcinol di (phenyl phosphate) (RDP) and studied the oxygen barrier properties of the organically modified clay composites and compared them with those of the unmodified clay composite. The O$_2$ barrier performance was explained using the work of adhesion (W$_a$) parameter obtained from contact angle measurements. W$_a$ essentially describes the strength of affinity between PLA and the clay sheets. Higher W$_a$ values, indicating strong affinity, were obtained for PLA/Cloisite 30B, and the lowest value was observed for the PLA/Cloisite-Na$^+$. The bulky tallow molecule in Cloisite 30B helped in forming exfoliated nanostructures in the PLA matrix that, in turn, demonstrated the best barrier performance of the three clays. On the other hand, the PLA/Cloisite-Na$^+$, with less interfacial interactions, was shown to be a poor barrier film. The authors obtained best fit of the experimental permeability data with Nielsen's model. The aspect ratios calculated from curve fitting of the Nielsen model were smaller than that observed by TEM, and the difference was attributed to the tilt angle (angle between platelet and the direction of diffusion).

Picard et al. [123] investigated the role of clay platelets (Nanofil 804; MMT modified with a dihydroxy methyl tallow quaternary ammonium cation) on the PLA crystallization, as well as the gas (O_2 and He) permeability and established a crystallization-permeability relationship in PLA/clay nanocomposites for the first time. Melt compounding of PLA and OMMT was carried out using a mini-extruder to prepare the nanocomposites, followed by compression molding to make 100 µm thick films. The nanocomposites showed improved gas barrier properties at two different filler concentrations. A ca. 15% to 25% reduction in permeability/diffusivity was observed –a change that was higher than that reported by Ray et al. [66–70] for several PLA/OMMT systems. Nielsen's tortuosity model was used to provide an accurate description of the experimental results for relative permeability and diffusivity of the nanocomposites. A mean clay aspect ratio of 24 calculated from the model curve was found to be in good agreement with that obtained from TEM micrographs. The presence of the OMMT platelets increased the crystallinity of the PLA by 46%, which decreased the O_2 permeability in the annealed nanocomposite films. This permeability decrease induced by the increase in crystallinity was described well by the Maxwell equation [122].

Li et al. [124] prepared PLA/OMMT nanocomposites by solution intercalation and, later, coagulation in water. The coagulated solid was dried, and compression molded to form films. Experimental results of relative permeability of CO_2 were found to follow the Nielsen model well at low clay loadings, but the model underestimated the permeability at higher clay content; the model predicted 60% reduction in the permeability of the clay-free material at a clay content of 3 wt%, whereas the corresponding experimental value was only 40%. This is because the model describes the system better in the dilute regime but is inaccurate in the semi-dilute regime. In their study, composites with >3 wt% OMMT loading belonged to the semi-dilute regime, and the theoretical permeability matched well with the measured permeability at 1 wt% and 3 wt% of clay loading. Nevertheless, the Cussler model still overestimated the permeability at higher clay loading of 7 wt% (theoretical permeability = 9% and measured permeability = 19%). The reason is the aggregation of silicate layers at higher clay concentrations leading to nonuniform dispersion of the layers/platelets and a decreased "effective" aspect ratio. According to Equation (40), a decrease in α will substantially increase the relative permeability. However, the Cussler model was found to give a better prediction of the system compared to other models. The Bharadwaj model fitting for $S' = 0$ was unsuitable at all clay loadings and was attributed to the uneven orientation of the silicate layers, corroborating with the observations from theTEM micrographs. The use of $S' = 1$ (platelets perpendicular to the diffusion direction) yielded the same description of the system as the Nielsen model.

PLA/poly(butylene succinate)/clay nanocomposites prepared by Bhatia et al. [125] by melt extrusion showed improved O_2 barrier property with increasing clay content. However, the formation of clay stacks and nonuniform dispersion at high clay loading (>3 wt%) led to reduced tortuosity, and further improvement in barrier properties was negligible. Nanocomposite films prepared by compression molding could be described by the Bharadwaj model only up to 3 wt% clay, beyond which deviations from the model occurred because of the aforementioned clay agglomerates/nonuniform clay dispersion.

Tenn et al. [126] investigated the effect of the clay platelet hydration on the barrier properties of PLA/OMMT nanocomposites. The relative water and oxygen permeability results were fitted to the Bharadwaj model. However, the fitting for the water permeability was quite unsuitable. It was concluded that, apart from the aspect ratio and orientation of the clays, other parameters, such as interaction between silicate layers and water molecules, rigidity of the polymer chains in the vicinity of the clay layers, degree of crystallinity and percolation effects at the clay–polymer interface can interfere with the tortuosity concept, and thus make the tortuosity models less useable for prediction purposes. Nevertheless, the Nielsen–Bharadwaj model was found to be the best model to fit for the relative oxygen permeability experimental data of the PLA/OMMT nanocomposites. It was suggested that the tortuosity concept can be applied in a straightforward way to a gas-polymer system when there is no interaction between the diffusing gas molecules and the polymer matrix, whereas in the case of water or organic species, in addition to the tortuosity, different physical phenomena and chemical interactions can play a large role during the course of permeation, which can result in significant deviation from the expected tortuosity-based results.

7. Conclusions

Two-dimensional platelet/disk-shaped fillers (e.g., nanoclays) have been identified as the most effective nano-filler for increasing the gas barrier properties of polymers. These nanoparticles not only improve barrier properties of the polymer, but also improve mechanical properties and, often, the thermal stability owing to interfacial interactions with the polymer matrix. In this article, the commonly followed preparation methods for PLA/organoclay nanocomposites were elaborated which are solution intercalation, melt processing and in situ polymerization. The melt processing method is the most preferred route because of ease of implementation in industry. The barrier performance of PLA/clay nanocomposites with different kinds of nanoclay and with a vast variety of modifiers were reviewed to highlight the structure-property relationship, which varied from case to case. In general, the extent of exfoliation and stacking orientation of the nanoclays was found to be the most important factor affecting the barrier properties of PLA, where improvement by one or two orders of magnitude can be observed for fully exfoliated platelets. The individual clay platelets act as blockages and create tortuosity to the diffusing permeant molecules, and thus extend the diffusion path length and time. In many cases, they also reduce the solubility of the permeating gas molecules. Best barrier performance was found to be obtained through the Lbl technique. Although it is successful on the laboratory scale, the future success of this technique will depend on industrial implementation. The ability of the Lbl prepared clays to impart delayed diffusion is most useful in packaging and coating applications.

Some important mathematical models for estimating the relative permeability of polymer/organoclay nanocomposites have been described. The commonality among the models is the dependence of relative permeability on three factors: clay aspect ratio, volume fraction and the clay platelet orientation with respect to the direction of diffusion. Experimental validation of the models on PLA/clay systems has been studied only by a few groups and the results were reviewed in this article. Most of the models, Nielsen's model in particular, were found to fit the data well at lower clay content. However, the models cannot be compared as the aspect ratios are different: some authors define aspect ratio as the width to thickness ratio while others define it as half width to thickness ratio. It can become more complicated because the degree of interaction between the polymer and the clay particles and the degree of delamination can be expected to vary. Nevertheless, with known aspect ratios of the clay, the simplest model proposed by Nielsen has proven to predict the relative permeability reasonably well. Another model which describes the tortuosity in polymer nanocomposites is the Fricke model that has been applied successfully in several composite systems, but has not so far been applied to PLA systems, although there is scope for in future studies [127].

Author Contributions: Conceptualization, M.S.H. and S.S.; writing—original draft preparation, S.S.; writing—review and editing, M.S.H. and S.S.; funding acquisition, M.S.H. All authors have read and agreed to the published version of the manuscript.

Funding: Knut och Alice Wallenbergs Stiftelse: 63235.

Acknowledgments: Knut and Alice Wallenberg foundation, through the Wallenberg Wood Science Center at KTH Royal Institute of Technology, is acknowledged for the financial support.

Conflicts of Interest: The authors declare no conflict of interest.

References

1. Jenck, J.F.; Agterberg, F.; Droescher, M.J. Products and processes for a sustainable chemical industry: A review of achievements and prospects. *Green Chem.* **2004**, *6*, 544–556. [CrossRef]
2. Kaplan, D.L. *Biopolymers from Renewable Resources*; Springer: Berlin, Germany, 1998.

3.   Mekonnen, T.; Mussone, P.; Khalil, H.; Bressler, D. Progress in bio-based plastics and plasticizing modifications. *J. Mater. Chem. A* **2013**, *1*, 13379–13398. [CrossRef]

4.   Kabir, E.; Kaur, R.; Lee, J.; Kim, K.-H.; Kwon, E.E. Prospects of biopolymer technology as an alternative option for non-degradable plastics and sustainable management of plastic wastes. *J. Clean. Prod.* **2020**, *258*, 120536. [CrossRef]

5.   Stefani, R.; Vinhal, G.L.R.R.B.; do Nascimento, D.V.; Pereira, M.C.S.; Pertuzatti, P.B.; da Silva Chaves, K. Smart Biopolymers in Food Industry. In *Industrial Applications for Intelligent Polymers and Coatings*; Hosseini, M., Makhlouf, A.S.H., Eds.; Springer International Publishing AG: Cham, Switzerland, 2016.

6.   Catalá, R.; López-Carballo, G.; Hernández-Muñoz, P.; Gavara, R. PLA and Active Packaging. In *Poly(lactic acid) Science and Technology: Processing, Properties, Additives and Applications*; Jiménez, A., Peltzer, M., Ruseckaite, R., Eds.; The Royal Society of Chemistry: Cambridge, UK, 2015.

7.   Fiori, S. Industrial uses of PLA. *RSC Polym. Chem. Ser.* **2015**, *2015*, 317–333.

8.   Lim, L.T.; Auras, R.; Rubino, M. Processing technologies for poly(lactic acid). *Prog. Polym. Sci.* **2008**, *33*, 820–852. [CrossRef]

9.   Castro-Aguirre, E.; Iñiguez-Franco, F.; Samsudin, H.; Fang, X.; Auras, R. Poly(lactic acid)—Mass production, processing, industrial applications, and end of life. *Adv. Drug Deliv. Rev.* **2016**, *107*, 333–366. [CrossRef] [PubMed]

10.   Farah, S.; Anderson, D.G.; Langer, R. Physical and mechanical properties of PLA, and their functions in widespread applications—A comprehensive review. *Adv. Drug Deliv. Rev.* **2016**, *107*, 367–392. [CrossRef]

11.   Armentano, I.; Bitinis, N.; Fortunati, E.; Mattioli, S.; Rescignano, N.; Verdejo, R.; Lopez-Manchado, M.A.; Kenny, J.M. Multifunctional nanostructured PLA materials for packaging and tissue engineering. *Prog. Polym. Sci.* **2013**, *38*, 1720–1747. [CrossRef]

12.   Maharana, T.; Mohanty, B.; Negi, Y.S. Melt-solid polycondensation of lactic acid and its biodegradability. *Prog. Polym. Sci.* **2009**, *34*, 99–124. [CrossRef]

13.   Kang, H.; Li, Y.; Gong, M.; Guo, Y.; Guo, Z.; Fang, Q.; Li, X. An environmentally sustainable plasticizer toughened polylactide. *RSC Adv.* **2018**, *8*, 11643–11651. [CrossRef]

14.   Mofokeng, J.P.; Luyt, A.S.; Tábi, T.; Kovács, J. Comparison of injection moulded, natural fibre-reinforced composites with PP and PLA as matrices. *J. Thermoplast. Compos. Mater.* **2012**, *25*, 927–948. [CrossRef]

15.   Anderson, K.S.; Schreck, K.M.; Hillmyer, M.A. Toughening polylactide. *Polym. Rev.* **2008**, *48*, 85–108. [CrossRef]

16.   Garlotta, D. A Literature Review of Poly(Lactic Acid). *J. Polym. Environ.* **2001**, *9*, 63–84. [CrossRef]

17.   Avérous, L. Polylactic acid: Synthesis, properties and applications. In *Monomer, Polymer and Composites from Renewable Resources*; Belgacem, M.N., Gandini, A., Eds.; Elsevier: London, UK, 2008.

18.   Wee, Y.J.; Kim, J.N.; Ryu, H.W. Biotechnological production of lactic acid and its recent applications. *Food Technol. Biotechnol.* **2006**, *44*, 163–172.

19.   Carothers, W.H.; Borough, G.L.; Natta, F.J. Studies of polymerization and ring formation. X. The reversible polymerization of six-membered cyclic esters. *J. Am. Chem. Soc.* **1932**, *54*, 761–772. [CrossRef]

20.   Albertsson, A.C.; Varma, I.K. Aliphatic polyesters: Synthesis, properties and applications. *Adv. Polym. Sci.* **2002**, *157*, 1–40.

21.   Okada, M. Chemical syntheses of biodegradable polymers. *Prog. Polym. Sci.* **2002**, *27*, 87–133. [CrossRef]

22.   Madhavan Nampoothiri, K.; Nair, N.R.; John, R.P. An overview of the recent developments in polylactide (PLA) research. *Bioresour. Technol.* **2010**, *101*, 8493–8501. [CrossRef]

23.   Guinault, A.; Menary, G.H.; Courgneau, C.; Griffith, D.; Ducruet, V.; Miri, V.; Sollogoub, C. The effect of the stretching of PLA extruded films on their crystallinity and gas barrier properties. *AIP Conf. Proc.* **2011**, *1353*, 826–831.

24.   Drieskens, M.; Peeters, R.; Mullens, J.; Franco, D.; Lemstra, P.J.; Hristova-Bogaerds, D.G. Structure versus properties relationship of poly(lactic acid). I. Effect of crystallinity on barrier properties. *J. Polym. Sci. Part B Polym. Phys.* **2009**, *47*, 2247–2258. [CrossRef]

25.   Guinault, A.; Sollogoub, C.; Domenek, S.; Grandmontagne, A.; Ducruet, V.J. Influence of crystallinity on gas barrier and mechanical properties of PLA food packaging films. *Int. J. Mater. Form.* **2010**, *3*, 603–606. [CrossRef]

26.   Sonchaeng, U.; Iñiguez-Franco, F.; Auras, R.; Selke, S.; Rubino, M.; Lim, L.T. Poly(lactic acid) mass transfer properties. *Prog. Polym. Sci.* **2018**, *86*, 85–121. [CrossRef]

27. Koros, W.J. Barrier Polymers and Structures. *Anal. Chem.* **1990**, *62*, 737A.

28. Hedenqvist, M.S. Barrier packaging materials. In *Handbook of Environmental Degradation of Materials*, 3rd ed.; Kutz, M., Ed.; Elsevier Inc.: Cambridge, MA, USA, 2018.

29. Majeed, K.; Jawaid, M.; Hassan, A.; Abu Bakar, A.; Abdul Khalil, H.P.S.; Salema, A.A.; Inuwa, I. Potential materials for food packaging from nanoclay/natural fibres filled hybrid composites. *Mater. Des.* **2013**, *46*, 391–410. [CrossRef]

30. Bee, S.L.; Abdullah, M.A.A.; Bee, S.T.; Sin, L.T.; Rahmat, A.R. Polymer nanocomposites based on silylated-montmorillonite: A review. *Prog. Polym. Sci.* **2018**, *85*, 57–82. [CrossRef]

31. Silvestre, C.; Duraccio, D.; Cimmino, S. Food packaging based on polymer nanomaterials. *Prog. Polym. Sci.* **2011**, *36*, 1766–1782. [CrossRef]

32. Sinha Ray, S.; Okamoto, M. Polymer/layered silicate nanocomposites: A review from preparation to processing. *Prog. Polym. Sci.* **2003**, *28*, 1539–1641. [CrossRef]

33. Nordqvist, D.; Hedenqvist, M.S. Transport Properties of Nanocomposites Based on Polymers and Layered Inorganic Fillers. In *Packaging Nanotechnology*; Mohanty, A.K., Misra, M., Eds.; American Scientific Publishers: Valencia, CA, USA, 2009.

34. Wokadala, O.C.; Ray, S.S.; Bandyopadhyay, J.; Wesley-Smith, J.; Emmambux, N.M. Morphology, thermal properties and crystallization kinetics of ternary blends of the polylactide and starch biopolymers and nanoclay: The role of nanoclay hydrophobicity. *Polymer* **2015**, *71*, 82–92. [CrossRef]

35. Hoidy, W.H.; Al-Mulla, E.A.J.; Al-Janabi, K.W. Mechanical and Thermal Properties of PLLA/PCL Modified Clay Nanocomposites. *J. Polym. Environ.* **2010**, *18*, 608–616. [CrossRef]

36. Ibrahim, N.; Jollands, M.; Parthasarathy, R. Mechanical and thermal properties of melt processed PLA/organoclay nanocomposites. *IOP Conf. Ser. Mater. Sci. Eng.* **2017**, *191*, 012005. [CrossRef]

37. Krishnamachari, P.; Zhang, J.; Lou, J.; Yan, J.; Uitenham, L. Biodegradable poly(Lactic Acid)/clay nanocomposites by melt intercalation: A study of morphological, thermal, and mechanical properties. *Int. J. Polym. Anal. Charact.* **2009**, *14*, 336–350. [CrossRef]

38. Hasook, A.; Tanoue, S.; Iemoto, Y.; Unryu, T. Characterization and mechanical properties of poly(lactic acid)/poly(ε-caprolactone)/organoclay nanocomposites prepared by melt compounding. *Polym. Eng. Sci.* **2006**, *46*, 1001–1007. [CrossRef]

39. Lai, S.M.; Wu, S.H.; Lin, G.G.; Don, T.M. Unusual mechanical properties of melt-blended poly(lactic acid) (PLA)/clay nanocomposites. *Eur. Polym. J.* **2014**, *52*, 193–206. [CrossRef]

40. Tabatabaei, S.H.; Ajji, A. Orientation, mechanical, and optical properties of poly (lactic acid) nanoclay composite films. *Polym. Eng. Sci.* **2011**, *51*, 2151–2158. [CrossRef]

41. Cele, H.M.; Ojijo, V.; Chen, H.; Kumar, S.; Land, K.; Joubert, T.; De Villiers, M.F.R.; Ray, S.S. Effect of nanoclay on optical properties of PLA/clay composite films. *Polym. Test.* **2014**, *36*, 24–31. [CrossRef]

42. Stloukal, P.; Pekařová, S.; Kalendova, A.; Mattausch, H.; Laske, S.; Holzer, C.; Chitu, L.; Bodner, S.; Maier, G.; Slouf, M.; et al. Kinetics and mechanism of the biodegradation of PLA/clay nanocomposites during thermophilic phase of composting process. *Waste Manag.* **2015**, *42*, 31–40. [CrossRef]

43. Castro-Aguirre, E.; Auras, R.; Selke, S.; Rubino, M.; Marsh, T. Impact of nanoclays on the biodegradation of poly(lactic acid) nanocomposites. *Polymers* **2018**, *10*, 202. [CrossRef]

44. Rovera, C.; Ghaani, M.; Farris, S. Nano-inspired oxygen barrier coatings for food packaging applications: An overview. *Trends Food Sci. Technol.* **2020**, *97*, 210–220. [CrossRef]

45. Choudalakis, G.; Gotsis, A.D. Permeability of polymer/clay nanocomposites: A review. *Eur. Polym. J.* **2009**, *45*, 967–984. [CrossRef]

46. Kotal, M.; Bhowmick, A.K. Polymer nanocomposites from modified clays: Recent advances and challenges. *Prog. Polym. Sci.* **2015**, *51*, 127–187. [CrossRef]

47. Zhu, T.T.; Zhou, C.H.; Kabwe, F.B.; Wu, Q.Q.; Li, C.S.; Zhang, J.R. Exfoliation of montmorillonite and related properties of clay/polymer nanocomposites. *Appl. Clay Sci.* **2019**, *169*, 48–66. [CrossRef]

48. Tan, B.; Thomas, N.L. A review of the water barrier properties of polymer/clay and polymer/graphene nanocomposites. *J. Memb. Sci.* **2016**, *514*, 595–612. [CrossRef]

49. Wolf, C.; Angellier-Coussy, H.; Gontard, N.; Doghieri, F.; Guillard, V. How the shape of fillers affects the barrier properties of polymer/non-porous particles nanocomposites: A review. *J. Memb. Sci.* **2018**, *556*, 393–418. [CrossRef]

50. Kalendova, A.; Merinska, D.; Gerard, J.F.; Slouf, M. Polymer/clay nanocomposites and their gas barrier properties. *Polym. Compos.* **2013**, *34*, 1418–1424. [CrossRef]

51. Cui, Y.; Kumar, S.; Rao Kona, B.; Van Houcke, D. Gas barrier properties of polymer/clay nanocomposites. *RSC Adv.* **2015**, *5*, 63669–63690. [CrossRef]

52. Raquez, J.M.; Habibi, Y.; Murariu, M.; Dubois, P. Polylactide (PLA)-based nanocomposites. *Prog. Polym. Sci.* **2013**, *38*, 1504–1542. [CrossRef]

53. Krikorian, V.; Pochan, D.J. Poly (l-Lactic Acid)/Layered Silicate Nanocomposite: Fabrication, Characterization, and Properties. *Chem. Mater.* **2003**, *15*, 4317–4324. [CrossRef]

54. Pavlidou, S.; Papaspyrides, C.D. A review on polymer-layered silicate nanocomposites. *Prog. Polym. Sci.* **2008**, *33*, 1119–1198. [CrossRef]

55. Yang, Z.; Peng, H.; Wang, W.; Liu, T. Crystallization behavior of poly(ε-caprolactone)/layered double hydroxide nanocomposites. *J. Appl. Polym. Sci.* **2010**, *116*, 2658–2667. [CrossRef]

56. Chen, G.X.; Kim, H.S.; Shim, J.H.; Yoon, J.S. Role of epoxy groups on clay surface in the improvement of morphology of poly(L-lactide)/clay composites. *Macromolecules* **2005**, *38*, 3738–3744. [CrossRef]

57. Jiang, L.; Liu, B.; Zhang, J. Properties of poly(lactic acid)/poly(butylene adipate-co-terephthalate)/ nanoparticle ternary composites. *Ind. Eng. Chem. Res.* **2009**, *48*, 7594–7602. [CrossRef]

58. Ozkoc, G.; Kemaloglu, S.; Quaedflieg, M. Production of poly(lactic acid)/organoclay nanocomposite scaffolds by microcompounding and polymer/particle leaching. *Polym. Compos.* **2010**, *31*, 674–683. [CrossRef]

59. Hasook, A.; Muramatsu, H.; Tanoue, S.; Iemoto, Y.; Unryu, T. Preparation of nanocomposites by melt compounding polylactic acid/polyamide 12/organoclay at different screw rotating speeds using a twin screw extruder. *Polym. Compos.* **2008**, *29*, 1–8. [CrossRef]

60. Katiyar, V.; Nanavati, H. In situ synthesis of high molecular weight poly(L-lactic acid) clay nanocomposites. *Polym. Eng. Sci.* **2011**, *51*, 2066–2077. [CrossRef]

61. Katiyar, V.; Nanavati, H. High molecular weight poly (L-lactic acid) clay nanocomposites via solid-state polymerization. *Polym. Compos.* **2011**, *32*, 497–509. [CrossRef]

62. Paul, M.A.; Delcourt, C.; Alexandre, M.; Degée, P.; Monteverde, F.; Rulmont, A.; Dubois, P. (Plasticized) polylactide/(organo-)clay nanocomposites by in situ intercalative polymerization. *Macromol. Chem. Phys.* **2005**, *206*, 484–498. [CrossRef]

63. Gorrasi, G.; Tammaro, L.; Vittoria, V.; Paul, M.A.; Alexandre, M.; Dubois, P. Transport properties of water vapor in polylactide/montmorillonite nanocomposites. *J. Macromol. Sci. Phys. Part B* **2004**, *43*, 565–575. [CrossRef]

64. Chang, J.H.; An, Y.U. Nanocomposites of polyurethane with various organoclays: Thermomechanical properties, morphology, and gas permeability. *J. Polym. Sci. Part B Polym. Phys.* **2002**, *40*, 670–677. [CrossRef]

65. Maiti, P.; Yamada, K.; Okamoto, M.; Ueda, K.; Okamoto, K. New polylactide/layered silicate nanocomposites: Role of organoclays. *Chem. Mater.* **2002**, *14*, 4654–4661. [CrossRef]

66. Ray, S.S.; Yamada, K.; Okamoto, M.; Fujimoto, Y.; Ogami, A.; Ueda, K. New polylactide/layered silicate nanocomposites. 5. Designing of materials with desired properties. *Polymer* **2003**, *44*, 6633–6646.

67. Sinha Ray, S.; Yamada, K.; Okamoto, M.; Ogami, A.; Ueda, K. New polylactide/layered silicate nanocomposites. 3. High-performance biodegradable materials. *Chem. Mater.* **2003**, *15*, 1456–1465. [CrossRef]

68. Sinha Ray, S.; Yamada, K.; Okamoto, M.; Ueda, K. New polylactide-layered silicate nanocomposites. 2. Concurrent improvements of material properties, biodegradability and melt rheology. *Polymer* **2002**, *44*, 857–866. [CrossRef]

69. Ray, S.S.; Yamada, K.; Okamoto, M.; Ogami, A.; Ueda, K. New polylactide/layered silicate nanocomposites, 4. Structure, properties and biodegradability. *Compos. Interfaces* **2003**, *10*, 435–450. [CrossRef]

70. Ray, S.S.; Yamada, K.; Ogami, A.; Okamoto, M.; Ueda, K. New polylactide/layered silicate nanocomposite: Nanoscale control over multiple properties. *Macromol. Rapid Commun.* **2002**, *23*, 943–947. [CrossRef]

71. Thellen, C.; Orroth, C.; Froio, D.; Ziegler, D.; Lucciarini, J.; Farrell, R.; D'Souza, N.A.; Ratto, J.A. Influence of montmorillonite layered silicate on plasticized poly(l-lactide) blown films. *Polymer* **2005**, *46*, 11716–11727. [CrossRef]

72. Lagarón, J.M.; Cabedo, L.; Cava, D.; Feijoo, J.L.; Gavara, R.; Gimenez, E. Improving packaged food quality and safety. Part 2: Nanocomposites. *Food Addit. Contam.* **2005**, *22*, 994–998. [CrossRef]

73. Rhim, J.W.; Hong, S.I.; Ha, C.S. Tensile, water vapor barrier and antimicrobial properties of PLA/nanoclay composite films. *LWT Food Sci. Technol.* **2009**, *42*, 612–617. [CrossRef]

74. Zenkiewicz, M.; Richert, J. Permeability of polylactide nanocomposite films for water vapour, oxygen and carbon dioxide. *Polym. Test.* **2008**, *27*, 835–840. [CrossRef]

75. Zenkiewicz, M.; Richert, J.; Rózański, A. Effect of blow moulding ratio on barrier properties of polylactide nanocomposite films. *Polym. Test.* **2010**, *29*, 251–257. [CrossRef]

76. Koh, H.C.; Park, J.S.; Jeong, M.A.; Hwang, H.Y.; Hong, Y.T.; Ha, S.Y.; Nam, S.Y. Preparation and gas permeation properties of biodegradable polymer/layered silicate nanocomposite membranes. *Desalination* **2008**, *233*, 201–209. [CrossRef]

77. Cabedo, L.; Feijoo, J.L.; Villanueva, M.P.; Lagarón, J.M.; Giménez, E. Optimization of biodegradable nanocomposites based on aPLA/PCL blends for food packaging applications. *Macromol. Symp.* **2006**, *233*, 191–197. [CrossRef]

78. Trifol, J.; Plackett, D.; Sillard, C.; Szabo, P.; Bras, J.; Daugaard, A.E. Hybrid poly(lactic acid)/nanocellulose/ nanoclay composites with synergistically enhanced barrier properties and improved thermomechanical resistance. *Polym. Int.* **2016**, *65*, 988–995. [CrossRef]

79. Trifol, J.; Plackett, D.; Sillard, C.; Hassager, O.; Daugaard, A.E.; Bras, J.; Szabo, P. A comparison of partially acetylated nanocellulose, nanocrystalline cellulose, and nanoclay as fillers for high-performance polylactide nanocomposites. *J. Appl. Polym. Sci.* **2016**, *133*, 43257. [CrossRef]

80. Singh, S.; Gupta, R.K.; Ghosh, A.K.; Maiti, S.N.; Bhattacharya, S.N. Poly(L-lactic acid)/layered silicate nanocomposite blown film for packaging application: Thermal, mechanical and barrier properties. *J. Polym. Eng.* **2010**, *30*, 361–375. [CrossRef]

81. Park, S.H.; Lee, H.S.; Choi, J.H.; Jeong, C.M.; Sung, M.H.; Park, H.J. Improvements in barrier properties of poly(lactic acid) films coated with chitosan or chitosan/clay nanocomposite. *J. Appl. Polym. Sci.* **2012**, *125*, E675–E680. [CrossRef]

82. Kim, H.K.; Kim, S.J.; Lee, H.S.; Choi, J.H.; Jeong, C.M.; Sung, M.H.; Park, S.H.; Park, H.J. Mechanical and barrier properties of poly(lactic acid) films coated by nanoclay-ink composition. *J. Appl. Polym. Sci.* **2013**, *127*, 3823–3829. [CrossRef]

83. Farmahini-Farahani, M.; Xiao, H.; Zhao, Y. Poly lactic acid nanocomposites containing modified nanoclay with synergistic barrier to water vapor for coated paper. *J. Appl. Polym. Sci.* **2014**, *131*, 40952. [CrossRef]

84. Zheng, W.; Beeler, M.; Claus, J.; Xu, X. Poly(lactic acid)/montmorillonite blown films: Crystallization, mechanics, and permeation. *J. Appl. Polym. Sci.* **2017**, *134*, 45260. [CrossRef]

85. Darie, R.N.; Pâslaru, E.; Sdrobis, A.; Pricope, G.M.; Hitruc, G.E.; Poiată, A.; Baklavaridis, A.; Vasile, C. Effect of nanoclay hydrophilicity on the poly(lactic acid)/clay nanocomposites properties. *Ind. Eng. Chem. Res.* **2014**, *53*, 7877–7890. [CrossRef]

86. Rhim, J.W. Effect of PLA lamination on performance characteristics of agar/κ-carrageenan/clay bio-nanocomposite film. *Food Res. Int.* **2013**, *51*, 714–722. [CrossRef]

87. Jalalvandi, E.; Majid, R.A.; Ghanbari, T.; Ilbeygi, H. Effects of montmorillonite (MMT) on morphological, tensile, physical barrier properties and biodegradability of polylactic acid/starch/MMT nanocomposites. *J. Thermoplast. Compos. Mater.* **2015**, *28*, 496–509. [CrossRef]

88. Othman, S.H.; Ling, H.N.; Talib, R.A.; Naim, M.N.; Risyon, N.P.; Saifullah, M. PLA/MMT and PLA/halloysite bio-nanocomposite films: Mechanical, barrier, and transparency. *J. Nano Res.* **2019**, *59*, 77–93. [CrossRef]

89. Mohsen, A.H.; Ali, N.A. Mechanical, Color and Barrier, Properties of Biodegradable Nanocomposites Polylactic Acid/Nanoclay. *J. Bioremediat. Biodegrad.* **2018**, *9*, 455. [CrossRef]

90. Busolo, M.A.; Fernandez, P.; Ocio, M.J.; Lagaron, J.M. Novel silver-based nanoclay as an antimicrobial in polylactic acid food packaging coatings. *Food Addit. Contam. Part A Chem. Anal. Control Expo. Risk Assess.* **2010**, *27*, 1617–1626. [CrossRef] [PubMed]

91. Şengül, B.; El-abassy, R.M.A.; Materny, A.; Dilsiz, N. Poly(lactic acid)/Organo-Montmorillonite Nanocomposites: Synthesis, Structures, Permeation Properties and Applications. *Polym. Sci. Ser. A* **2017**, *59*, 891–901. [CrossRef]

92. Chowdhury, S.R. Some important aspects in designing high molecular weight poly(L-lactic acid)–clay nanocomposites with desired properties. *Polym. Int.* **2008**, *57*, 1326–1332. [CrossRef]

93. Jorda-Beneyto, M.; Ortuño, N.; Devis, A.; Aucejo, S.; Puerto, M.; Gutiérrez-Praena, D.; Houtman, J.; Pichardo, S.; Maisanaba, S.; Jos, A. Use of nanoclay platelets in food packaging materials: Technical and cytotoxicity approach. *Food Addit. Contam. Part A Chem. Anal. Control Expo. Risk Assess.* **2014**, *31*, 354–363. [CrossRef]

94. Svagan, A.J.; Åkesson, A.; Cárdenas, M.; Bulut, S.; Knudsen, J.C.; Risbo, J.; Plackett, D. Transparent films based on PLA and montmorillonite with tunable oxygen barrier properties. *Biomacromolecules* **2012**, *13*, 397–405. [CrossRef]

95. Carosio, F.; Colonna, S.; Fina, A.; Rydzek, G.; Hemmerlé, J.; Jierry, L.; Schaaf, P.; Boulmedais, F. Efficient gas and water vapor barrier properties of thin poly(lactic acid) packaging films: Functionalization with moisture resistant Nafion and clay multilayers. *Chem. Mater.* **2014**, *26*, 5459–5466. [CrossRef]

96. Crank, J.; Crank, E.P.J. *The Mathematics of Diffusion*; Oxford University Press: Oxford, UK, 1979.

97. Wijmans, J.G.; Baker, R.W. The solution-diffusion model: A review. *J. Memb. Sci.* **1995**, *107*, 1–21. [CrossRef]

98. Crank, J. A theoretical investigation of the influence of molecular relaxation and internal stress on diffusion in polymers. *J. Polym. Sci.* **1953**, *11*, 151–168. [CrossRef]

99. Crank, J.; Park, G.S. *Diffusion in Polymers*; Academic Press: London, UK; New York, NY, USA, 1968.

100. Barrer, R.M. Diffusivities in glassy polymers for the dual mode sorption model. *J. Memb. Sci.* **1984**, *18*, 25–35. [CrossRef]

101. Meares, P. Transient permeation of organic vapors through polymer membranes. *J. Appl. Polym. Sci.* **1965**, *9*, 917–932. [CrossRef]

102. Hedenqvist, M.S.; Gedde, U.W. Parameters affecting the determination of transport kinetics data in highly swelling polymers above T(g). *Polymer* **1999**, *40*, 2381–2393. [CrossRef]

103. Vieth, W.R. *Diffusion in and Through Polymers: Principles and Applications*; Hanser Publications: Cincinnati, OH, USA, 1991.

104. Dhoot, S.N.; Freeman, B.D.; Stewart, M.E. Barrier Polymers. In *Encyclopedia of Polymer Science and Technology*; John Wiley and Sons Inc.: New York, NY, USA, 2002.

105. Ziegel, K.D.; Frensdorff, H.K.; Blair, D. Measurement of Hydrogen Isotope Transport in Poly(Vinyl Fluoride) Films By Permeation-Rate Method. *J. Polym. Sci. Phys.* **1969**, *7*, 809–819. [CrossRef]

106. Hedenqvist, M.; Angelstok, A.; Edsberg, L.; Larssont, P.T.; Gedde, U.W. Diffusion of small-molecule penetrants in polyethylene: Free volume and morphology. *Polymer* **1996**, *37*, 2887–2902. [CrossRef]

107. Webb, J.A.; Bower, D.I.; Ward, I.M.; Cardew, P.T. The effect of drawing on the transport of gases through polyethylene. *J. Polym. Sci. Part B Polym. Phys.* **1993**, *31*, 743–757. [CrossRef]

108. Paternak, R.; Schimscheimer, J.F.; Heller, J. Dynamic Approach To Diffusion and Permeation Measurements. *J. Polym. Sci. Part A 2 Polym. Phys.* **1970**, *8*, 467–479. [CrossRef]

109. Hedenqvist, M.S.; Ritums, J.E.; Condé-Brana, M.; Bergman, G. Sorption and desorption of tetrachloroethylene in fluoropolymers: Effects of the chemical structure and crystallinity. *J. Appl. Polym. Sci.* **2003**, *87*, 1474–1483. [CrossRef]

110. Hiltner, A.; Liu, R.Y.F.; Hu, Y.S.; Baer, E. Oxygen transport as a solid-state structure probe for polymeric materials: A review. *J. Polym. Sci. Part B Polym. Phys.* **2005**, *43*, 1047–1063. [CrossRef]

111. Cohen, M.H.; Turnbull, D. Molecular transport in liquids and glasses. *J. Chem. Phys.* **1959**, *31*, 1164–1169. [CrossRef]

112. Nilsson, F.; Hedenqvist, M.S.; Gedde, U.W. Small-molecule diffusion in semicrystalline polymers as revealed by experimental and simulation studies. *Macromol. Symp.* **2010**, *298*, 108–115. [CrossRef]

113. Nilsson, F.; Gedde, U.W.; Hedenqvist, M.S. Penetrant diffusion in polyethylene spherulites assessed by a novel off-lattice Monte-Carlo technique. *Eur. Polym. J.* **2009**, *45*, 3409–3417. [CrossRef]

114. Mogri, Z.; Paul, D.R. Gas sorption and transport in side-chain crystalline and molten poly(octadecyl acrylate). *Polymer* **2001**, *42*, 2531–2542. [CrossRef]

115. Nielsen, L.E. Models for the Permeability of Filled Polymer Systems. *J. Macromol. Sci. Part A Chem.* **1967**, *1*, 929–942. [CrossRef]

116. Cussler, E.L.; Hughes, S.E.; Ward, W.J.; Aris, R. Barrier membranes. *J. Memb. Sci.* **1988**, *38*, 161–174. [CrossRef]

117. Brydges, W.T.; Gulati, S.T.; Baum, G. Permeability of glass ribbon-reinforced composites. *J. Mater. Sci.* **1975**, *10*, 2044–2049. [CrossRef]

118. Lape, N.K.; Nuxoll, E.E.; Cussler, E.L. Polydisperse flakes in barrier films. *J. Memb. Sci.* **2004**, *236*, 29–37. [CrossRef]

119. Fredrickson, G.H.; Bicerano, J. Barrier properties of oriented disk composites. *J. Chem. Phys.* **1999**, *110*, 2181–2188. [CrossRef]

120. Gusev, A.A.; Lusti, H.R. Rational design of nanocomposites for barrier applications. *Adv. Mater.* **2001**, *13*, 1641–1643. [CrossRef]

121. Bharadwaj, R.K. Modeling the barrier properties of polymer-layered silicate nanocomposites. *Macromolecules* **2001**, *34*, 9189–9192. [CrossRef]

122. Guo, Y.; Yang, K.; Zuo, X.; Xue, Y.; Marmorat, C.; Liu, Y.; Chang, C.C.; Rafailovich, M.H. Effects of clay platelets and natural nanotubes on mechanical properties and gas permeability of Poly (lactic acid) nanocomposites. *Polymer* **2016**, *83*, 246–259. [CrossRef]

123. Picard, E.; Espuche, E.; Fulchiron, R. Effect of an organo-modified montmorillonite on PLA crystallization and gas barrier properties. *Appl. Clay Sci.* **2011**, *53*, 58–65. [CrossRef]

124. Li, Y.; Ren, P.; Zhang, Q.; Shen, T.; Ci, J.; Fang, C. Properties of Poly(lactic acid)/Organo-Montmorillonite Nanocomposites Prepared by Solution Intercalation. *J. Macromol. Sci. Part B Phys.* **2013**, *52*, 1041–1055. [CrossRef]

125. Bhatia, A.; Gupta, R.K.; Bhattacharya, S.N.; Choi, H.J. Analysis of Gas Permeability Characteristics of Poly(Lactic Acid)/ Poly (Butylene Succinate) Nanocomposites. *J. Nanomater.* **2012**, *2012*, 249094. [CrossRef]

126. Tenn, N.; Follain, N.; Soulestin, J.; Crétois, R.; Bourbigot, S.; Marais, S. Effect of Nanoclay Hydration on Barrier Properties of PLA/Montmorillonite Based Nanocomposites. *J. Phys. Chem. C* **2013**, *117*, 12117–12135. [CrossRef]

127. Hedenqvist, M.; Gedde, U.W. Diffusion of small-molecule penetrants in semicrystalline polymers. *Prog. Polym. Sci.* **1996**, *21*, 299–333. [CrossRef]

MDPI
St. Alban-Anlage 66
4052 Basel
Switzerland
Tel. +41 61 683 77 34
Fax +41 61 302 89 18
www.mdpi.com

Polymers Editorial Office
E-mail: polymers@mdpi.com
www.mdpi.com/journal/polymers